T0324690

LASERS FOR SCIENTISTS AND ENGINEERS

LASERS FOR SCIENTISTS AND ENGINEERS

L. Wilmer Anderson • John B. Boffard

University of Wisconsin-Madison, USA

NEW JERSEY • LONDON • SINGAPORE • BEIJING • SHANGHAI • HONG KONG • TAIPEI • CHENNAI • TOKYO

Published by

World Scientific Publishing Co. Pte. Ltd.
5 Toh Tuck Link, Singapore 596224
USA office: 27 Warren Street, Suite 401-402, Hackensack, NJ 07601
UK office: 57 Shelton Street, Covent Garden, London WC2H 9HE

Library of Congress Cataloging-in-Publication Data
Names: Anderson, L. W. (Louis Wilmer), 1933– author. | Boffard, John B., author.
Title: Lasers for scientists and engineers / L. Wilmer Anderson, John B. Boffard,
University of Wisconsin--Madison, USA.
Description: Singapore ; Hackensack, NJ : World Scientific, [2017] |
Includes bibliographical references and index.
Identifiers: LCCN 2017016095| ISBN 9789813224285 (hardcover ; alk. paper) |
ISBN 9813224282 (hardcover ; alk. paper) | ISBN 9789813224292 (pbk. ; alk. paper) |
ISBN 9813224290 (pbk. ; alk. paper)
Subjects: LCSH: Lasers.
Classification: LCC TA1675 .A495 2017 | DDC 621.36/6--dc23
LC record available at https://lccn.loc.gov/2017016095

British Library Cataloguing-in-Publication Data
A catalogue record for this book is available from the British Library.

Desk Editor: Christopher Teo

Typeset by Stallion Press
Email: enquiries@stallionpress.com

Printed in Singapore

This book is dedicated to L.W.A.'s wife M.G.A. and three children M.M.A., L.C.A., and E.B.A., and to our colleague C.C.L.

Preface

This book discusses the principles of laser operation and the details of how selected lasers operate. Lasers are now of great importance in many applications. Since the invention of the laser the variety of lasers and their uses has grown at a phenomenal rate. Scientists and engineers have at their disposal an enormous array of sophisticated laser equipment with the possibility of carrying out experiments that were inconceivable only a few decades ago. No scientist or engineer can afford to remain ignorant of the revolutionary developments made possible by lasers and by apparatus that utilizes lasers. The needed training in the principles and uses of lasers has been met in colleges and universities by courses on lasers and their applications. This book is suitable for a one semester course for senior or first year graduate level students in physics, chemistry, biology, astronomy, and the various fields of engineering. The background needed for this book are junior level courses in optics and modern physics including elementary quantum mechanics. No prior knowledge of lasers is assumed. The subject of lasers is treated without the use of quantum field theory.

The subject of lasers and their applications is immense. A complete discussion of all aspects of lasers and their applications is not attempted in this book. Instead the object of this text is to acquaint the reader with the general principles of laser operation and the details of selected lasers. Little attempt is made to cover laser applications. It is not necessary for most scientists or engineers to understand all the specialized applications of lasers in order to use lasers for their research or work. The exclusive goal of this book is to enable scientists and engineers to obtain a basic understanding of lasers that is adequate for their careers.

The decision to write a book requires some explanation. It is a common experience that many students have a difficult time beginning research or work using sophisticated lasers. Although there are other books on this subject, they often cover the entire field of lasers and their applications. The objective of this book is to provide a short easily read book that enables the reader to thoroughly grasp the subject. The reader of this book will find discussions that begin at an elementary level and lead to a complete understanding of lasers. In addition to college and university students, scientists and engineers who have completed their formal education and are working with lasers, but lack adequate training in this area, will find the information they need in order to understand the operation of lasers.

The first seven chapters contain a detailed and reasonably complete treatment of the principles of laser operation and optical cavities. These chapters contain treatments of stimulated emission and the criterion for laser action, laser thresholds, line shapes, saturation, optical cavities, and Gaussian optical beams. Emphasis is placed on a clear understanding of theses topics so that the reader is able to pursue successfully other more specialized topics after reading these chapters. The next seven chapters contain material on the details of particular lasers and other specialized topics. Chapter 8 is on diode lasers. Because some students have not had a course on solid state physics this chapter includes a treatment of semiconductor physics. Chapter 9 contains a discussion of lasers that utilize ions in a solid state or glass matrix as the active laser medium. Chapter 10 contains a discussion of the Helium-Neon laser. Although HeNe lasers are not widely used at the present time they can be analyzed in detail in a reasonably complete manner at the level of this text and this will acquaint the reader with knowledge of how complicated a through treatment of a laser really is. Later chapters treat the subject of tuneable lasers including the free electron laser. The last two chapters discuss briefly the subjects of the generation of coherent beams of light using non-linear optics and quantum optics.

This book contains somewhat more material than can be covered in a one semester course on lasers. Even so in order to keep the length of the book reasonably short a great deal of material that might be appropriate in a course on lasers has been omitted. Most instructors will want to cover the material contained in the first seven chapters carefully. They can then select material from the final seven chapters, and if they desire provide supplementary material to fit the needs of the particular students they are teaching. Solving problems is essential for understanding any scientific subject. A problem set is included at the end of each chapter. Some problems are merely drills, but other problems illuminate and significantly extend the material in the text.

Finally it is a pleasure to thank Professor James E. Lawler for many discussions of the properties of lasers in general and especially his insights on the operation of free electron lasers and other modern optical developments. We, also thank Professor Thad Walker for many helpful discussions and especially for his discussion with one of us (LWA) on why the spring model of the atom works so well, and Professor Charles Goebel for many valuable comments on the text and basic science. In addition we acknowledge numerous helpful discussions with Professors Chun C. Lin, Mark Saffman, and Deniz Yavuz. Of course inadequacies in the text are entirely the responsibility of the authors.

L. Wilmer Anderson
John B. Boffard
Madison, Wisconsin, January 2017

Contents

Chapter 1

An Introduction to Lasers

The word **LASER** is an acronym for **L**ight **A**mplification by **S**timulated **E**mission of **R**adiation. The name indicates that the process of stimulated emission makes a laser work. Stimulated emission is studied in detail in this book. The name does not describe the special characteristics of laser light that makes lasers so useful. Compared to conventional sources of light such as an incandescent light, the light from a laser may be quite intense, very monochromatic, and may be emitted in a beam with an angular divergence limited by diffraction. The monochromaticity and high degree of collimation of a laser beam are manifestations of the coherence of the laser beam. If one repeatedly samples the electromagnetic field in a light beam at different points in space either longitudinally along the beam axis or transversely across the beam axis and finds that definite phase differences are maintained, then the light is said to be coherent. These properties imply that laser light is nearly a perfect classical sinusoidal electromagnetic wave.

Lasers have many applications in scientific or engineering research, in telecommunications, in industry, and in medicine. For example in scientific or engineering research variable wavelength lasers are used for spectroscopy experiments, for the ultra-sensitive detection of atoms or molecules, for atom trapping experiments, for laser isotope separation experiments, for measurements of the time dependence of physical or chemical reactions, for studies of laser induced chemical reactions, and for many other research projects. In commercial or industrial applications lasers are used for the transmission of communication signals, for reading bar codes, for cutting of precise patterns in metal tooling, and for many other operations. In medicine lasers are used for the treatment of detached retinas, for sculpting corneas, for the treatment of ulcers, for use as a precision scalpel, and for many other applications. The remarkable properties of lasers are the result of stimulated emission. This book discusses how stimulated emission is utilized to produce laser action, and how a number of lasers operate. Because the applications of lasers is so enormous no effort is made to discuss the applications of lasers. The purpose of Chapter 1 is to introduce the concepts and notation used in this book, to provide a brief review of the theories of electricity and magnetism and atomic physics, that are involved in the operation of lasers, and to describe the basics of how a laser works.

1.1 Electromagnetic Waves

Light is an electromagnetic wave. The light generated by a laser must satisfy Maxwell's equations just as any electromagnetic wave does. Maxwell's equations are

$$\begin{aligned} \nabla \cdot \boldsymbol{D} &= \rho \\ \nabla \cdot \boldsymbol{B} &= 0 \\ \nabla \times \boldsymbol{E} &= -\frac{\partial \boldsymbol{B}}{\partial t} \\ \nabla \times \boldsymbol{H} &= j + \frac{\partial \boldsymbol{D}}{\partial t} \end{aligned} \tag{1.1}$$

where $\boldsymbol{D}$, $\boldsymbol{E}$, $\boldsymbol{B}$, and $\boldsymbol{H}$ are the displacement, the electric field, the magnetic field, and the magnetic intensity vectors respectively. The charge density is ρ and the current density is j. An electromagnetic wave consists of orthogonal electric and magnetic fields. For many purposes, the light from a laser can be described as an ideal plane wave. The electric field in a plane wave propagating in a vacuum in the direction of the wave vector $\boldsymbol{k}$ can be represented by

$$\boldsymbol{E} = \boldsymbol{E}_0 \, e^{i(\boldsymbol{k}\cdot\boldsymbol{r} - \omega t + \phi)} \tag{1.2}$$

where $k = 2\pi/\lambda$ is the magnitude of the wave vector, λ is the wavelength of the light, $\nu = c/\lambda$ is the frequency of the electromagnetic wave, c is the speed of light in a vacuum, $\omega = 2\pi\nu$ is the angular frequency, $\boldsymbol{r}$ is the position, t is the time, and ϕ is the phase constant. The complex notation is convenient for describing the field of an electromagnetic wave. The actual electric field is obtained by taking the real part of the expression in Equation 1.1. The vector $\boldsymbol{E}_0$ is orthogonal to the vector $\boldsymbol{k}$. In the International System of units (SI) the electric field, $\boldsymbol{E}$, has units of Volts per meter (V/m). The magnetic field associated with this plane electromagnetic wave is given by

$$\boldsymbol{B} = \boldsymbol{B}_0 \, e^{i(\boldsymbol{k}\cdot\boldsymbol{r} - \omega t + \phi)} \tag{1.3}$$

The magnetic field, $\boldsymbol{B}$, has units of Tesla (T) or Webers per square meter (W/m^2). The vector $\boldsymbol{B}_0$ is orthogonal to both the vectors $\boldsymbol{E}_0$ and $\boldsymbol{k}$. Maxwell's equations require that $B_0 = E_0/c$ (in SI units) where B_0 and E_0 are the magnitudes of $\boldsymbol{B}_0$ and $\boldsymbol{E}_0$ respectively. If an electromagnetic wave with frequency ν is propagating in a medium with an index of refraction, n, then the speed of light in the medium (i.e. the phase velocity of the wave) is given by $v = c/n$, and in the expression for the electromagnetic fields one should replace c with v. The index of refraction of the medium is a function of the frequency of the electromagnetic wave. The frequency of an electromagnetic wave in a medium with index n is the same as in vacuum, and the wavelength in a medium is equal to the wavelength in vacuum divided by the index of refraction of the medium. The electric and magnetic fields of an electromagnetic wave are orthogonal, and the vector product of the electric and magnetic fields, $\boldsymbol{E}_0 \times \boldsymbol{B}_0$, is in the direction of the propagation of the electromagnetic wave, i.e. in the direction of $\boldsymbol{k}$. The relation between the speed of light in a medium, the wave vector, and the frequency of the wave is $v = c/n = \lambda\nu = \omega/k$. Lasers with wavelengths in the infrared, visible, ultraviolet and even the X-ray regions of the electromagnetic spectrum have been constructed. Because the light generated by a laser satisfies Maxwell's equations, just as any electromagnetic wave must, laser beams obey the ordinary laws of optics including the laws for reflection, refraction, and diffraction.

The light generated by a laser may be very coherent. A light beam is *coherent* if there a fixed relationship between the phase at different locations of the beam. The coherence of the laser light is due to its generation by stimulated emission. Laser beams can have large longitudinal coherence lengths. The *longitudinal* coherence length is the distance parallel to the direction of propagation, over which the laser light wave maintains its coherence. This is often expressed as a *coherence time* where the coherence time is equal to the longitudinal coherence length divided by the velocity of propagation of the wave. Laser beams can also have large transverse coherence lengths. The *transverse* coherence length is the distance orthogonal to the direction of propagation over which the laser beam is coherent. The coherence time, the longitudinal coherence length, and the bandwidth of the laser are all simply related. The bandwidth of a laser is the spread in frequencies in the laser light. For example if the laser operates on average for 10^{-3} s without a phase slip or disruption, then the coherence time is 10^{-3} s and the longitudinal coherence length is $(10^{-3}c) = 3 \times 10^5$ m. A Michelson interferometer with a path difference up to 3×10^5 m will produce sharp fringes using the laser as a light source. The relationship of the coherence time of the laser to the laser bandwidth is analyzed carefully in a Chapter 7 of this book where it is shown that the laser bandwidth is given by the inverse of π times the coherence time. Thus a beat wave experiment with a stable source shows that a laser with a coherence time of 10^{-3} s has a bandwidth of $(\pi 10^{-3})^{-1}$ s$^{-1} = 318$ Hz.

In later chapters it is shown that the light emitted by a laser can be diffraction limited so that the light can emerge into a small solid angle. This means that a laser beam can be very bright i.e. it can have a large value of the power per solid angle. A laser beam can be focused into a very small diffraction limited spot. At the focus the laser beam can have an extremely high intensity i.e. a high power per area.

1.2 Photons

In Section 1.1 we briefly reviewed the wave properties of light. The wave character of light is shown by interference or diffraction experiments. Light also may be thought of as consisting of packets of energy or particles called photons. The photon behavior of light arises because of the quantization of the electromagnetic field. Each photon with frequency ν carries an energy given by $h\nu = \hbar\omega$ where $h = 6.626 \times 10^{-34}$ J s is Planck's constant, and where $\hbar = h/2\pi$. The quantum or particle character of light is apparent in experiments such as the Compton scattering of photons. The energy density, w, in an electromagnetic wave is given by the expression

$$w = \frac{\epsilon\, \boldsymbol{E}_{\mathrm{RMS}}^2}{2} + \frac{\boldsymbol{B}_{\mathrm{RMS}}^2}{2\mu_0} \tag{1.4}$$

where $\boldsymbol{E}_{\mathrm{RMS}}$ and $\boldsymbol{B}_{\mathrm{RMS}}$ are the root mean square values of the fields, $\epsilon = K\epsilon_0$ is the electric permitivity of the material in which the wave is traveling, $\epsilon_0 = 8.854 \times 10^{-12}$ C^{2}N^{-1}m^{-2} is the permitivity of free space, K is the dielectric constant of the material at the frequency ν, and $\mu_0 = 4\pi \times 10^{-7}$ TA^{-1}m is the permeability of free space. In Eq. 1.4 it is assumed that the relative permeability of the material is one, i.e., that the magnetic susceptibility is zero. The dielectric constant and the index of refraction are related by $K = n^2$. Since the energy of a photon is quantized it follows that the energy density in an electromagnetic wave with a given frequency and wave vector is also given by

$$w = n_{\mathrm{ph}}\, h\nu \tag{1.5}$$

where $n_{\rm ph}$ is the density[1] of photons with a given frequency and wave vector in the electromagnetic field. Only photons with particular allowed frequencies and wave vectors can exist in an optical cavity as is discussed in later chapters. A photon with one of the given allowed frequencies and wave vectors is said to be in one of the allowed modes of the cavity. Thus $n_{\rm ph}$ is the photon density in a mode with the particular frequency and wave vector. The intensity, I_ν , of a light beam with a given frequency and wave vector is given by

$$I_\nu = \frac{wc}{n} = \frac{n_{\rm ph}\, h\nu c}{n} \tag{1.6}$$

where I is measured in Watts per square meter (W/m^2). In discussing lasers it is more convenient to treat some problems using the wave picture whereas it is more convenient to treat other problems using the particle or photon picture. In many situations it is convienent to think of a photon as a wave packet that has a length equal to the longitudinal coherence length and a transverse size equal to the transverse coherence length.

1.3 Spontaneous and Stimulated Emission

Atoms and molecules have quantized energy levels. It is found experimentally that isolated atoms (or molecules) in an excited level, labeled u, can decay radiatively to a lower level, labeled l, with the emission of a photon. The radiative decay of an isolated atom is called spontaneous emission. The frequency of the photon emitted in spontaneous decay is related to the energy difference between the atomic energy levels by

$$\nu_0 = \frac{\Delta W}{h} = \frac{W_u - W_l}{h} \tag{1.7}$$

where W_u is the energy of the upper level u and W_l is the energy of the lower level l. It is found experimentally that when an ensemble of atoms in an excited level, such as level u, decays radiatively by spontaneous emission the number density of atoms in the upper level decreases exponentially with time provided that there is no source of production of atoms into the upper level. It is also found experimentally that an atom can absorb a photon from the electromagnetic field. The absorption of a photon with frequency ν_0 excites an atom initially in lower level l into the upper level u. It might be thought that absorption and spontaneous emission are adequate

[1]The number *density* of photons is the number of photons per unit volume per unit frequency. In general this text uses the term density to mean number density for atoms, molecules, photons and modes in the electromagnetic field.

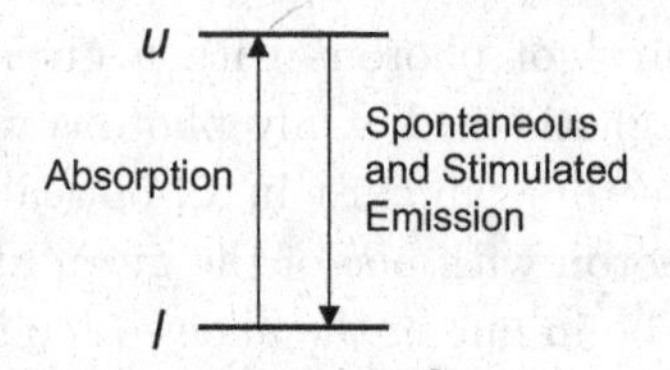

Fig. 1.1 Two atomic energy levels with absorption and spontaneous and stimulated emission indicated.

to explain the level populations of atoms exposed to radiation fields. This is, however, not the case. A third process called stimulated emission is necessary. Stimulated emission occurs when a photon interacts with an atom in an excited level, and the atom decays to a lower level with the emission of a second photon. Figure 1.1 shows two atomic energy levels with absorption and spontaneous and stimulated emission occurring. In stimulated emission the photon that is emitted is identical in every way with the photon that causes the stimulated emission so that the photon that is emitted has the same frequency, the same phase, the same polarization, and the same direction of propagation as the original photon. Einstein analyzed the level populations of atoms in the presence of electromagnetic radiation, and he demonstrated the necessity for considering absorption, spontaneous emission, and stimulated emission in determining the level populations of atoms exposed to radiation fields.

Stimulated emission is the subject of much of this text book, so that here we present only a brief discussion as to why stimulated emission is necessary. In order to understand this consider a simple physical system composed of an ensemble of atoms that is maintained in thermal equilibrium with a container at the temperature T. The atoms are therefore exposed to blackbody radiation characteristic of the temperature T. That segment of the blackbody spectrum at frequencies corresponding to the energy differences between different levels of the atoms can be absorbed by the atoms. Atoms in excited levels decay both by spontaneous and induced emission. In order to see that induced emission is necessary, consider the situation that would arise if there were no stimulated emission. In this case, the population of any excited level would be, in the steady state, determined by equating the rate at which the excited atoms are produced by absorption, to the rate at which the excited atoms decay by spontaneous emission. The number of blackbody photons per unit frequency and unit volume increases monotonically with increasing temperature, and hence the absorption rate per atom increases monotonically with the temperature. The spontaneous decay rate per excited atom is independent of the temperature. When the temperature is large enough that the absorption rate per atom exceeds the spontaneous decay rate per atom the population in an upper level would exceed the population in a lower level if there were no mechanism other than spontaneous emission for the decay of atoms in the upper level to the lower level. This is, however, not physically reasonable because in thermal equilibrium the populations of the various energy levels in an atom must be given by the Boltzmann distribution, which predicts that the lower level must have a higher population than the upper level. Therefore some process leading to the depopulation of the upper level other than spontaneous emission must occur. That process is the stimulated emission of radiation. When one calculates the populations of atomic levels by equating the absorption rate per atom into the upper level with the sum of the decay rates per atom due to spontaneous emission plus stimulated emission one finds that the population of the atomic levels satisfies the expected Boltzmann distribution. The qualitative discussion presented in this section is put on a quantitative basis in Chapter 2 where the process of stimulated emission is analyzed in detail and in Chapter 3 where stimulated emission is shown to lead to laser action.

One final comment on stimulated emission may be appropriate here. Although our discussion has focused on transitions in atoms where the energy levels are quantized, any process that radiates spontaneously can be stimulated to radiate. Stimulated emission is not an intrinsically quantum mechanical effect, and classical systems that radiate can be stimulated to radiate. The free electron laser is an example of a classical system that utilizes the stimulated emission of radiation.

1.4 Lasers

The concept of a laser was proposed by Schawlow and Townes. A typical laser is indicated schematically in Figure 1.2. It is made up of an active "lasing" medium and an *optical cavity.* Chapter 7 describes the requirements of the optical cavity and various designs used in some lasers. Some properties of the lasing medium are addressed in Chapters 5 and 6, along with complete laser systems in Chapters 8-9.

Although a substantial part of this book is devoted to the discussion of particular lasers a short discussion of some particular lasers may be in order here as an introduction. The most important lasers are the solid state

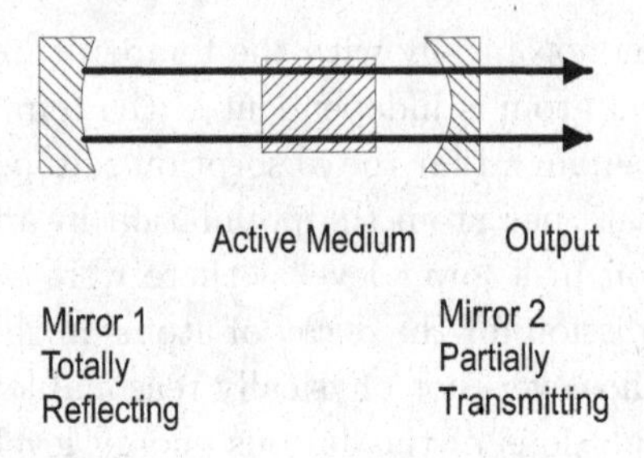

Fig. 1.2 A schematic diagram of a laser.

lasers. Solid state lasers are very reliable and some can be tuned to a particular wavelength and some can produce very high output power. A common type of solid state laser is the semiconductor diode laser, for instance a GaAlAs-GaAs-GaAlAs sandwich that lases in the near infrared region of the spectrum and widely used for optical data storage (CD-ROM). Semiconductor lasers are also used extensively for communications, for readout of bar codes, and for numerous other applications. Semiconductor lasers also have extensive uses in scientific research where they are used as narrow band width lasers for absorption spectroscopy and for many other scientific purposes. These lasers can operate continuously. This is called cw (continuous wave) operation. Semiconductor lasers usually are operated at low output powers, but some can be operated with very high output powers. The intensity of semiconductor lasers can be very rapidly changed as a function of the time. This is called modulation of the laser intensity and is used in high-speed optical communication. Glass or silica fiber lasers are often used to amplify the signals from diode lasers.

A type of cw gas discharge laser is the helium-neon (He-Ne) laser. This gas discharge laser operates with a wave length of 632.8 nm and with typical output powers of a few mW. The HeNe laser is sometimes used for college demonstrations. The Ar ion laser is also a gas discharge laser. The Ar ion laser can have an optical output power in the visible of 10's of Watts. Another gas discharge laser is the CO_2 laser, which operates in the infra-red with a wavelength of 10.6 microns (μm) and can have high output power levels. An application of the CO_2 laser is for machine tooling. Gas lasers are not used as often as they were in the past due primarily to their short lifetimes and lack of reliability. The operating wavelengths of various lasers are shown in Figure 1.3.

Some lasers are typically operated in a pulsed mode rather than cw.

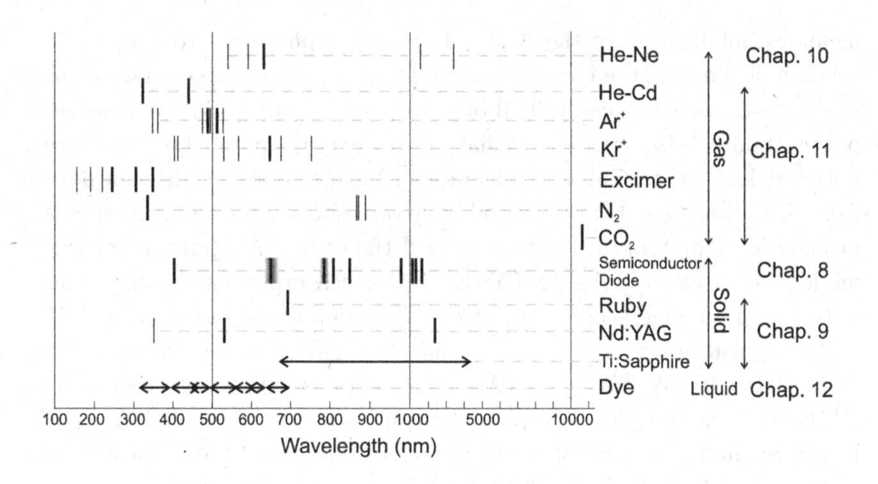

Fig. 1.3 Wavelengths of some common lasers.

These lasers can operate with instantaneous peak powers that are very high. Lasers that can operate in a pulsed mode include Nd:YAG lasers, Nd:Glass lasers, titanium-sapphire (Ti:Sapph) lasers, and excimer lasers. Common uses for these lasers are for the pumping of high power pulsed tunable lasers or for other applications such as irradiating the photo resist material in the manufacture of semiconductor chips.

All lasers are tunable over some spread in wavelength. For many lasers the range over which the laser can be tuned is, however, very small and these lasers are called fixed frequency lasers. An example of a fixed frequency laser is a HeNe laser. For other lasers the operating wavelength can be varied over a substantial range, sometimes hundreds of nanometers. These lasers are called tunable lasers. Tunable lasers are used for application where it is desireable to set the laser to a particular desired wavelength such as scientific applications like spectroscopy and other applications. Titanium-sapphire lasers, diode lasers, and free electron lasers are examples of tunable lasers. One often hears the statement "Lasers can be very powerful." While this statement is true it requires some explanation. Even a powerful cw laser such as the Ar ion laser which has an output optical power of 15 W emits only a few times as much optical power as a 100 W incandescent light bulb. It should be noted that a 100 W incandescent light uses 100 W of power from the electric power grid but only about 3-6% of that power or about 3-6 W is converted into visible photons. Of course the Ar ion laser has a much higher brightness than the incandescent light since the Ar ion laser emits photons into a very small solid angle whereas the incandescent light emits photons into all directions. In a pulsed laser, one can talk about both the average power and the peak power. Some pulsed lasers may have a low average power (i.e., by having a long delay 'off' period between pulses), but extremely high peak power. Indeed, by limiting the 'on' period to a very short pulse duration one can produce very high optical powers even if the energy per pulse is relatively modest. Consider a pulsed Nd:YAG laser which produces a string of light pulses with an energy per pulse of $E = 0.1$ J in optical radiation and with a pulse duration of $t = 10^{-8}$ s. The peak power of this laser is about $P_{\mathrm{p}} = E/t = 10^7$ W. If the Nd:YAG laser operates with a repetition rate of 20 Hz then the laser has an average output of power of $P_{\mathrm{av}} = 20E = 2$ W. In this example the ratio of the peak output power and the average output power is $P_{\mathrm{p}}/P_{\mathrm{av}} = 5 \times 10^6$. This is rather typical. Pulsed lasers can have very high peak output powers but usually have moderate to low average output powers; cw lasers typically have moderate to low output powers. There are only few lasers such as the CO_2 laser that can operate with cw output powers of 1 kW or more.

This book discusses how a number of different lasers operate in detail, but only briefly touches on a few scientific or engineering applications since the variety of applications is simply too vast to cover adequately. The discussion in this chapter is intended only to introduce the reader to the many types of lasers and to familarize the reader with the notation used in this text.

Summary of Key Ideas

- Light is emitted from transitions between the quantized energy levels of atoms and molecules, $\Delta W = W_u - W_l = h\nu = hc/\lambda$.
- Excited levels of atoms or molecules decay by either spontaneous or stimulated emission.
- Lasers utilize stimulated emission to produce coherent light beams.
- Lasers come in many varieties: pulsed or cw, operating with a fixed wavelength or tunable over a range of wavelengths (Fig. 1.3), and utilize a number of different mediums (gas discharges, solid crystals, *etc.*..).

Suggested Additional Reading

Charles H. Townes, *How the Laser Came to Happen: Adventures of a Scientist*, Oxford University Press (2002).

William T. Silfvast, *Laser Fundamentals*, Cambridge University Press (1996).

Donald C. O'Shea, W. Russell Callen and William T. Rhodes, *An Introduction to Lasers and their Applications*, Addison Wesley Publishing Co. (1977).

Frank S. Crawford, Jr., *Waves - The Berkeley Physics Course - Volume 3*, McGraw Hill (1965).

A. Einstein, "Zur Quantentheorie der Strahlung [On the Quantum Theory of Radiation]" *Phys. Z.* **18**, 121 (1917).

A. L. Schawlow and C. H. Townes, "Infrared and Optical Masers" *Phys. Rev.* **112**, 1940 (1958).

A. L. Schawlow, *Lasers and Light*, W. H. Freeman and Co. (1969).

J. D. Jackson, *Classical Electordynamics 3rd Ed.*, John Wiley and Sons (1998).

P. W. Milonni and J. H. Eberly, *Laser Physics 2nd Ed.*, John Wiley and Sons (2010).

Problems

1. An electromagnetic wave is traveling in a vacuum in the direction from $-x$ toward $+x$. The amplitude of the electric field in the wave is 0.1 V/m and the frequency of the wave is 5×10^{14} Hz. The wave is polarized along the y axis i.e. the electric field lies parallel or anti parallel to the y axis. (a) What is the wavelength? (b) Write an expression for the electric field of the electromagnetic wave if the electric field is zero at $x = 0$ and $t = 0$. (c) What is the magnitude and direction of the wave vector of the electromagnetic wave? (d) What is the magnitude and direction of the magnetic field in the electromagnetic wave?

2. An electromagnetic wave traveling in free space with $\lambda = 500$ nm has a coherence time of 5×10^{-4} s. (a) What is the longitudinal coherence length? Express your answer in terms of both meters and wavelengths. (b) What is the bandwidth of the laser?

3. A photon traveling in free space has a wavelength of 500 nm. What is the energy of the photon? Give your answer both in joules and in electron volts.

4. Show that a photon with a wavelength λ in nanometers has an energy, E, in electron volts of $E = 1239.85/\lambda$. (It will be useful to remember this relationship.)

5. A single mode electromagnetic wave in a vacuum is 1 cm in diameter, has a wavelength of 500 nm, and an intensity of 0.1 W/m^2. (a) What is the energy density in the wave? (b) What is the electric field in the wave? (c) What is the magnetic field in the wave? (d) What is the photon density in the wave?

6. If the light beam described in Problem 5 is focused to a spot 100 microns in diameter what are the intensity, the energy density, and the electric field at the focus?

7. An atom has an upper energy level 2.1 eV above the ground level. If the atom spontaneously makes a transition from the upper level to the ground level what is the wavelength and frequency of the emitted radiation?

8. If an electromagnetic wave with wavelength $\lambda = 500$ nm in vacuum passes from a vacuum into a material with an index of refraction $n = 1.5$, what are the wavelength and frequency inside the material?

9. A pulsed laser has a repetition rate of 20 pulses per second and an average power of 10 mW. If each pulse is 5 nsec in duration what is the peak instantaneous power emitted during a pulse?

10. For the laser of Problem 9 is the coherence time necessarily 5 nsec? Explain your answer.

Chapter 2

Stimulated Emission

2

In this chapter the concepts of spontaneous emission, absorption, and stimulated emission are discussed in a quantitative manner. Blackbody radiation is analyzed in terms of both Planck's and Einstein's derivations. Using Einstein's analysis the relationships between spontaneous emission, absorption, and stimulated emission are derived. Both classical and semi-classical treatments are presented for absorption and stimulated emission.

2.1 Electromagnetic Modes of a Cavity

In order to understand blackbody radiation it is necessary to understand the modes of a cavity for electromagnetic radiation. In our analysis we assume that the reader has a knowledge of Maxwell's equations for electromagnetism. Consider the modes of a cubic cavity with perfectly conducting walls (i.e. walls with an infinite conductivity) located at $x = 0$ and $x = L$, $y = 0$ and $y = L$, and $z = 0$ and $z = L$. For a cavity with walls having an infinite conductivity the component of the electric field tangential to the walls must be zero at the walls. The electric field can be written, as was discussed in Chapter 1, either in a complex notation where the true electric field is the real part of the complex electric field or written directly in a real notation. For clarity in this discussion the electric field is written using a real notation. The general expressions for the components of the electric field of a sinusoidal standing electromagnetic wave in a perfectly conducting cavity are the following;

$$E_x = A\cos(k_x x)\sin(k_y y)\sin(k_z z)\sin(\omega t + \phi) \tag{2.1}$$

$$E_y = B\sin(k_x x)\cos(k_y y)\sin(k_z z)\sin(\omega t + \phi) \tag{2.2}$$

$$E_z = C\sin(k_x x)\sin(k_y y)\cos(k_z z)\sin(\omega t + \phi) \tag{2.3}$$

where A, B, and C are constants representing the magnitudes of the field components and ϕ is the phase constant for the standing electromagnetic wave. In order that the tangential component of the electric field be zero at the walls it is necessary that $k_x = N_x\pi/L$, $k_y = N_y\pi/L$, and $k_z = N_z\pi/L$ where N_x, N_y, and N_z are positive integers. From Maxwell's equations it straightforward to show that the electric field satisfies the wave equation

$$\frac{\partial^2 \boldsymbol{E}}{\partial x^2} + \frac{\partial^2 \boldsymbol{E}}{\partial y^2} + \frac{\partial^2 \boldsymbol{E}}{\partial z^2} = \left(\frac{n}{c}\right)^2 \frac{\partial^2 \boldsymbol{E}}{\partial t^2} \tag{2.4}$$

By substitution into Equation 2.4 one finds that the electric field specified by Equations 2.1, 2.2, and 2.3 satisfies the wave equation provided

$$k^2 = k_x^2 + k_y^2 + k_z^2 = \left(\frac{n}{c}\right)^2 \omega^2 \,. \tag{2.5}$$

Since N_x, N_y, and N_z are positive integers, the allowed values of k form a lattice in the first octant of "k space". For each set of values of N_x, N_y, and N_z (i.e. for each unit of volume in "N space") there are two cavity modes with orthogonal polarizations. A unit volume in "N space" corresponds to a volume $(\pi/L)^3$ in "k space". Thus for each volume $(\pi/L)^3$ in "k space" there are two cavity modes with orthogonal polarizations. Provided k is much greater than π/L, the number of modes dN between k and $k + dk$ is equal to two times the volume of the first octant in "k space" of a spherical shell of thickness dk divided by the volume in "k space" for two cavity modes, $(\pi/L)^3$. This leads to

$$dN = \frac{\pi k^2\, dk}{\left(\frac{\pi}{L}\right)^3} \,. \tag{2.6}$$

The number of modes per unit frequency is given by

$$\frac{dN}{d\nu} = \frac{L^3\, k^2}{\pi^2} \frac{dk}{d\nu} = \frac{k^2}{\pi^2} \frac{2\pi\, n}{c} L^3 , \tag{2.7}$$

where n is the index of refraction. The number density of modes per unit frequency (i.e. number of modes per unit frequency per volume) is obtained by dividing the number of modes per unit frequency by the volume of the cavity L^3. The result is

$$\frac{1}{L^3} \frac{dN}{d\nu} = \frac{8\pi\, \nu^2\, n^3}{c^3} \,. \tag{2.8}$$

The number density of modes per unit frequency is independent of the size of the system provided k is much greater than π/L i.e. provided λ is much less than L. This expression will prove quite useful in later chapters when the number of possible modes, p, of a laser cavity within the $\Delta\nu$ bandwidth of the laser is calculated. Assuming the index of refraction is one,

$$p = \frac{8\pi\,\nu^2}{c^3}\,\Delta\nu\,V\,, \tag{2.9}$$

where V is the volume of the cavity.

Perhaps a numerical example may be useful at this point. If one considers a cavity with length $L = 10$ cm and a cavity mode with $N_y = N_z = 0$ then for a wavelength $\lambda = 500$ nm one finds that $N_x = 2L/\lambda = 4\times10^5$. The fractional change in the wavelength when N_x changes by 1 is equal to $\Delta\lambda/\lambda = -\lambda/2L = -2.5\times10^{-6}$. Thus for visible wavelengths the integer N_x is very large and the change in the wavelength when N_x is changed by 1 is very small. The number density of modes per unit frequency is equal to $1/L^3\,dN/d\nu = 8\pi\nu^2\,n^3/c^3 = 8\pi\,n/\lambda^2\,c$. For an index of refraction of one and for a wavelength $\lambda = 500$ nm one finds that $1/L^3\,dN/d\nu = 0.3$ modes/(cm^3 Hz).

The discussion of the modes of a cavity has focused on a cavity with perfectly conducting walls. A cavity need not have perfectly conducting walls. For example it is common to use an open laser cavity consisting of two spherical mirrors with dielectric coatings that have a reflectivity very near one. The boundary conditions for electric and magnetic fields in a cavity with dielectric walls are different from those for a cavity with perfectly conducting walls. The number density of modes per unit frequency can, however, be shown to be essentially independent of the type of walls. In the short wavelength (high frequency) limit all systems including any laser cavity or even the universe have the same average number density of modes per unit frequency.

Because the discussion has focused on an idealized cavity with perfectly reflecting walls the cavity modes are discreet and a mode with a given frequency is separated from another mode with a different frequency by a region in frequency space with no allowed modes. For a more realistic situation with walls that are not perfectly reflecting an individual cavity mode has a finite width in frequency which is given by $\Delta\nu_c = 1/(2\pi\,\tau_c)$ where τ_c is the $1/e$ lifetime of an electromagnetic wave in the mode. The modes overlap if the separation of the modes is less than $\Delta\nu_c$.

The magnetic field associated with this standing electromagnetic wave in a perfectly reflecting cavity can be obtained using Maxwell's equations. For example by using the x component of

$$\nabla\times\boldsymbol{E} = -\frac{\partial\boldsymbol{B}}{\partial t} \tag{2.10}$$

one obtains

$$B_x = \frac{(C\,k_y - B\,k_z)\,\sin(k_x\,x)\,\cos(k_y\,y)\,\cos(k_z\,z)\,\cos(\omega t - \phi)}{\omega}\,. \quad (2.11)$$

Similar results can be obtained for B_y and B_z. From Equation 2.11 it is seen that B_x vanishes at $x = 0$ and at $x = L$. In a similar way one can show that B_y vanishes at $y = 0$ and at $y = L$ and that B_z vanishes at $z = 0$ and at $z = L$. Thus as expected the normal component of magnetic field is zero at the walls of a perfectly conducting cavity.

2.2 Planck's Derivation of Blackbody Radiation

Planck explained the energy density per unit frequency for blackbody radiation in the following way. In thermal equilibrium the rates of emission and absorption are the same. The radiation field present in a cavity in thermal equilibrium is called blackbody radiation. Blackbody radiation does not depend in any way on the material from which the walls of the cavity are made but depends only on the temperature of the walls. The energy density per unit frequency, $\rho(\nu)$, of blackbody radiation can be calculated by multiplying the density of modes per unit frequency by the average energy per mode. The average energy per mode at the temperature T is given by the the probability of n photons in a mode times the energy of n photons summed over all the values of n

$$\overline{W} = \frac{h\nu\,\exp(-h\nu/kT) + 2h\nu\,\exp(-2h\nu/kT) + \cdots}{1 + \exp(-h\nu/kT) + \exp(-2h\nu/kT) + \cdots}\,. \quad (2.12)$$

The ratio of the two infinite series is given exactly by

$$\overline{W} = \frac{h\nu}{\exp(h\nu/kT) - 1} = h\nu\,N_\nu \quad (2.13)$$

where N_ν is the average number of photons in a mode at the frequency ν and at the temperature T. The average number of photons in a mode is also called the photon occupation number. The photon occupation number for blackbody radiation is given by

$$N_\nu = \frac{1}{\exp(h\nu/kT) - 1}\,. \quad (2.14)$$

Multiplying the average energy per mode times the density of modes per unit frequency yields the energy density per unit frequency at the frequency ν in the blackbody spectrum,

$$\rho(\nu) = \frac{8\pi\,h\nu^3}{c^3}\,\frac{1}{\exp(h\nu/kT) - 1} = \frac{8\pi\,h\nu^3}{c^3}\,N_\nu \quad (2.15)$$

where it is assumed that the index of refraction of the material in the cavity is equal to one so that the speed of light in the cavity is the speed of light in a vacuum. It should be noted that the black body expression for N_ν is valid only when the radiation field is in thermal equilibrium with its surroundings.

It is interesting to ask at what temperature is the photon occupation number for blackbody radiation equal to one photon per mode for visible wave length photons? If the photon occupation number is to be one then $1/[\exp(h\nu/kT) - 1] = 1$. If $\lambda = 620$ nm then the photon energy is equal to 2 eV. At the temperature $T = 3.4\times10^4$ K the photon occupation number is one. For black body radiation at room temperature and for wavelengths in the visible the photon occupation number is extremely small. For blackbody radiation the occupation number for modes with almost the same energy is nearly the same. In contrast to blackbody radiation it will be shown in later chapters that for a laser the photon occupation number for a particular mode or a few particular modes can be very large while the occupation number for other modes with almost the same energy is nearly zero.

2.3 Einstein's Treatment of Stimulated Emission

Einstein discovered an alternate derivation of the energy density per unit frequency in the blackbody radiation field in terms of absorption, spontaneous emission, and stimulated emission. Einstein's derivation of the energy density per unit frequency in the blackbody radiation field assumes the universality of blackbody radiation, i.e., that the blackbody spectrum does not depend on the nature of the walls of the cavity but depends only on their temperature. He made use of his analysis together with Planck's calculation of the energy density per unit frequency for blackbody radiation to determine relationships between the absorption rate, the spontaneous emission rate, and the stimulated emission rate.

Einstein made use of the concept of "detailed balance" in his analysis of blackbody radiation. Consider a large number or ensemble of atoms (or molecules) in thermal equilibrium at a temperature T and interacting with the bath of blackbody radiation characteristic of the temperature T. Our attention is focussed on only two energy levels in the atom as shown in Figure 2.1. The upper level, labeled u, has energy W_u, angular momentum J_u, and degeneracy $g_u = 2J_u + 1$. Similarly the lower level, labeled l, has energy W_l, angular momentum J_l, and degeneracy $g_l = 2J_l + 1$. The energy difference between the upper and lower levels is $W_u - W_l = h\nu_0$, where ν_0

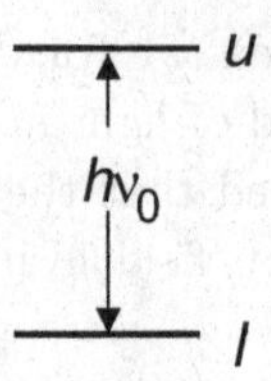

Fig. 2.1 Two energy levels, denoted by u and l for upper and lower, and separated by a difference in energy of $h\nu_0$.

is the frequency of the transition between the energy levels. In thermal equilibrium the ratio of the populations or number of atoms in each level is given by

$$\frac{N_u}{N_l} = \frac{g_u}{g_l} \exp\left(-h\nu_0/kT\right). \tag{2.16}$$

Although in thermal equilibrium the populations are constant, atoms in the system are undergoing transitions constantly. The transitions do not alter the populations because there are as many atoms leaving a given level as are entering that level. Einstein reasoned that the absorption rate from level l to level u must be proportional to the blackbody energy density per unit frequency times the number of atoms in level l. The absorption rate is given by

$$N_l \, R_{lu} = N_l \, B_{lu} \, \rho(\nu_0) \tag{2.17}$$

where R_{lu} is the absorption rate per atom and B_{lu} is a proportionality constant, which depends only on the strength of the transition. It is found experimentally that an atom in an excited level decays spontaneously to a lower level by the emission of radiation even in the absence of any other radiation. The rate of this spontaneous emission must depend only on the properties of the atom since it does not require that any other radiation be present to occur. The spontaneous rate from level u to level l is given by

$$N_u \, R_{ul\ \mathrm{spont}} = N_u \, A_{ul}. \tag{2.18}$$

$R_{ul\ \mathrm{spont}}$ is the spontaneous emission rate per atom from level u to level l per atom and where A_{ul} is a constant that depends only on atomic factors that determine the strength of the transition. Einstein and his contemporaries understood both absorption and spontaneous emission quite well in terms of a classical model, which is discussed in section 2.6 of this chapter.

Einstein realized that at high temperatures $\rho(\nu_0)$ becomes large and the absorption rate becomes larger than the spontaneous emission rate resulting in a population in the upper level u larger than the population in the lower level l unless there is a de-excitation mechanism other than spontaneous emission. The concept of "detailed balance" requires that excitation and de-excitation be balanced for each transition. The concept suggests that the needed de-excitation mechanism is the time reversed excitation mechanism. Thus Einstein proposed that stimulated emission must occur with a rate given by

$$N_u R_{ul\ \mathrm{stim}} = N_u B_{ul}\, \rho(\nu_0) \tag{2.19}$$

where $R_{ul\ \mathrm{stim}}$ is the rate of stimulated emission per atom and where B_{ul} is a constant that depends only on atomic factors that determine the strength of the transition. Detailed balance requires that the absorption rate must equal the sum of the rates for spontaneous emission plus stimulated emission so that

$$N_l B_{lu}\, \rho(\nu_0) = N_u \left[A_{ul} + B_{ul}\, \rho(\nu_0)\right] . \tag{2.20}$$

Solving Equation 2.20 for $\rho(\nu_0)$ yields

$$\rho(\nu_0) = \frac{A_{ul}}{B_{lu}\frac{N_l}{N_u} - B_{ul}} = \frac{A_{ul}}{B_{lu}\frac{g_l}{g_u}\exp\left(h\nu_0/kT\right) - B_{ul}} . \tag{2.21}$$

The relationships between the B coefficients and the A coefficient are derived by equating Planck's and Einstein's expressions for $\rho(\nu_0)$. These two expressions are identical only if both

$$g_l B_{lu} = g_u B_{ul} \tag{2.22}$$

and

$$\frac{A_{ul}}{B_{ul}} = \frac{8\pi\, h\nu_0^3}{c^3} . \tag{2.23}$$

Equations 2.22 and 2.23 must be true for any quantum mechanical or classical system. It should be mentioned that Equation 2.22 can also be derived with the use of time reversal without using Planck's expression for the energy density per unit frequency in blackbody radiation. Stimulated emission becomes absorption under time reversal i.e. stimulated emission and absorption are really the same process but viewed with direction of the time reversed. This implies that the photon produced in stimulated emission is identical in every way to the incident photon. In order to see this consider a mode with two photons. If one of these photons is absorbed then one photon remains in the mode. In the time reversed process one photon is incident and two photons emerge after stimulated emission. Obviously the stimulated photon is identical to the incident photon in this time reversed picture.

The expression for the total transition rate per atom from level u to level l, R_{ul}, can be written as

$$\begin{aligned} R_{ul} &= R_{ul\,\mathrm{spont}} + R_{ul\,\mathrm{stim}} \\ &= A_{ul} + B_{ul}\,\rho(\nu_0) \\ &= A_{ul}\,(1 + N_{\nu_0}) \end{aligned} \tag{2.24}$$

where Equations 2.15 and 2.23 are used to eliminate $\rho(\nu_0)$. In a similar manner one can show that the absorption rate per atom is

$$R_{lu} = B_{lu}\,\rho(\nu_0) = \frac{g_u}{g_l}\,A_{ul}\,N_{\nu_0}\ . \tag{2.25}$$

An ensemble of atoms does not in reality emit only single frequency photons, but instead emits photons with a spread in frequencies centered about ν_0. The probability of a given frequency being emitted as a function of the frequency is called the *line shape*, and the spread in frequencies about ν_0 is called the *line width.* The origin of line shapes is discussed in Chapter 4. Based upon Equation 2.24, one can write $A_{ul} = R_{ul}/(1 + N_{\nu_0})$ which can be interpreted as follows. The rate of spontaneous emission into a particular mode is the total rate of spontaneous emission per atom, A_{ul}, divided by the number of modes interacting with the two levels u and l. For blackbody radiation all the modes within a small frequency range such as an atomic line width have essentially the same photon occupation number. Usually the spontaneous rate of decay for nearby modes is the same. The total emission rate into a particular mode, which is equal to the spontaneous rate plus the stimulated rate into the particular mode, is equal to the spontaneous emission rate into that mode times $1 + N_{\nu_0}$. When one sums up the total emission rate into all the modes within the line width one obtains the result shown in Equation 2.24. In order to calculate the number of modes interacting with the two atomic levels one must understand the line shape of the transition as is discussed in Chapter 4.

It is important to recognize that the rates of stimulated emission and the spontaneous emission are related by the factor $1 + N_\nu$. Thus if the spontaneous emission decay rate into a particular mode is 10^3 s^{-1} then the stimulated emission rate into that mode when there are 100 photons in that mode is equal to 10^5 s^{-1} and the total decay rate is equal to 1.01×10^5 s^{-1}.

The reader should note that the Einstein A and B coefficients are constants that depend only on the strength of the transition and are applicable for all systems not just systems that are in thermal equilibrium and interacting with the blackbody spectrum. In general in non-thermal equilibrium situations the photon occupation number for a mode, N_ν, is not equal to the photon occupation number characteristic of thermal equilibrium i.e. blackbody radiation. In lasers the photon occupation numbers are very different from the occupation numbers for thermal equilibrium.

As a final comment on blackbody radiation one should observe that the total energy density in the blackbody spectrum is

$$w = \int_0^\infty \rho(\nu)\, d\nu = \int_0^\infty \frac{8\pi\, h\nu^3}{c^3}\left(\frac{1}{\exp\left(h\nu/kT\right)} - 1\right) d\nu\,. \tag{2.26}$$

If one makes the change of variables from ν to $u = h\nu/kT$ then the total energy in the blackbody spectrum can be written as

$$w = (kT)^4 \int_0^\infty \left(\frac{8\pi\, u^3}{h^3\, c^3}\right)\left(\frac{1}{\exp(u) - 1}\right) du\,. \tag{2.27}$$

Thus as is well known the total energy density in the blackbody spectrum is proportional to T^4 and the temperature independent integral times k^4 is equal to the Stefan-Boltzmann constant.

Before leaving the subject the relation between the two quantities N_ν, the number of photons in a given mode, and n_{ph} the density of photons in a given mode is presented. The relation between these quantities is

$$N_\nu = \int n_{\text{ph}}\, dV \tag{2.28}$$

where the integration is over the volume of the cavity. The energy density in a given mode is equal to $n_{\text{ph}}\, h\nu$.

2.4 A Semi-classical Analysis of the Einstein *B* Coefficients

Section 2.3 introduced the Einstein A and B coefficients. This section presents a semi-classical calculation for an atom interacting with blackbody radiation to determine the Einstein A and B coefficients. In this analysis the atom is treated quantum mechanically, but the electromagnetic radiation field is treated classically. Consider the absorption from a lower level, l, into an upper level, u. It is assumed that the lower level has a total angular momentum J_l and a degeneracy $g_l = 2J_l + 1$ corresponding to $2J_l + 1$ states with different values of the z-component of the total angular momentum, m_l; and it is assumed that the upper level has a total angular momentum J_u and a degeneracy $g_u = 2J_u + 1$ corresponding to $2J_u + 1$ states with different values of the z-component of the total angular momentum, m_u. In the semi-classical theory of absorption one uses time dependent perturbation theory.

The Hamiltonian is taken as $H = H_0 + H_1$ where H_0 is the Hamiltonian for the atom alone and H_1 is the time dependent perturbation that results from the interaction of the electromagnetic field with the atom. Schrödinger's Equation is

$$H\,\Psi = (H_0 + H_1)\,\Psi = -\frac{h}{2\pi i}\frac{\partial \Psi}{\partial t} \tag{2.29}$$

where Ψ is the wave function for the atom. The time dependent state of the atom is expanded in terms of the stationary states of the atom in the absence of the electromagnetic field so that

$$\Psi = \sum_n a_n(t)\,\psi_n \exp\left(\frac{-i\,2\pi\,E_n t}{h}\right) \tag{2.30}$$

where $a_n(t)$ is the time dependent coefficient for the stationary state ψ_n, E_n is the energy of the stationary state labeled by n, and the summation is over all the atomic states n. It is assumed that the wave functions ψ_n are orthogonal and normalized. Using the expression for Ψ and Schrödinger's Equation yields

$$\sum_n a_n(t)\,H_0\psi_n \exp\left(\frac{-i2\pi\,E_n t}{h}\right) + \sum_n a_n(t)\,H_1\psi_n \exp\left(\frac{-i2\pi\,E_n t}{h}\right) = -\frac{h}{2\pi i}\sum_n \dot{a}_n(t)\,\psi_n \exp\left(\frac{-i2\pi\,E_n t}{h}\right) + \sum_n a_n(t)\,E_n\psi_n \exp\left(\frac{-i2\pi\,E_n t}{h}\right). \tag{2.31}$$

Since H_0 is the Hamiltonian for the atom it follows that $H_0\psi_n = E_n\psi_n$ so that the first term on the left side of the equation is identical to the second term on the right side of the equation. This leads to the result that

$$\sum_n a_n(t)\,H_1\psi_n \exp\left(\frac{-i2\pi\,E_n t}{h}\right) = -\frac{h}{2\pi i}\sum_n \dot{a}_n(t)\,\psi_n \exp\left(\frac{-i2\pi\,E_n t}{h}\right). \tag{2.32}$$

If one multiplies both sides of the above equation by $\psi_k^* \exp(i2\pi\,E_k/h)$ where the subscript k indicates a particular atomic state and integrates over all space then because the states are orthonormal one obtains

$$\dot{a}_k(t) = -\frac{2\pi i}{h}\sum_n a_n(t)\exp\left(\frac{i2\pi\,E_k t}{h}\right)\exp\left(\frac{-i2\pi\,E_n t}{h}\right)\int \psi_k^*\,H_1\,\psi_n\,dV \tag{2.33}$$

where V is the volume.

In the electric dipole approximation the classical time dependent interaction Hamiltonian, H_1, is given by $H_1 = -\boldsymbol{p}\cdot\boldsymbol{E} = -(p_x\,E_x + p_y\,E_y + p_z\,E_z)$

where $\boldsymbol{p}$ is the electric dipole moment of the atom and $\boldsymbol{E}$ is the oscillatory electric field. The electric dipole approximation is valid when the wavelength of the electromagnetic radiation is much larger than the atomic radius, i.e., when $ka \ll 1$ where $k = 2\pi/\lambda$ is the wave number and a is the atomic radius. In the electric dipole approximation the electric field is nearly constant over the atomic radius. Let us consider the term $-p_x E_x$ in the interaction Hamiltonian. The x component of the electric field in an infinitely conducting cavity is given $E_x = A \cos k_x x \sin k_y y \sin k_z z \sin(\omega t)$ where the zero of time is taken to be such that the phase factor in Equation 2.1 is $\phi = 0$. The x-component of the electric field is taken as $E_x = E_{0x}(\nu) \sin \omega t$ where the quantity $E_{0x}(\nu)$ contains the spatial variation of the electric field. It should be noted that $E_{0x}(\nu)$ is a function of the frequency (or wavelength) of the radiation since it depends on k_x, k_y, and k_z. E_{0x} is regarded as a constant over the atomic radius. In this approximation it is found that

$$\begin{aligned}
\dot{a}_k(t) &= \frac{2\pi i}{h} \sum_n a_n(t) \int \psi_k^* e^{\frac{i2\pi E_k t}{h}} H_1 \psi_n e^{\frac{-i2\pi E_n t}{h}} dV \\
&= \frac{2\pi i}{h} \sum_n a_n(t) \left(\frac{e^{i\omega t} - e^{-i\omega t}}{2i} \right) e^{\frac{i2\pi(E_k t - E_n t)}{h}} E_{0x} \int \psi_k^* p_x \psi_n \, dV \\
&= \frac{2\pi i}{h} \sum_n a_n(t) \left(\frac{e^{i\omega t} - e^{-i\omega t}}{2i} \right) e^{\frac{i2\pi(E_k t - E_n t)}{h}} E_{0x} M_{kn\,x} \qquad (2.34)
\end{aligned}$$

where the relation $\sin \omega t = [\exp(i\omega t) - \exp(-i\omega t)]/2i$ is used and where

$$M_{kn\,x} = \int \psi_k^* p_x \psi_n \, dV \qquad (2.35)$$

is the matrix element of the x component of the electric dipole moment between the final state k and the initial state n.

Thus far the analysis has been a general treatment of time dependent perturbation theory for electric dipole absorption or stimulated emission. The theory is now specialized to treat absorption. Consider the state labeled by n to be the state in the lower level, l, with magnetic quantum number m_l so that $\psi_n = \psi(l, m_l)$ where l indicates all the quantities upon which the wave function depends other than the magnetic quantum number m_l. This means that l includes the representations of the position vectors for each of the electrons in the atom when the atom is in the state ψ_n, the total angular momentum of the lower level, J_l, and all other quantities upon which the state ψ_n depends except m_l. In a similar way the state labeled k is considered to be the state in the upper level, u, with magnetic quantum number m_u so that $\psi_k = \psi(u, m_u)$ where u indicates all the quantities upon which the wave function depends other than the magnetic quantum number m_u. It is further assumed that the atom is in lower level state ψ_n at the time $t = 0$. This assumption implies that the initial conditions for the differential equations are $a_n(t=0) = 1$ and $a_k(t=0) = 0$. Integrating the differential equation with respect to the time from time $t = 0$ to $t = T$ gives

$$a_k = \frac{i\pi\, E_{0x}\, M_{kn\;x}}{h}\left(\frac{1 - e^{i(\omega_0+\omega)T}}{\omega_0 + \omega} - \frac{1 - e^{i(\omega_0-\omega)T}}{\omega_0 - \omega}\right) \tag{2.36}$$

where $\omega_0 = 2\pi(E_k - E_n)/h$. For angular frequencies in the radiation field ω near ω_0 the second term is much larger than the first term since the denominator in the second term is very small if ω is near ω_0. Since only frequencies near the atomic frequency are being considered the first term is ignored. With this approximation the coefficient a_k at the time T is given by

$$a_k = -\frac{i\pi\, E_{0x}\, M_{kn\;x}}{h}\, e^{i(\omega_0-\omega)T/2}\, \frac{\sin(\omega_0 - \omega)T/2}{(\omega_0 - \omega)/2}\,. \tag{2.37}$$

The probability, P_k, that the atom is in the state k at the time T is called the transition probability. The transition probability is equal to the absolute square of the coefficient $a_k(T)$ averaged over all frequencies. Thus the contribution to P_k due to E_{0x} is given by $\int |a_k|^2\, d\nu$ which is equal to

$$\begin{aligned}\int_0^\infty |a_k|^2\, d\nu = \int_0^\infty a_k^* \cdot a_k\, d\nu &= \frac{\pi^2\, E_{0\;x}^2\, M_{kn\;x}^2\, T^2}{h^2}\int_0^\infty \frac{\sin^2(\omega_0 - \omega)T/2}{[(\omega_0 - \omega)/2]^2}\, d\nu \\ &= \frac{\pi^2\, E_{0\;x}^2\, M_{kn\;x}^2\, T}{h^2}\end{aligned} \tag{2.38}$$

where the lower limit of integration has been extended from 0 to $-\infty$ since the function being integrated is small and oscillatory in the range from $-\infty$ to 0 and it makes a negligible contribution to the integral over that range. For blackbody radiation the x, y, and z components of the electric field are independent of one another and have the same time average. The terms $-p_y E_{0y}$ and $-p_z E_{0z}$ make similar contributions to P_k. The final result is that P_k is given by

$$\begin{aligned}P_k &= \frac{\pi^2\left(E_{0\;x}^2\, M_{kn\;x}^2 + E_{0\;y}^2\, M_{kn\;y}^2 + E_{0\;z}^2\, M_{kn\;z}^2\right)T}{h^2} \\ &= \frac{\pi^2\, E_{0\;x}^2\left(M_{kn\;x}^2 + M_{kn\;y}^2 + M_{kn\;z}^2\right)T}{h^2}\end{aligned} \tag{2.39}$$

where use is made of the fact that for blackbody radiation $E_{0x}^2 = E_{0y}^2 = E_{0z}^2$. The transition probability per unit time, P_k/T, is equal to $B_{nk}\rho(\nu)$. Thus it is only needed to express E_{0x} in terms of $\rho(\nu)$ in order to obtain B_{nk}. The energy density in an electromagnetic field is given by

$$\rho(\nu) = \overline{\frac{\epsilon_0 E^2}{2} + \frac{B^2}{2\mu_0}} = \epsilon_0 \overline{E^2} = \frac{\epsilon_0 E_0^2}{2} = \frac{3}{2}\epsilon_0 E_{0\,x}^2 \tag{2.40}$$

where the overline indicates a time average. Combining the expressions for P_k and for $\rho(\nu)$ yields

$$B_{nk} = \frac{2\pi^2}{3\epsilon_0\, h^2}\left(M_{kn\ x}^2 + M_{kn\ y}^2 + M_{kn\ z}^2\right) = \frac{2\pi^2}{3\epsilon_0\, h^2} M_{kn}^2 \tag{2.41}$$

where the absolute square of the total dipole matrix element between states k and n is denoted by $M_{kn}^2 = (M_{kn\,x}^2 + M_{kn\,y}^2 + M_{kn\,z}^2)$.

The Einstein B_{nk} is a *state to state* absorption coefficient, i.e., it is the Einstein absorption coefficient from a lower state n with magnetic quantum number m_l to the upper state k with magnetic quantum number m_u. The lower level has $g_l = 2J_l + 1$ degenerate states with magnetic quantum numbers ranging from m_l=$-J_l$ to $m_l = J_l$, and the upper level has $g_u = 2J_u + 1$ degenerate states with magnetic quantum numbers ranging from m_u=$-J_u$ to m_u=J_u. In order to obtain the Einstein absorption coefficient B_{lu} for a transition from a *lower level* l to an *upper level* u it is necessary sum over all the states with different magnetic quantum numbers in the final upper level and average over all the states with different magnetic quantum numbers in the lower initial level. This leads to the result that

$$B_{lu} = \frac{2\pi^2}{3\epsilon_0\, h^2}\frac{1}{g_l}\sum_{m_u}\sum_{m_l}\left|\int \psi^*(u, m_u)\, \boldsymbol{p}\, \psi(l, m_l)\, dV\right|^2 \tag{2.42}$$

where the wave function for the lower level, n, as $\psi(l, m_l)$ and the wave function for the upper level, k, as $\psi(u, m_u)$ are explicitly indicated. In the wave functions, as was previously discussed, the symbols l and u denote all the other quantities upon which the wave function depends except the magnetic quantum numbers. It should be noted that it is common to put the initial state wave function on the right and the final state wave function on the left in the matrix element, and it is common practice to put the initial level on the left and the final level on the right in the subscript for the Einstein B coefficient. In a similar manner

$$B_{ul} = \frac{2\pi^2}{3\epsilon_0\, h^2}\frac{1}{g_u}\sum_{m_l}\sum_{m_u}\left|\int \psi^*(l, m_l)\, \boldsymbol{p}\, \psi(u, m_u)\, dV\right|^2 = \frac{g_l}{g_u} B_{lu}\,. \tag{2.43}$$

The Einstein A_{ul} coefficient is obtained by using the relationship between the Einstein A_{ul} and B_{ul} coefficients with the result

$$A_{ul} = \frac{8\pi\, h\nu^3}{c^3}\, B_{ul} = \frac{16\pi^3\,\nu^3}{3\epsilon_0\, hc^3}\,\frac{1}{g_u}\sum_{m_l}\sum_{m_u}\left|\int \psi^*(l,m_l)\,\boldsymbol{p}\,\psi(u,m_u)\,dV\right|^2 \quad (2.44)$$

$$= \frac{4}{3}\,\frac{\omega^3}{4\pi\epsilon_0\,\frac{h}{2\pi}c^3}\,\frac{1}{g_u}\sum_{m_l}\sum_{m_u}\left|\int \psi^*(l,m_l)\,\boldsymbol{p}\,\psi(u,m_u)\,dV\right|^2 .$$

The atomic electric dipole moment is given by

$$\boldsymbol{p} = \sum_i e\,\boldsymbol{r}_i \quad (2.45)$$

where $\boldsymbol{r}_i$ is the vector position of the ith electron in the atom. It should be noted that an atom has an electric dipole moment only during a transition between two levels. An atom in a stationary state has no electric dipole moment. For a one electron atom $\boldsymbol{p} = e\,\boldsymbol{r}$. Thus for a one electron atom the Einstein B coefficient is given by

$$B_{lu} = \frac{2\pi^2\,e^2}{3\epsilon_0\,h^2}\,\frac{1}{g_l}\sum_{m_u}\sum_{m_l}\left|\int \psi^*(u,m_u)\,\boldsymbol{r}\,\psi(l,m_l)\,dV\right|^2 = \frac{g_u}{g_l}\,B_{ul}\,, \quad (2.46)$$

and the Einstein A coefficient is given by

$$A_{ul} = \frac{8\pi\, h\nu^3}{c^3}\, B_{ul} = \frac{16\pi^3\,\nu^3\,e^2}{3\epsilon_0\, hc^3}\,\frac{1}{g_u}\sum_{m_l}\sum_{m_u}\left|\int \psi^*(l,m_l)\,\boldsymbol{r}\,\psi(u,m_u)\,dV\right|^2$$

$$= \frac{4}{3}\,\frac{\omega^3\,e^2}{4\pi\epsilon_0\,\frac{h}{2\pi}c^3}\,\frac{1}{g_u}\sum_{m_l}\sum_{m_u}\left|\int \psi^*(l,m_l)\,\boldsymbol{r}\,\psi(u,m_u)\,dV\right|^2 . \quad (2.47)$$

Thus both the Einstein A and B coefficients depend on the square of the matrix element of the position vector.

Although a semi-classical treatment of the Einstein A and B coefficients in this section has been presented one should note that the concept of stimulated emission is not an intrinsically quantum mechanical one. Stimulated emission occurs in classical physics. In fact the free electron laser can be treated in an entirely classical manner.

2.5 Absorption and Stimulated Emission Cross Sections

The Einstein B coefficients can be used to calculate absorption and stimulated emission rates per atom for a system immersed in a bath of radiation with a spectral distribution that is much broader than the atomic transition line width. When atoms are interacting with a monochromatic beam of light such as a laser beam the concept of a cross section is more useful. In a typical situation a laser beam is incident on a target material. The absorption cross section for a thin target is defined as the number of photons absorbed per second divided by the product of the number of incident photons per second and the target thickness. The target thickness is defined as the line integral of the target number density along the direction of the incident beam. If the target is constant in number density then the target thickness is equal to the product of the target number density times the length of the target parallel to the beam axis. Atomic or molecular absorption or stimulated emission cross sections are usually rapidly varying functions of the laser frequency with a sharp peak at or near $\nu = \nu_0$. The absorption and stimulated emission cross sections are related to the Einstein A and B coefficients for the transition. The absorption rate per atom is given by

$$R_{lu} = B_{lu}\,\rho(\nu_0) = \frac{g_u}{g_l}\,\frac{c^3}{8\pi\,h\nu_0^3}\,A_{ul}\,\rho(\nu_0)\,. \tag{2.48}$$

This rate can also be calculated from the absorption cross section as

$$R_{lu} = \int_0^\infty \frac{\rho(\nu)\,c}{h\nu}\,\sigma_{lu}(\nu-\nu_0)\,d\nu \tag{2.49}$$

where the frequency dependence of the *absorption cross section*, $\sigma_{lu}(\nu - \nu_0)$, is shown explicitly. Equation 2.49 follows from the definition of the absorption cross section if one recognizes that $\rho(\nu)c\,d\nu/h\nu$ is equal to the number of photons per second per unit area with a frequency between ν and $\nu + \Delta\nu$ passing through the target.

If the spectral width (i.e. the width in frequency) of $\rho(\nu)/\nu$ is much greater than the spectral width of $\sigma_{lu}(\nu - \nu_0)$ then the absorption rate per atom can be written as

$$R_{lu} = \frac{\rho(\nu)\,c}{h\nu_0}\int_0^\infty \sigma_{lu}(\nu-\nu_0)\,d\nu\,. \tag{2.50}$$

Combining the expressions for R_{lu} one finds

$$\int_0^\infty \sigma_{lu}(\nu-\nu_0)\,d\nu = \frac{g_u}{g_l}\,\frac{\lambda_0^2}{8\pi}\,A_{lu} = \frac{h\nu_0}{c}\,B_{lu}\,. \tag{2.51}$$

The absorption cross section is usually written as

$$\sigma_{lu}(\nu-\nu_0) = \frac{g_u}{g_l}\,\frac{\lambda_0^2}{8\pi}\,A_{ul}\,g(\nu-\nu_0) \tag{2.52}$$

where $g(\nu - \nu_0)$ is called the *line shape* function. The line shape function is relatively narrow and is normalized so that

$$\int_0^\infty g(\nu-\nu_0)\,d\nu = 1\,. \tag{2.53}$$

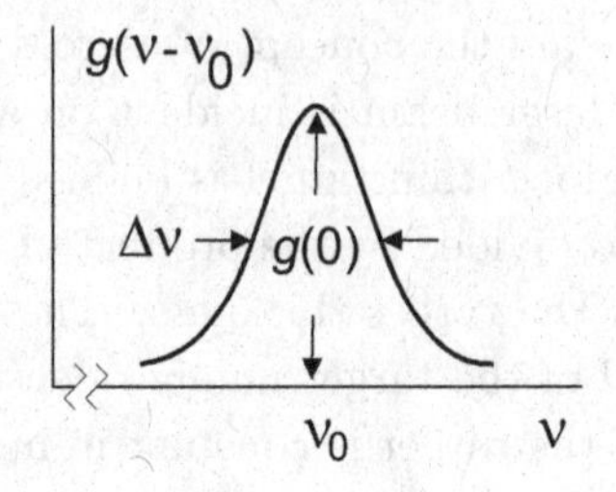

Fig. 2.2 A normalized line shape $g(\nu - \nu_0)$ as a function of the frequency ν. For a normalized line shape $g(0) \approx 1/\Delta\nu$.

The line shape function is not yet specified, but in Chapter 4 the various mechanisms that broaden a spectral line are discussed and the line shape functions for different mechanisms are presented. Figure 2.2 shows a possible line shape as a function of the frequency ν. Since there is a unit area under the line shape function it is seen that $g(0) \approx 1/\Delta\nu$ where $g(0)$ is the value of the line shape function at $\nu = \nu_0$ and $\Delta\nu$ is the full width at half maximum of $g(\nu - \nu_0)$.

The stimulated emission cross section can be derived in a similar manner yielding

$$\sigma_{ul} = \frac{\lambda_0^2}{8\pi} A_{ul}\, g(\nu - \nu_0) \; . \tag{2.54}$$

The notation g is used for both the line shape function, $g(\nu - \nu_0)$, and the statistical weights of the upper and lower levels, g_u and g_l. These particular notations are widely used, and hence they are used in this text. In order to minimize any possible confusion, however, a parenthesis and the frequency difference are used with the line shape function and a subscript with the statistical weight of a level are always used.

Before leaving this material there is one important consideration not yet discussed. In the analysis of the Einstein A and B coefficients and of the absorption and stimulated emission cross sections, as presented in this chapter, the approximation that the index of refraction is one has been used. For a laser this is usually not a good approximation. The index of refraction of a material varies rapidly near an atomic transition if either the upper or lower level of that transition is populated. This leads to many interesting effects. Among these effects is the result that a laser operates at a frequency that is "pulled" away from the cavity frequency in the absence of the excited laser medium and toward the atomic transition frequency of the laser. This effect and other effects due to the variation of the index of refraction from one are discussed in later chapters.

2.6 Classical Treatment of the Absorption Cross Section

The derivation of the absorption cross section in the previous section utilized Einstein's analysis of stimulated emission and the relationship between the Einstein A and B coefficients. There is an alternative derivation of the absorption cross section based on a classical model of an atom as an electron with charge, $-e$, and mass, m, which is bound to a massive nucleus by a simple harmonic potential with a spring constant k and a damping constant γ. As will be seen this classical analysis provides a very satisfactory treatment of the absorption cross section. This may seem surprising at first since the true force between an electron and the nucleus is the Coulomb force rather than a simple harmonic spring force. The reason that the classical analysis is successful is discussed in the final section of this chapter.

This classical system obeys an equation of motion

$$m\frac{d^2x}{dt^2} + m\gamma\frac{dx}{dt} + kx = -e\text{Re}(E_0 e^{i\omega t}) \tag{2.55}$$

when the charge $-e$ is bound to the atom and the atom is in the electric field of an electromagnetic wave with angular frequency ω. It is assumed that the electromagnetic wave is polarized in the x direction. It is also assumed that the force due to the wave's magnetic field can be ignored. As in Chapter 1 a complex phasor notation is used. The notation Re indicates that one must take the real part of the complex function to obtain the real function. The constant E_0 may be complex. This allows the electric field to have an arbitrary phase. The electric field is taken arbitrarily to be along the x axis. Since the electric field and the motion of the electron are along the x axis we treat both the electric field and the position of the electron as scalars rather than vectors. The steady state sinusoidal solution to Equation 2.55 is

$$x = \text{Re}(x_0 e^{i\omega t}) \tag{2.56}$$

where x_0, which is complex, is given by

$$x_0 = \frac{-\frac{e}{m}E_0}{\frac{k}{m} - \omega^2 + i\omega\gamma} . \tag{2.57}$$

The oscillating induced dipole moment of the atom is given by $-ex$. The steady state "velocity" of the electron is given by

$$\frac{dx}{dt} = \text{Re}(i\omega\, x_0\, e^{i\omega t}) . \tag{2.58}$$

The power, P, absorbed by the bound electron is calculated by taking the time average over a period of the product of velocity times the force yielding

$$P = \frac{1}{T_0}\int_0^{T_0} \mathrm{Re}(i\omega\, x_0\, e^{i\omega t})\,\mathrm{Re}(-e\, E_0\, e^{i\omega t})\, dt$$
$$= -\frac{e}{2}\mathrm{Re}(i\omega\, x_0\, E_0^*) \tag{2.59}$$

where E_0^* is the complex conjugate of E_0. Combining Equations 2.57 and 2.59 one obtains

$$P = \frac{e^2}{2m}\left|E_0\right|^2 \frac{\omega^2\,\gamma}{\left(\omega^2-\omega_0\right)^2+\omega^2\,\gamma^2} \tag{2.60}$$

where $\omega_0^2 = k/m$. In this classical analysis of atomic absorption, where it is assumed that the atom can be represented as an electron bound by a one dimensional simple harmonic potential, the resonant frequency, ω_0, of the simple harmonic oscillator is taken as corresponding to the angular frequency of the transition from level l to level u. The incident power in the electromagnetic wave is $c\,\epsilon_0\,|E_0|^2/2$, and the power absorbed from the wave by the atom during the transition from level l to level u is in this model given by

$$P_{lu} = \frac{c\,\epsilon_0}{2}\left|E_0\right|^2 \sigma_{lu}\,. \tag{2.61}$$

Combining Equations 2.60 and 2.61 yields

$$\sigma_{lu} = \frac{e^2}{m\,c\,\epsilon_0}\,\frac{\omega^2\,\gamma}{(\omega^2-\omega_0^2)+\omega^2\,\gamma^2}\,. \tag{2.62}$$

If the resonance is narrow then the absorption cross section is large only near line center where $\omega \approx \omega_0$. Near line center where $\omega \approx \omega_0$ it is a good approximation to take $\omega + \omega_0 \approx 2\omega$ and $(\omega^2-\omega_0^2)^2 = 4\omega^2(\omega-\omega_0)^2$. With this approximation the absorption cross section can be written as

$$\sigma_{lu} = \pi\left(\frac{e^2}{4\pi\,\epsilon_0\,mc^2}\right) c\,\frac{\gamma}{\left(\omega-\omega_0\right)^2+\left(\frac{\gamma}{2}\right)^2}$$
$$= \pi\, r_e\, c\, g(\nu-\nu_0) \tag{2.63}$$

where the classical radius of the electron is

$$r_e = \frac{e^2}{4\pi\,\epsilon_0\,mc^2} = 2.82\times10^{-13}\ \mathrm{cm} \tag{2.64}$$

and where the normalized line shape function is

$$g(\nu-\nu_0) = \frac{(\gamma/4\pi^2)}{\left(\nu-\nu_0\right)^2+(\gamma/4\pi)^2} = \frac{\Delta\nu/2\pi}{\left(\nu-\nu_0\right)^2+(\Delta\nu/2)^2} \tag{2.65}$$

where $\Delta\nu = \gamma/2\pi$ is the full width at half maximum of the line shape. This is called a Lorentzian line shape.

It is customary to preserve the simple expression for the absorption cross section using it for atoms or other quantum mechanical systems together with a correction factor called the oscillator strength, f_{lu}. The oscillator strength incorporates the quantum mechanical effects into a dimensionless constant. The quantum mechanical absorption and stimulated emission cross sections are written respectively as

$$\sigma_{lu} = \pi\, r_e\, c\, f_{lu}\, g(\nu - \nu_0) \tag{2.66}$$

and

$$\sigma_{ul} = \pi\, r_e\, c\, f_{ul}\, g(\nu - \nu_0)\,. \tag{2.67}$$

The statistical weights g and oscillator strengths f are related by

$$g_u\, f_{ul} = g_l\, f_{lu} = g\, f\,. \tag{2.68}$$

The values of gf or $\log(gf)$ are the quantities typically reported in compilations of atomic data. In general f values are of the order of unity for strongly allowed transitions and are much smaller for forbidden transitions. It can be shown that

$$A_{ul} = 2\,\omega_0^2\,\frac{r_e}{c}\, f_{ul}\,. \tag{2.69}$$

In the expressions given in Equations 2.52 and 2.54 for the absorption and stimulated emission cross sections the line shape function is always a normalized function of the frequency, but the line shape does not need to be the Lorentzian line shape calculated in the classical model. The variety of line shapes that can occur are discussed in Chapter 4.

A numerical calculation may help make the discussion of the relationships between oscillator strengths, Einstein A coefficients, and the absorption cross sections more concrete. For the $6^2\mathrm{S}_{1/2} \rightarrow 6^2\mathrm{P}_{1/2}$ transition in Cs the wavelength is $\lambda = 894$ nm and the lifetime of the $6^2\mathrm{P}_{1/2}$ level is $\tau_u = 3.8\times 10^{-8}$ s. The absorption oscillator strength of the transition can be found from $f_{lu} = (\lambda^2 A_{ul})/(8\pi^2 r_e c) = (\lambda^2)/(8\pi^2 r_e\, c\,\tau_u) = 0.32$. The absorption cross section at line center is given by $\sigma_{lu} = (\lambda^2 A_{ul}/8\pi) g(0) \approx (\lambda^2 A_{ul}/8\pi)(1/\Delta\nu)$. If the line width for the transition is $\Delta\nu = 10^9$ Hz then the absorption cross section at line center is $\sigma_{lu} = 8.4\times 10^{-12}$ cm^2.

A model for spontaneous emission that is similar to the model for absorption can be developed. Even in the absence of a driving force an electron bound by a harmonic potential will oscillate about the center of force thereby producing an oscillating dipole moment. Since the electron is accelerated it radiates at the frequency of oscillation. The radiation carries away energy and thereby damps the oscillator. The damping constant γ that results from the radiation is interpreted as being due to the spontaneous radiation. As the energy of excitation is radiated the amplitude of oscillation decreases leading to an exponential decrease of the radiated power with the time. This results in a Lorentzian line shape for the emitted radiation. The spontaneous emission line shape has the same line width and center frequency as the line shape for absorption. In general the rate of spontaneous emission per atom (or molecule) from an excited level u to a lower level l and into modes with frequencies between $\nu - d\nu$ and $\nu + d\nu$ is taken as $A_{ul}\, g(\nu - \nu_0)d\nu$ where $g(\nu - \nu_0)$ is the normalized line shape. Integrating over all frequencies gives the total rate of spontaneous emission per atom for the transition as A_{ul}, which is generally correct even for situations where the line shape is not the Lorentzian line shape of the classical model. In the discussion of absorption the equation of motion was taken as that of a damped harmonic oscillator with a damping constant γ. The general damping constant can include damping effects other than those due to spontaneous emission as well as the damping due to spontaneous emission. Since the damping constant is partly due to spontaneous emission one can regard spontaneous emission as being included in the analysis of absorption. Thus both absorption and spontaneous emission can be understood using the same damped harmonic oscillator model, the only difference being that the driving force is zero for the analysis of spontaneous emission.

The discussion of spontaneous radiation just presented assumes the harmonically bound electron is excited and decays in the absence of driving forces. If the electron is driven by a continuous and monochromatic light wave with angular frequency ω then the response of the electron in the steady state is a sinusoidal motion with angular frequency ω. An electron moving in a sinusoidal motion with angular frequency ω is accelerated, and hence it must radiate a monochromatic electromagnetic wave with angular frequency ω. This corresponds to the scattering of the electromagnetic wave. Thus in addition to understanding both absorption and spontaneous emission using the simple classical model of an electron bound to a nucleus by a harmonic oscillator potential one can also understand the scattering of an electromagnetic wave using this model.

2.7 Why the Classical Spring Model of an Atom Works

In section 2.6 of this chapter a classical analysis of the absorption cross section, which used as a model of an atom an electron bound to a massive nucleus by a spring, was shown to be very successful. Since the actual force between the electron and the nucleus is the Coulomb force it is, at first sight, surprising that the model works so well. This section discusses why the spring model of the atom works. In order to make the discussion concrete, consider atomic hydrogen. When the electric field in an electromagnetic wave interacts with a hydrogen atom the atomic electron cloud is displaced anti-parallel to the electric field and the nucleus is displaced parallel to the electric field so that an atomic electric dipole moment is formed. If the electric field is moderate in magnitude then the nucleus is displaced by a length, r, that is only a small fraction of the radius of the atomic electron cloud. In this situation that part of the electric field at the nucleus due to the electron cloud is produced only by the part of the electron cloud in a sphere whose radius is equal to the displacement of the nucleus from the center of the electron cloud. When the center of the electron cloud and the nucleus are separated by a vector $\boldsymbol{r}$, the electric field, $\boldsymbol{E}$, at the nucleus due to the atomic electron cloud can be calculated by the use of Gauss's law applied to a sphere of radius r at the center of the electron cloud, $4\pi r^2 E = (\rho/\epsilon_0)(4/3)\pi r^3$. The final result for the electric field is $\boldsymbol{E} = (\rho/3\epsilon_0)\boldsymbol{r}$ where ρ is the charge density in the atom and where $\boldsymbol{r}$ is directed from the nucleus toward the center of the electron cloud. The force on the nucleus due to the electron is given by $\boldsymbol{F} = e\boldsymbol{E} = (\rho e/3\epsilon_0)\boldsymbol{r}$ where $+e$ is the nuclear charge The force on the electron due to the nucleus, which is equal and opposite to the force on the nucleus due to the electron, is $\boldsymbol{F} = -e\boldsymbol{E} = -(\rho e/3\epsilon_0)\boldsymbol{r}$. The restoring force is therefore a spring force with a spring constant $k = \rho e/3\epsilon_0$. Thus it has been shown that for the small displacements of an atomic electron cloud in a moderate electric field that the spring model is appropriate.

It now remains to investigate how the resonant spring frequency is related to the atomic frequencies. The spring constant divided by the electron mass, m_e, is equal to square of the resonant angular frequency. Thus $\omega^2 = \rho e/3m_e\epsilon_0$. The charge density in an atom is approximately given by $\rho = e/(4/3)\pi a^3$ where a is the radius of the atom. Thus the $\omega^2 = e^2/4\pi\epsilon_0 m_e a^3$. The radius of the atom is taken as $a = \zeta a_0$ where a_0 is the Bohr radius and ζ is a constant. The Bohr radius of a hydrogen atom is given by $a_0 = 4\pi\epsilon_0/m_e e^2 = 0.051$ nm. One must now estimate what is a reasonable value of ζ. In the n=1 level of atomic hydrogen the atomic radius of a hydrogen atom is $a = a_0$ and in the n=2 level $a = 4a_0$. When a ground level hydrogen atom is placed into an electric field, the atom becomes polarized and has an electric dipole moment. An atom in an atomic eigen state has no dipole moment. In order for an atom to have an electric dipole moment the atom must be in a state that is a linear superposition of two eigen states of opposite parity such as the 1S and 2P states of atomic hydrogen. Thus one might reasonably take ζ=2, a value between the values of a for the 1S and 2P states. With this value of ζ it is found that $\omega^2 = e^2/4\pi\epsilon_0 m_e a^3 = (m_e 2e^8)/[8(4\pi\epsilon_0)^4\hbar^6]$. This value of ω^2 is nearly equal to the value of the square of the energy difference between the 2P and 1S levels of atomic hydrogen divided by $\hbar$ i.e. $\omega \approx (E_{2P} - E_{1S})/\hbar$. For this reason one expects the spring model for an atom to work well provided one associates the resonant spring frequency with the atomic resonant frequency. Of course one does not expect the spring model of the atom to work perfectly and hence the oscillator strength is introduced as a constant that enables one to correct the absorption cross section obtained using the classical spring model of the atom to yield the true cross section. In this treatment the very small parity violations in atomic eigenstates have been ignored.

The classical spring model of the atom is very useful in a wide variety of calculations for things such as the frequency dependence of the index of refraction of a gas and frequency dependence of many other properties of matter.

Summary of Key Ideas

- A reflecting cavity such as a laser cavity has discrete modes. The blackbody energy density, Eq. 2.15, was calculated by both Planck and Einstein. The number of modes p of a cavity within the $\Delta\nu$ bandwidth is equal to

$$p = \frac{8\pi\,\nu^2}{c^3}\,\Delta\nu\,V \qquad (2.9)$$

- The relations between spontaneous emission (A_{ul}), stimulated emission (B_{ul}) and absorption (B_{lu})are:

$$g_l\,B_{lu} = g_u\,B_{ul} \qquad (2.22) \qquad\qquad \frac{A_{ul}}{B_{ul}} = \frac{8\pi\,h\nu_0^3}{c^3} \qquad (2.23)$$

- The stimulated emission cross section is (Eqs. 2.54 and 2.67)
$$\sigma_{ul}(\nu-\nu_0) \;=\; \frac{\lambda_0^2}{8\pi}\,A_{ul}\,g(\nu-\nu_0) \;=\; \pi\,r_e\,c\,f_{ul}\,g(\nu-\nu_0)$$
where $g(\nu-\nu_0)$ is the normalized line shape and f_{ul} is the oscillator strength. Similarly, Eqs. 2.52 and 2.66 give the absorption cross section:
$$\sigma_{lu}(\nu-\nu_0) \;=\; \frac{g_u}{g_l}\,\frac{\lambda_0^2}{8\pi}\,A_{ul}\,g(\nu-\nu_0) \;=\; \pi\,r_e\,c\,f_{lu}\,g(\nu-\nu_0)$$
- Oscillator strengths and Einstein coefficients are linked by
$$A_{ul} = 2\,\omega_0^2\,\frac{r_e}{c}\,f_{ul} \qquad (2.69)$$
while emission and absorption oscillator strengths are related by
$$g_u\,f_{ul} = g_l\,f_{lu} = g\,f \qquad (2.68)$$
where g_u and g_l are statistical weights. For strongly allowed transitions f values are on the order of one.

Suggested Additional Reading

Max Planck, *The Theory of Heat Radiation* [M. Masius (transl.) (2nd Ed.)], P. Blakiston's Son & Co. (1914).

Max Planck, *Acht Vorlesungen uber Theoretische Physik*, Verlag von S. Hirzel (1910).

A. Einstein, "Zur Quantentheorie der Strahlung [On the Quantum Theory of Radiation]" *Phys. Z.* **18**, 121 (1917).

Donald C. O'Shea, W. Russell Callen and William T. Rhodes, *An Introduction to Lasers and their Applications*, Addison Wesley Publishing Co. (1977).

William T. Silfvast, *Laser Fundamentals*, Cambridge University Press (1996).

Linus Pauling and E. Bright Wilson, *Introduction to Quantum Mechanics*, McGraw Hill (1935).

L. I. Schiff, *Quantum Mechanics*, North Holland (1968).

A. E. Siegman, *An Introduction to Lasers and Masers*, McGraw Hill (1971).

A. E. Siegman, *Lasers*, University Science Books (1986).

A. E. Ruark and H. C. Urey, *Atoms Molecules and Quanta*, McGraw Hill (1930).

Problems

1. (a) For a perfectly conducting cubic cavity with $L = 20$ cm what is the fractional change in the frequency when N_x is changed by one if the wavelength is near 500 nm assuming that N_y and N_z are zero? (b) Assuming that $N_x = N_y = N_z$?

2. What is the blackbody occupation number i.e. the average number of photons per mode for $\lambda = 500$ nm at T=5000 K? For T=1000 K? For T=100 K?

3. What is the average energy per mode for the conditions given in Problem 2?

4. What is the energy density per unit frequency in the blackbody radiation for the conditions given in Problem 2?

5. Estimate the number of modes per unit volume within a line of width 2×10^9 Hz if the line has a wavelength of 600 nm.

6. For the atomic transition of Problem 5 estimate the spontaneous decay rate into a single mode if $A_{ul} = 3\times10^8$ s^{-1} if the volume of the laser cavity is 10 cm^3.

7. An atomic transition has λ_0=500 nm and A_{ul}=10^8 s^{-1}. The full width of the line at half maximum is equal to 10^9 Hz. Estimate the stimulated emission cross section at line center.

8. If the transition of Problem 7 is between an upper level with J_u=1 and a lower level with J_l=0 estimate the absorption cross section of the transition at line center.

9. An atomic transition has $g_u = g_l = 1$ and f_{lu}=0.3. If the line width of the transition is 5×10^8 Hz estimate the absorption and stimulated emission cross sections at line center.

10. For the transition of Problem 9 the wavelength is λ_0=800 nm. What is the Einstein A_{ul} for the transition?

Chapter 3

The Criterion for Laser Action

In Chapter 2 the relationships between the Einstein A and B coefficients were derived. In addition the absorption and stimulated emission cross sections were derived. In this chapter use is made of these concepts to analyze the gain or loss of a light beam as it passes through a material, and the criterion for laser action is determined.

3.1 The Laser Oscillation Criterion

In order to obtain the laser criterion consider a light beam traveling through a material. This beam will normally be absorbed. If, however, a sufficiently large population inversion exists the beam may increase in intensity. In this section the gain or loss of the intensity for a light beam as it passes through a material is analyzed, and the condition for laser oscillation is obtained.

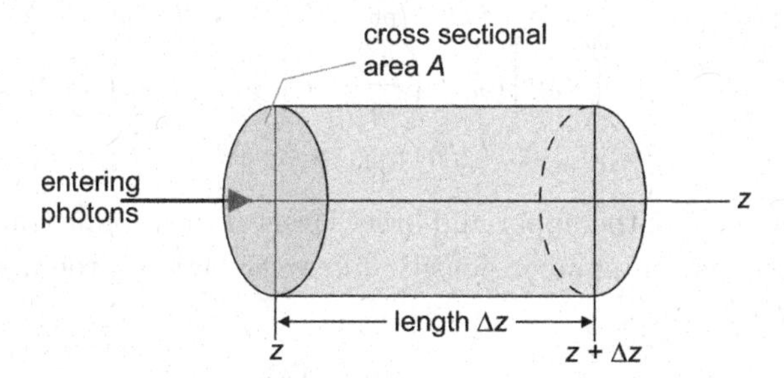

Fig. 3.1 Light enters a sample volume of material with length Δz, cross sectional area A, and index of refraction n. The material consists of a mix of atoms in both the lower level l and upper level u.

Consider the rate of change of the photon number density in a volume that has the shape of a right circular cylinder as shown in Figure 3.1. The axis of the cylinder is collinear with the z axis, and the cylinder has a cross sectional area A and lies between z and $z + \Delta z$ so that the cylinder has a length Δz. A photon beam with a particular frequency, phase, and wave vector (i.e. a beam in a particular mode) is propagating along the z axis. The differential equation for the photon number density in the light beam as a function of the position z is obtained by considering the change in the photon number density in the cylindrical volume. The photon number density in the cylindrical volume can change from photons entering or leaving the cylinder, from photons being created or lost in the cylinder by stimulated emission or absorption on an atomic transition with a frequency near the frequency of the light beam, or from photons in the cylinder being lost from the mode due to other loss mechanisms such as scattering. The net number of photons that exit from the cylindrical volume in a time t is given by the number of photons that leave the end of the cylinder at $z + \Delta z$ in a time t minus the number that enter the cylinder at z in a time t. The net number of photons leaving the cylindrical volume in a time Δt is therefore given by

$$[n_{\text{ph}}(z + \Delta z) - n_{\text{ph}}(z)]\, A\,(c/n)\,\Delta t = \frac{\partial n_{\text{ph}}}{\partial z}\,(c/n)\,A\,\Delta z\,\Delta t\,. \tag{3.1}$$

The net number of photons that are created in the cylindrical volume in a time Δt is equal to the number created by stimulated emission minus the number absorbed. The net number of photons created in the cylindrical volume in a time Δt is therefore given

$$\begin{aligned}
\Delta n_{\text{ph creation}} &= [\sigma_{ul}\, n_u\,(c/n)\,n_{\text{ph}} - \sigma_{lu}\, n_l\,(c/n)\,n_{\text{ph}}]\; A\,\Delta z\,\Delta t \\
&= \left(\sigma_{ul}\left[n_u - \left(\frac{g_u}{g_l}\right) n_l\right](c/n)\,n_{\text{ph}}\right) A\,\Delta z\,\Delta t \\
&= [\sigma_{ul}\,\Delta n\,(c/n)\,n_{\text{ph}}]\; A\,\Delta z\,\Delta t
\end{aligned} \tag{3.2}$$

where u and l denote the upper and lower levels of the atomic transition and where the population number density difference between the upper level u and the lower level l is

$$\Delta n = n_u - \left(\frac{g_u}{g_l}\right) n_l\,. \tag{3.3}$$

In this expression for the net number of photons created in the cylindrical volume in a time Δt spontaneous emission into the particular mode is ignored since the rate for spontaneous emission into a particular mode is usually very small. In other words absorption followed by spontaneous emission results in almost all the spontaneously emitted photons going into modes other than the particular mode being considered and hence being lost from that mode. Finally the number of photons lost from the cylindrical volume in the time Δt due to other loss mechanisms is given by

$$\Delta n_{\text{ph loss}} = \left(\frac{n_{\text{ph}}}{\tau_{\text{ph}}}\right) A\,\Delta z\,\Delta t \tag{3.4}$$

where τ_{ph} is a phenomenological constant called the *photon lifetime.* The photon lifetime represents the $1/e$ lifetime of a photon in the mode due to losses from all processes *except* absorption on the atomic transition. The total increase in the number of photons in the cylindrical volume, $\Delta n_{\text{ph}}\,A\Delta z$, is equal to the net number of photons created in the cylinder minus the sum of net number of photons leaving the cylinder and the number of photons lost from the cylinder by other processes. This leads to the result that the total rate of change in the photon number density is given by

$$\frac{dn_{\text{ph}}}{dt} = -\,(c/n)\,\frac{\partial n_{\text{ph}}}{\partial z} + \sigma_{ul}\,\Delta n\,(c/n)\,n_{\text{ph}} - \frac{n_{\text{ph}}}{\tau_{\text{ph}}}\,. \tag{3.5}$$

In the steady state $dn_{\text{ph}}/dt = 0$ so that the fractional change in the photon number density per unit length is given by

$$-\,\frac{1}{n_{\text{ph}}}\,\frac{\partial n_{\text{ph}}}{\partial z} = -\sigma_{ul}\,\Delta n + \frac{1}{(c/n)\,\tau_{\text{ph}}}\,. \tag{3.6}$$

It should be noted that the photon number density at a given position z can be constant in time so that $dn_{\text{ph}}/dt = 0$ at every position z even though the light beam increases or decreases as a function of z.

The fractional change in the photon number density per unit length is equivalent to the fractional change in the intensity of the light beam per unit length since the intensity of the light beam is related to the photon number density by $I_\nu = h\nu\,n_{\text{ph}}\,c$. Thus one finds that the fractional change in intensity per unit length for a light beam in a single mode as it passes through a material is expressed mathematically as

$$\begin{aligned} -\,\frac{1}{I_\nu}\,\frac{dI_\nu}{dz} &= -\,\frac{1}{n_{\text{ph}}}\,\frac{\partial n_{\text{ph}}}{\partial z} = -\sigma_{ul}\,\Delta n + \frac{1}{(c/n)\,\tau_{\text{ph}}} \\ &= \sigma_{lu}\,n_l - \sigma_{ul}\,n_u + \frac{1}{(c/n)\,\tau_{\text{ph}}}\,. \end{aligned} \tag{3.7}$$

In this expression, as was mentioned earlier, the fractional change of the intensity of the light beam per unit length due to the spontaneous emission of a photons into the mode is ignored since this is usually a very small effect. For a laser, however, the effect of spontaneous emission into the laser mode can, even though small, have important ramifications in some situations. Consequently spontaneous emission into a laser mode is treated in Chapter 6. If the population densities n_l and n_u are independent of the position z then solution for I_ν is

$$I_\nu = I_{\nu 0} \exp(-\alpha z) \tag{3.8}$$

where $I_{\nu 0}$ is the intensity of the light beam at $z = 0$. The value of α is

$$\begin{aligned} \alpha &= \sigma_{lu}\, n_l - \sigma_{ul}\, n_u + \frac{1}{(c/n)\, \tau_{\text{ph}}} \\ &= \sigma_{ul} \left(\frac{g_u}{g_l}\, n_l - n_u \right) + \frac{1}{(c/n)\, \tau_{\text{ph}}} = -\sigma_{ul}\, \Delta n + \frac{1}{(c/n)\, \tau_{\text{ph}}} \end{aligned} \tag{3.9}$$

where α is called the *absorption coefficient* if α is positive or the *gain coefficient* if α is negative.

The light beam's intensity decreases exponentially as the beam passes through the material if α is positive and increases exponentially if α is less than zero. In order for α to be positive so that the light intensity decreases exponentially, Δn must be less than $1/[\sigma_{ul}(c/n)\tau_{\text{ph}}]$. In order for α to be negative so that the light intensity increases exponentially, Δn must be greater than $1/[\sigma_{ul}(c/n)\tau_{\text{ph}}]$.

If all the atoms or molecules in the material are in the ground level then n_u=0 and Δn is negative. If Δn is negative then obviously Δn is less than $1/[\sigma_{ul}(c/n)\tau_{\text{ph}}]$ which is positive. Thus if n_u=0 then α is positive. In this situation where all the atom or molecules in the material are in their ground level the light beam intensity decreases exponentially. This exponential decrease of the intensity of a light beam as it passes through a material where all the atoms are in the ground level is called Beer's law.

The intensity of the light beam is constant as a function of the position z if α=0. In this situation one photon is stimulated into the beam on the u to l transition for every photon that is lost from the beam by absorption on the l to u transition or is lost due to other processes. In order for α to be equal to zero there must be a population inversion. For constant intensity cw laser operation with α=0 the population number density difference is given by

$$\begin{aligned} \Delta n_{\text{th}} &= \left(n_u - \frac{g_u}{g_l}\, n_l \right)_{\text{th}} \\ &= \frac{1}{\sigma_u l\, (c/n)\, \tau_{\text{ph}}} = \frac{8\pi}{\lambda^2 A_{ul}\, g(\nu - \nu_0)\, (c/n)\, \tau_{\text{ph}}}\,. \end{aligned} \tag{3.10}$$

The population number density difference needed for constant intensity cw operation is called the threshold population number density difference and the subscript "th" is used to denote the threshold population number density difference. The threshold population number density difference depends on $\nu - \nu_0$, the difference between the frequency of the light beam and the line center for the atomic transition of the material. Typically the normalized line shape function, $g(\nu - \nu_0)$, has a maximum value of $g(0)$ at line center. The full width of the line at half maximum is denoted by $\Delta\nu$. Since $g(\nu - \nu_0)$ is normalized

$$1 = \int_0^\infty g(\nu - \nu_0)\, d\nu \approx g(0)/\Delta\nu. \tag{3.11}$$

Thus the peak value of $g(\nu - \nu_0)$ is approximately given by $g(0) \approx 1/\Delta\nu$. The minimum value of the threshold population number density difference occurs at line center and is given by

$$\Delta n_{\text{th}} = \frac{8\pi}{\lambda_0^2} \frac{\Delta\nu}{A_{ul}\,(c/n)\,\tau_{\text{ph}}} = \frac{8\pi\,\nu_0^2\,\Delta\nu}{(c/n)^3} \frac{1}{A_{ul}\,\tau_{\text{ph}}}\,. \tag{3.12}$$

The discussion of radiation has focused on an upper level labeled u and a lower level labeled l. The spontaneous transition probability from level u to level l is A_{ul}. Level u may be able to decay to other lower levels as well as level to l. The total transition probability out of the upper level u is given by $A_u = \sum_j A_{uj}$ where A_{uj} is the Einstein A coefficient from upper level u to a lower level j and where the summation is over all levels j (including l) of the atom lower than level u. The spontaneous radiative lifetime of level u is given by $\tau_u^{-1} = \sum_j A_{uj}$. *If* A_{ul} is much larger than the sum of all the Einstein A's from level u to all other lower levels so that $\sum_j A_{uj} \approx A_{ul}$ then

$$\tau_u^{-1} = \sum_j A_{uj} \approx A_{ul}\,. \tag{3.13}$$

It is clear that if l is the ground level and if level u is the lowest excited level above the ground level then $\tau_u^{-1} = A_{ul}$ and there are many other situations where an excited level decays primarily to a single lower level so that $\tau_u^{-1} = A_{ul}$. If $\tau_u = A_{ul}^{-1}$ then the threshold population number density difference is given by

$$\Delta n_{\text{th}} = \frac{8\pi\,\nu_0^2\,\Delta\nu}{(c/n)^3} \frac{\tau_u}{\tau_{\text{ph}}} = -\frac{8\pi\,\Delta\lambda}{\lambda_0^4} \frac{\tau_u}{\tau_{\text{ph}}} \tag{3.14}$$

where $\Delta\lambda = -(c/n)\Delta\nu/\nu^2$ is the line width expressed in terms of wavelength rather than frequency.

From expression for Δn_{th} it can be seen that for the following conditions a small value of the threshold population density difference is obtained:

(i) If $\Delta\lambda$ is small then Δn_{th} is small.
(ii) If τ_{ph} is long then Δn_{th} is small. One of the important functions of the highly reflective mirrors used for most laser cavities is to assure that the photons in the cavity bounce repeatedly between the mirrors before they escape from the cavity so that the photon lifetime is long.
(iii) If the spontaneous radiative lifetime of the upper level is short then Δn_{th} is small.
(iv) If the wavelength of the laser is long then Δn_{th} is small.

A transition with a long wavelength usually has a long radiative lifetime since the Einstein A coefficient varies as ν^3. Hence it may be difficult to find a transition for which conditions (iii) and (iv) are both satisfied.

3.2 Example: Threshold Population Difference for a KrF Laser

A numerical calculation of the threshold population difference density for a particular laser may aid in understanding the subjects covered in this Chapter. A KrF excimer laser is analyzed. An excimer laser operates on a transition between a bound upper level of a molecule and an unbound lower level. The molecular potential energy curves for KrF as a function of the internuclear separation, R, are shown in Figure 3.2.

The KrF excimer laser can operate as follows. A gas mixture containing argon (or helium or neon), krypton, and nitrogen trifluoride (or molecular fluorine) is excited using an intense pulsed electron beam or an intense pulsed glow discharge. The optical cavity can be very lossy (i.e. the photon

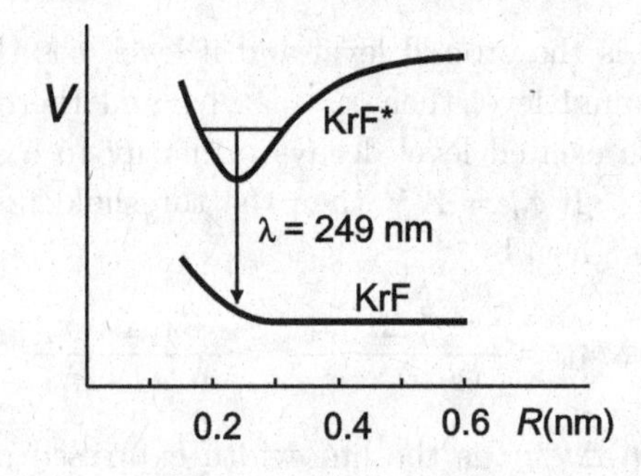

Fig. 3.2 The potential curves for KrF* and KrF as functions of the internuclear distance. The KrF laser is an example of an excimer laser.

lifetime can be very short) since the excimer laser has a very high gain. The cavity might consist of one mirror that has a reflectivity near one and a second mirror that has a reflectivity of about one half. With these reflectivities a photon has a probability of 0.5 of escaping from the cavity for each round trip in the cavity. This means that after two round trips a photon has only a 25% probability of remaining in the cavity. The photon lifetime is approximately equal to the time for two round trips in the cavity. In Section 3.3.2 of this chapter a rigorous expression for the photon lifetime in a cavity is obtained. For the KrF excimer laser analysis consider an active medium (the excited gas) that is 0.5 m long and has a cross sectional area of 2 $\mathrm{cm}^2 = 2\times10^{-4}$ m^2. For the laser length of 0.5 m, with n=1 and τ_{ph} equal to the time required for two round trips then $c\tau_{\mathrm{ph}}$=2 m and τ_{ph}=7 ns.

For excitation with a pulsed electron beam suitable proportions for the gas mixture and total pressure are Ar:Kr:$\mathrm{NF_3}$ = 1300:130:1 and 2.25 atmospheres respectively. The population inversion is produced by a complicated series of chemical reactions initiated by the pulsed electron beam excitation of the gas mixture. With simplification two of the important reaction pathways that produce the KrF excimer molecules are indicated below. The reactions for the first pathway are represented as follows:

1. $\mathrm{Kr^+}$ is produced by the reactions
   i. $\mathrm{e^-} + \mathrm{Kr} \rightarrow \mathrm{Kr^+} + 2\mathrm{e^-}$
   or
   ii. $\mathrm{e^-} + \mathrm{Ar} \rightarrow \mathrm{Ar^+} + 2\mathrm{e^-}$
   $\mathrm{Ar^+} + \mathrm{Kr} \rightarrow \mathrm{Kr^+} + \mathrm{Ar}$
2. $\mathrm{F^-}$ is produced by the reaction
   $\mathrm{e^-} + \mathrm{NF_3} \rightarrow \mathrm{F^-} + \mathrm{NF_2}$
3. The excited KrF* molecule is produced by the reaction
   $\mathrm{Kr^+} + \mathrm{F^-} + X \rightarrow \mathrm{KrF^*} + X$

where X is a third body which is an Ar atom in most cases, and where the asterisk (*) is used to indicate an excited molecule or atom. The third body is required for the conservation of energy and momentum so that the recombination of $\mathrm{Kr^+}$ and $\mathrm{F^-}$ to form KrF* can occur. The reactions for the second or alternate reaction pathway for producing the KrF excimers are represented (after the $\mathrm{Kr^+}$ ions are formed) as follows:

1. Kr* is produced by the reactions
   i. $\mathrm{Kr^+} + \mathrm{Kr} + X \rightarrow \mathrm{Kr_2^+} + X$
   ii. $\mathrm{e^-} + \mathrm{Kr_2^+} \rightarrow \mathrm{Kr^*} + \mathrm{Kr}$
2. The excited KrF* molecule is produced by the reaction
   $\mathrm{Kr^*} + \mathrm{NF_3} \rightarrow \mathrm{KrF^*} + \mathrm{NF_2}$

The principal function of the Ar is to provide a high enough atomic number density that the electron beam energy is primarily deposited in the gas thereby producing Ar^+. It is not possible to use higher pressure Kr because the formation of polymers such as Kr_2F reduces the concentration of KrF^*. There are also other important reaction pathways that occur in the KrF excimer laser so that the discussion has been much simplified.

The KrF molecular electronic potential curves for the excited level and the ground level are shown schematically as functions of the internuclear separation in Figure 3.2. The KrF^* potential corresponds to Kr^+ and F^- ions in the separated atom limit, and the KrF potential curve corresponds to ground level Kr and F atoms in the separated atom limit. The excited KrF^* molecular potential has a bound vibrational level, whereas the ground potential is repulsive and has no bound vibrational levels. Because the ground potential is repulsive the ground level population density is always near zero since the KrF^* molecule in the ground electronic level dissociates rapidly. The excited level involved in the laser action has a vibrational quantum number v=0. The radiation from the excited level has a wavelength centered at λ=249 nm, and a line width of about 4 nm. The lifetime of the upper level is about 20 ns. The 4 nm line width occurs primarily because the wave function in the excited level is essentially a harmonic oscillator function, and the radiation occurs with the distribution of inter-nuclear separations predicted by the harmonic oscillator wave function. The energy difference between the potential curves and hence the wavelength of the radiation depends on the inter-nuclear separation. Thus the distribution of internuclear separations maps into the line width.

The stimulated emission cross section (at line center) for this case can be obtained from Equation 2.54,

$$\begin{aligned} \sigma_{ul} &= \frac{\lambda_0^2}{8\pi} A_{ul}\, g(0) \approx \frac{\lambda_0^2}{8\pi} \frac{1}{\tau_u} \frac{1}{\Delta\nu} \\ &= \frac{\lambda_0^4}{8\pi} \frac{1}{c\,\tau_u\,\Delta\lambda} = 6.4\times 10^{-21}\ \mathrm{m}^2 \end{aligned} \tag{3.15}$$

where the index of refraction is taken to be one. The threshold population number density difference is given by

$$\Delta n_{\mathrm{th}} = \frac{1}{\sigma_{ul}\, c\, \tau_{\mathrm{ph}}} = 7.8\times 10^{19}\ \frac{\mathrm{molecules}}{\mathrm{m}^3} \tag{3.16}$$

where again the index of refraction of the gas is taken as one. For the KrF^* excimer laser since there is zero population in the lower level the number of excited molecules that must be produced is equal to the threshold population density difference i.e. $n_u = \Delta n_{\text{th}}$ since n_l=0. For the total volume of the laser V=1×10^{-4} m^3 there must be 7.8×10^{15} excited molecules produced in order to obtain laser action. The energy per pulse that is emitted by the laser if the threshold is just reached is about $\Delta n_{\text{th}}\, V\, h\nu_0 = 6\times10^{-3}$ J. One can estimate a lower bound on the energy input per pulse as $\Delta n_{\text{th}}\, V\, E_I$ where E_I is the ionization energy of the Ar atom, which is 15.8 eV or 2.5×10^{-18} J. The minimum value of the energy input is thus $\Delta n_{\text{th}}\, V\, E_I = 1.6\times10^{-2}$ J per pulse. In order to obtain laser action it is necessary to drive the laser with an peak input power that exceeds $\Delta n_{\text{th}}\, V\, E_I/\tau_u = 9\times10^5$ W. Bhaumik *et al.* have operated a KrF excimer laser with an energy output of 1.5 J per pulse and with a pulse duration of 125 ns so that the peak output power is 1.2×10^7 W. Their laser had an output energy per pulse about 250 times the threshold output energy per pulse. This of course required an energy input per pulse greatly in excess of that required just to reach threshold.

The KrF* excimer laser typically does not lase on a single mode with a unique wavelength and wave vector but instead lases on many modes with a broad spread in wavelength and wave vectors. The laser medium is simply amplifying the spontaneous emission emitted in a nearly axial direction by the KrF* molecules. This process is called Stimulated Amplification of Spontaneous Emission or SASE. The amplification of the spontaneous emission occurs for a time duration approximately equal to the photon lifetime i.e. for a time corresponding to a few round trips through the amplifying medium. A very high gain system such as the KrF laser can lase with an optical cavity that has a short photon lifetime. Although the optical cavity mentioned in this discussion might have sounded crude so that a very poor optical beam quality would result, it is possible using an unstable resonator to produce a very high quality optical beam from an excimer laser even though the photon lifetime for the cavity is very short. This is discussed in the chapter on laser optical cavities. It is possible to produce an excimer laser beam in nearly a single mode. Excimer lasers are used for applications where a high peak power in a short pulse is required such as the exposure of the photo-resist masks for making integrated circuit wafers. For the exposure of photo-resist masks it is necessary that the excimer laser beam is in nearly a single mode. It should be noted that many of the lasers discussed in this book use optical cavities with long photon lifetimes.

3.3 Alternate Discussions of the Laser Threshold Criterion

In this section two instructive alternate derivations of the laser threshold population number density difference are discussed. The first of these is based on a derivation given by Schawlow and Townes, the second considers the amplitude of the electric field of the beam as it completes a round trip in the laser cavity.

3.3.1 *Schawlow and Townes Derivation*

The spontaneous decay rate per atom from an upper level u into a lower level l is A_{ul}. The spontaneous decays go into the large number of modes, p, that lie within the line width of the transition, i.e., into a number of modes given by

$$p = \frac{8\pi\,\nu_0^2\,\Delta\nu}{(c/n)^3}\,V = \frac{8\pi\,\Delta\lambda}{\lambda_0^4}\,V \tag{3.17}$$

where V is the volume of the cavity. The spontaneous rate per atom into an individual mode is A_{ul}/p, and the induced rate per atom into an individual mode is $N_\nu\,A_{ul}/p$ where N_ν is the number of photons in the individual mode. The induced decay rate from level u into level l with the emission of a photon into a particular mode minus the total absorption rate from level l into level u with the loss of a photon from the same mode yields a net emission into the mode that is given by$N_\nu\,A_{ul}\,\Delta n\,V/p$ where $\Delta n = n_u - (g_u/g_l)n_l$. The loss rate from the mode due to all processes other than absorption on the l to u transition is N_ν/τ_{ph}. Equating the net emission into the mode with the loss from the mode from all processes except absorption on the l to u transition yields

$$\Delta n_{\mathrm{th}} = \frac{p}{V}\,\frac{1}{A_{ul}\,\tau_{\mathrm{ph}}} = \frac{8\pi\nu_0^2\,\Delta\nu}{(c/n)^3}\cdot\frac{1}{A_{ul}\,\tau_{\mathrm{ph}}} \tag{3.18}$$

which is the same result for the threshold population number density difference as was obtained previously (Eq. 3.12).

3.3.2 *Round Trip Cavity Approach*

The second alternative derivation of the threshold condition for a laser analyzes the electric field as the laser beam makes a round trip in the laser cavity as is shown in Figure 3.3. The intensity of an electromagnetic wave of frequency ν is given by

$$I_\nu = \frac{c}{n}\left(\frac{\varepsilon\,\boldsymbol{E}_{\mathrm{RMS}}^2}{2} + \frac{\boldsymbol{B}_{\mathrm{RMS}}^2}{2\mu_0}\right) = \frac{c\,\varepsilon\,\boldsymbol{E}_0^2}{2n} \tag{3.19}$$

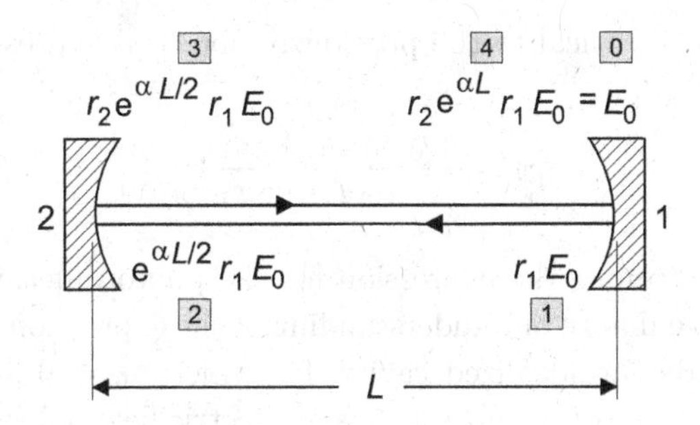

Fig. 3.3 A laser beam making a round trip in a cavity of length L. The electric field is indicated just before and just after each reflection.

where $\boldsymbol{E}_{\text{RMS}} = \boldsymbol{E}_0/2^{1/2}$ is the RMS value of the electric field in the laser beam, $\boldsymbol{B}_{\text{RMS}} = \boldsymbol{B}_0/2^{1/2}$ is the RMS value of the magnetic field, and n is the index of refraction. In the last part of Equation 3.19 the relationship between the electric and magnetic fields in an electromagnetic wave has been employed to eliminate $\boldsymbol{B}_{\text{RMS}}$ in the expression for the laser intensity. Solving for the magnitude of the electric field leads to

$$E_0 = \left(\frac{2n\, I_\nu}{c\,\varepsilon}\right)^{1/2} . \tag{3.20}$$

In order for a laser to oscillate in the steady state it is necessary that the electric field to be unchanged after a round trip in the cavity. This means that after a round trip both the amplitude and the phase of the field are unchanged. The condition that the phase be unchanged after a round trip in the optical cavity determines the frequency at which the laser oscillates and is discussed in Chapter 7. At this point the focus is on the condition that the amplitude be unchanged after a round trip in the optical cavity. The electric field reflection coefficients for the two mirrors that make up the laser cavity are taken to be r_1 and r_2 and L is taken to be the length of the cavity. The condition that the electric field amplitude be unchanged after the wave makes a round trip in the cavity is

$$\left| r_1\, r_2 \exp\left(2L\left[\frac{(\lambda_0^2/8\,\pi)\, A_{ul}\, g(\nu-\nu_0)\,\Delta n_{\text{th}} - \alpha_0}{2}\right]\right)\right| = 1 \tag{3.21}$$

where α_0 is a term that represents all losses other than losses due to the reflection coefficients being less than one or losses due to absorption from level l to level u. Solving Equation 3.21 for the threshold population difference yields

$$\Delta n_{\text{th}} = \frac{8\pi\, n^2\, \nu_0^2}{c^2}\, \frac{1}{A_{ul}\, g(\nu-\nu_0)} \left(\alpha_0 - \frac{\ln r_1\, r_2}{L}\right) . \tag{3.22}$$

This equation is identical to the previously obtained expressions for Δn_{th} provided that

$$\tau_{\text{ph}} = \frac{n\,L}{c}\,\frac{1}{\alpha_0\,L - \ln r_1\,r_2}\;. \tag{3.23}$$

Now let us examine the expression for the photon lifetime in a simple limit in order to enhance our understanding of the expression for the photon lifetime. Consider an idealized cavity, for which $\alpha_0 = 0$ and made with mirrors such that one of the mirrors has an electric field reflection coefficient r_1=1 and the other has an electric field reflection coefficient r_2 that is only a little less than 1. At the second mirror $E_{\text{refl}} = r_2\,E_{\text{inc}}$ for each reflection where E_{refl} and E_{inc} are respectively the magnitudes of reflected and incident electric fields in the electromagnetic wave. We denote the intensity reflection coefficient by R_2 so that $I_{\text{refl}} = R_2\,I_{\text{inc}}$ where I_{refl} and I_{inc} are the reflected and incident intensities. It is clear that $R_2 = r_2^2$ since $I = c\,\varepsilon_0 E^2_{\text{RMS}}$. If the absorption at mirror 2 is zero and the losses due to diffraction are zero then $R_2 = 1 - T_2$ where T_2 is the transmission coefficient for the mirror. It follows that $r_2 = R_2^{1/2} = (1 - T_2)^{1/2}$. For this situation the photon lifetime can be expressed as

$$\tau_{\text{ph}} = \frac{n\,L}{-c\,\ln r_2} = \frac{n\,L}{-c\,\ln\,(1 - T_2)^{1/2}} \approx \frac{2n\,L}{c\,T_2}\;. \tag{3.24}$$

The question now arises, how many bounces from the mirrors are required to reduce the intensity of the wave to $1/e$ of the incident intensity? After N reflections from the second mirror and in the absence of absorption or stimulated emission between levels u and l the ratio of the intensity of a wave in the cavity to its initial intensity is given by

$$\frac{I_\nu}{I_{\nu 0}} = (1 - T_2)^N = \left(1 - \frac{N\,T_2}{N}\right)^N \to \exp(-N\,T_2) \text{ as } N \to \infty \tag{3.25}$$

where the approximation that T_2 is small and N is large is used. After $N = 1/T_2$ reflections from mirror 2 the wave is decreased by a factor of $1/e$. The photon lifetime is $N = 1/T_2$ times the time for a round trip time for the cavity so that $\tau_{\text{ph}} = (2nL/c)/T_2$. This is the same result as obtained in Equation 3.24. It is hoped that by calculating the photon lifetime in two ways the reader has obtained a good feeling for the mathematical expression for the photon lifetime, $\tau_{\text{ph}} = (nL/c)/[\alpha_0\,L - \ln\,r_1\,r_2]$.

As a numerical example consider a cavity for which L = 30 cm, T_2=0.01, $n = 1$, and α_0=0. The photon lifetime for this cavity is $\tau_{\text{ph}} = (2L/c)/T_2 = 2\times10^{-7}$ s.

Summary of Key Ideas

- To have a laser, the gain of the active medium must overcome all the losses including those due to due to absorption, refection, and diffraction.
- At line center, the threshold population difference required for lasing is
$$\Delta n_{\rm th} = \frac{8\pi}{\lambda_0^2}\,\frac{\Delta\nu}{A_{ul}\,(c/n)\,\tau_{\rm ph}} = \frac{8\pi\,\nu_0^2\,\Delta\nu}{(c/n)^3}\,\frac{1}{A_{ul}\,\tau_{\rm ph}} \tag{3.12}$$
- The photon lifetime represents the $1/e$ lifetime of the photon in the cavity due to all losses except absorption,
$$\tau_{\rm ph} = \frac{n\,L}{c}\,\frac{1}{\alpha_0\,L - \ln r_1\,r_2} \tag{3.23}$$
where $r_{1,2}$ are the electric field reflection coefficients and α_0 represents all losses other than reflection and absorption losses. In terms of the intensity reflection (R) and transmission coefficients (T), $r_x = R_x^{1/2} = (1 - T_x)^{1/2}$.

Suggested Additional Reading

A. L. Schawlow and C. H. Townes, "Infrared and Optical Masers" *Phys. Rev.* **112**, 1940 (1958).

A. L. Schawlow, *Lasers and Light*, W. H. Freeman and Co. (1969).

William T. Silfvast, *Laser Fundamentals*, Cambridge University Press (1996).

A. E. Siegman, *Lasers*, University Science Books (1986).

M. L. Bhaumik, R. S. Bradford Jr. and E. R. Ault, "High-efficiency KrF excimer laser" *Appl. Phys. Lett.* **28**, 23 (1976).

D. L. Huestis, Chapter 1 in *Appl. Atomic Coll. Phys. Vol. 3: Gas Lasers*, (Eds: E. W. McDaniel and W. L. Nighan) Academic Press (1982).

A. Yariv, *Quantum Electronics 3rd Ed.*, John Wiley and Sons (1989).

B. E. A. Saleh and M. C. Teich, *Fundamentals of Photonics*, John Wiley and Sons (1991).

P. W. Milonni and J. H. Eberly, *Laser Physics 2nd Ed.*, John Wiley and Sons (2010).

Problems

1. Atomic sodium vapor with an atomic density of 2×10^{14} cm^{-3} and a temperature of 2000 K absorbs light at a wavelength of 589.6 nm in making transitions from the $3^2S_{1/2}$ level to the $3^2P_{1/2}$ level. The oscillator strength of the transition is approximately 1/3. Calculate the following:
(a) What is the Einstein A coefficient and the radiative lifetime for the $3^2P_{1/2}$ level of sodium?
(b) Estimate the absorption cross section at line center if the line width is equal to 3.5×10^9 Hz.
(c) Estimate the absorption coefficient, α, at line center. You may ignore all loss processes except absorption by the $3^2S_{1/2} \to 3^2P_{1/2}$ transition, and you may assume that all the sodium atoms are in the ground level.
(d) At what atomic density will the absorptance be 0.1 for a path length of 5 cm.?

2. For the sodium atoms of Problem 1 estimate how many modes per unit volume lie under the line width? What is the spontaneous decay rate per atom in the $3^2P_{1/2}$ level into a given mode at line center? Take the volume as 10 cm^3.

3. It has been found that one can make a pulsed Xe excimer laser by bombarding high pressure gaseous Xe with an intense pulsed electron beam. With simplification the reactions that lead to emission by a Xe_2^* eximer molecule are the following:

i. $e^-(E > 12.127 \text{ eV}) + Xe \to Xe^+ + 2e^-$
ii. $Xe^+ + 2Xe \to Xe_2^+ + Xe$
iii. $Xe_2^+ + Xe^- \to Xe^{**} + 2Xe$
iv. $Xe^{**} + 2Xe \to Xe_2{}^{**} + Xe$
v. $Xe_2^{**} \to Xe^* + Xe$
vi. $Xe^* + 2Xe \to Xe_2^* + Xe$
vii. $Xe_2^* \to 2Xe + h\nu$

The notation uses a single asterisk to denote a low lying excited level and two asterisks to denote a high lying excited level. At the densities of high pressure gaseous Xe, the reactions ii to vi occur very rapidly. Excited Xe^* atoms are also produced by the reaction $e^-(E > 8.31 \text{ eV}) + Xe \to Xe^* + e^-$. The laser transition is between the lowest vibrational level of the bound level of the Xe_2^* molecule and the unbound repulsive singlet level formed by two ground level Xe atoms as indicated in reaction vii. The radiative

transition has a center wavelength of 173 nm and a line width of 20 nm. A volume of Xe that is 10 cm long and 1 cm^2 in cross sectional area is excited by electron bombardment. The lifetime of the excited Xe_2^* level is about 20 ns. The upper laser level is the lowest excited level of the Xe_2^* eximer molecule. No mirrors are used for the laser cavity. Find the following:
(a) An estimate of the minimum population inversion density necessary for laser action.
(b) A lower bound on the deposited energy per pulse necessary for laser action.
(c) A lower limit on the power that must be deposited into the Xe for laser action if the energy is deposited in the Xe in a time equal to $0.1\times$ the spontaneous lifetime of the excited level.
(d) An estimate of the stimulated emission cross section at line center.

4. Experiments attempting to produce laser action by He_2^* eximers have been carried out using pulsed electron beam excited liquid He. The reactions leading to the He_2^* eximer are similar to those outlined for Xe in Problem 3, although the ionization energy of He (24.59 eV) is much higher than that of Xe (12.13 eV). The proposed laser action occurs with a center wavelength of about 80 nm and a line width of about 20 nm. The spontaneous lifetime of the excited level is about 10 ns. Assume that the electron accelerator deposits the energy into a volume that is 10 cm long and 2 cm^2 in cross sectional area. Find the following:
(a) An estimate of the minimum population inversion density necessary for laser action.
(b) An estimate of the stimulated emission cross section at line center.
(c) The number of modes within the line width.
(d) The spontaneous decay rate per atom and into a given mode at line center.
(e) An estimate of the minimum energy per pulse and peak power the electron beam must deposit for laser action to occur.
(f) Experiments indicated that the He could not be made to lase as an excimer laser. This is probably because the 80 nm wavelength photons photoionized the eximer molecules by the reaction $h\nu + \text{He}_2^* \rightarrow \text{He}_2^+ + \text{e}^-$. Estimate the minimum size of the photoionizaton cross section that would prevent laser action.

5. A laser cavity has two mirrors one with an intensity reflection coefficient of 1.00 and the other with an intensity reflection coefficient of 0.99. The mirrors are separated by 0.5 m. If the index of refraction of the material in the cavity is one, and if the only important loss is by transmission through the second mirror what is the photon lifetime in the cavity?

6. How many photons must be in a mode in order that the total emission rate per atom into a given mode is 200 times the spontaneous rate per atom into the same mode?

7. Explain carefully how a system can have a population inversion but can still have a positive absorption coefficient.

8. Two different laser systems have the same Einstein A coefficients, the same line widths $\Delta\lambda$, and optical cavities with the same photon lifetime. The wavelength for one laser is 600 nm and the wavelength for the other is 900 nm. What is the ratio of the threshold population number density differences for the two lasers?

9. An optical cavity is to be constructed using mirrors with intensity reflection coefficients of 1.0000 and R. The cavity is to be 25 cm in length and is filled with a material with an index of refraction near one. If the photon lifetime is to be 1 μs what value of R is required?

10. A light beam passing through a material with an atomic density of 10^{15} cm^{-3} and a length of 30 cm is attenuated to 0.20 of its initial intensity. What is absorption cross section for the material at the wavelength of the light beam? You may assume that the only significant loss is by absorption on a single atomic transition.

Chapter 4

Line Shapes

4

Line shapes are an important concept in the study of lasers. The line shape of a laser transition defines the possible emission frequencies at which the laser can operate. Line shapes are also important for understanding the absorption of light, either from a broadband source , or from a narrow band source such as a laser. Fundamental to the operation of lasers, the 'type' of the line shape determines how the gain of an active medium varies with the operating wavelength of the laser. This is addressed in detail in Chapter 5 on saturation. Most of this chapter is devoted to line shapes in gases. Line shapes in liquids and solids are discussed briefly, but these line shapes are much more dependent on the particular material. As a result, most of the details on line shapes for liquid or solid lasers will be found in the later chapters where particular laser systems are discussed.

4.1 Line Shapes in Atomic Gases

Although gas lasers are not as widely used as they were in the past it is still desirable to treat gas lasers in a text book on lasers. The normalized line shape that occurs in the absorption or stimulated emission cross sections is a very important and basic quantity. The same line shape also usually occurs in spontaneous emission. There are a number of different sources of line broadening, and these different sources of broadening give rise to a variety of different line shapes. The most basic distinction between different broadening mechanisms is whether the resulting line shape is *homogeneously* or *inhomogeneously* broadened. A homogeneous line is one for which the broadening mechanism produces the same distribution of absorption (or emission) frequencies for all atoms (or molecules) of an ensemble. Both radiative broadening and collisional broadening produce homogeneous line shapes. On the other hand an inhomogeneous line is one, for which the broadening mechanism produces a different distribution of frequencies for different atoms in an ensemble of atoms. The term inhomogeneous line shape was initially applied to line shapes for nuclear magnetic resonance where the primary line broadening mechanism arose from the inhomogeneous magnetic field that caused the absorption frequencies to depend on the position of the absorber in the magnet. The term inhomogeneous broadening is now used for any broadening mechanism that produces different distributions of frequencies for different atoms in an ensemble. For example Doppler broadening produces an inhomogeneous line shape.

In this section line shapes for a number of broadening mechanisms for absorbing or radiating atoms in gases are discussed. For gases homogeneous broadening mechanisms are discussed first, and then the inhomogeneous broadening mechanisms are discussed. It is stressed that for most situations in the gas phase the same line shape usually occurs for absorption, for spontaneous emission, and for stimulated emission. In high pressure gases, however, rapid non-radiative relaxation of the atoms or molecules in excited levels to lower excited levels can change the excited level populations, which results in a different line shape for absorption and emission.

4.1.1 *Natural Line Shapes*

The natural line width of a transition is the spread in frequencies that results from the radiative lifetimes of the levels. In order to understand the natural line width that is due to the radiation emitted by an atom or molecule it is necessary to discuss the time and hence the frequency dependence of the intensity of the light emitted. The simple case where an atom or molecule is in an excited level labeled u and is radiating a photon during a transition to the ground level labeled l is treated. An atom in excited level u has an oscillating electric dipole moment as it makes a transition to a lower level. The intensity, I, of radiation emitted by an atom in excited level u as observed at a particular position as a function of the time, t, is given by

$$I(t) = \begin{cases} I_0 \exp\left[-t/\tau_u\right] & \text{for } t \geq 0 \\ 0 & \text{for } t < 0 \end{cases} \tag{4.1}$$

where I_0 is the intensity at $t = 0$ and τ_u is the radiative lifetime of the excited level. In Equation 4.1 it is assumed that the atom is excited into the level u at time t=0. The intensity is proportional to the square of the

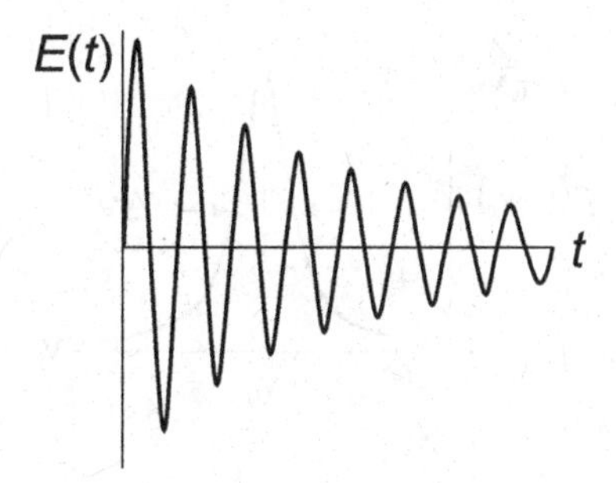

Fig. 4.1 An exponentially damped oscillating electric field as a function of the time.

electric field since $I = (c/n)\,\epsilon_0 \boldsymbol{E}^2_{\mathrm{RMS}}$. Therefore the electric field produced by a radiating atom as observed at a particular position is given by

$$\boldsymbol{E}(t) = \begin{cases} \boldsymbol{E}_0 \exp\left[i(\omega_0\, t + \phi) - \frac{t}{2\tau_u}\right] & \text{for } t \geq 0 \\ 0 & \text{for } t < 0 \end{cases} \tag{4.2}$$

where $h\omega_0/2\pi$ is the energy difference between the upper level u and the lower level l, where ϕ is a phase constant, and where $I_0 = \epsilon_0 \boldsymbol{E}_0^2 c/2$. Of course the electric field is the real part of the complex expression given in Equation 4.2. Figure 4.1 shows such an electric field where the phase constant ϕ is taken arbitrarily as $-\pi/2$. In order to determine the frequencies present in the emitted radiation one must evaluate the Fourier transform of the electric field. The Fourier transform of $\boldsymbol{E}(t)$ is given by

$$\boldsymbol{E}(\omega) = \int_{-\infty}^{\infty} \boldsymbol{E}(t)\, e^{-i\omega t}\, dt = \frac{\boldsymbol{E}_0\, e^{i\phi}}{i(\omega - \omega_0) + \frac{1}{2\tau_u}} \,. \tag{4.3}$$

The energy per unit frequency in the electromagnetic wave passing through a unit area as a function of the angular frequency is called the *fluence*, $F(\omega)$. The fluence represents the frequency distribution of the energy in the electromagnetic wave per unit area. The fluence is directly proportional to the absolute value of the square of $\boldsymbol{E}(\omega)$ so that

$$F(\omega) \propto \boldsymbol{E}(\omega) \cdot \boldsymbol{E}^*(\omega) = \frac{\boldsymbol{E}_0^2}{(\omega - \omega_0)^2 + \left(\frac{1}{2\tau_u}\right)^2} \,. \tag{4.4}$$

The normalized line shape, $g(\nu - \nu_0)$, is directly proportional to $F(\omega)$. The normalized line shape is

$$g(\nu - \nu_0) = \frac{\frac{\Delta\nu_L}{2\pi}}{(\nu - \nu_0)^2 + \left(\frac{\Delta\nu_L}{2}\right)^2} \tag{4.5}$$

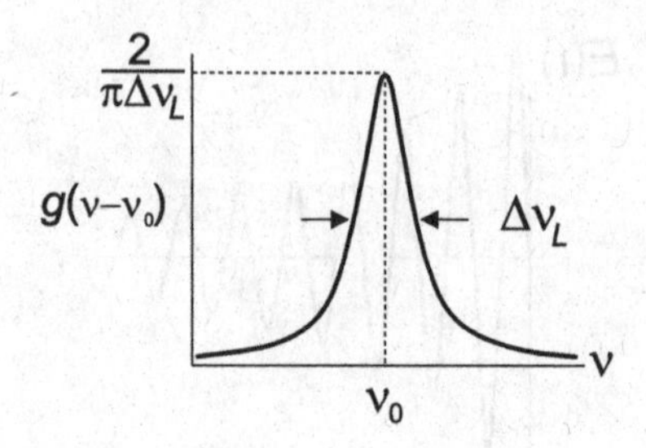

Fig. 4.2 A normalized Lorentzian line shape as a function of the frequency.

where line width, $\Delta\nu_L = (2\pi\,\tau_u)^{-1}$, is the full width at half maximum of the line shape. The constant, $\Delta\nu_L/2\pi$, in Equation 4.5 has been selected so that the line shape is normalized. This is called a Lorentzian line shape. Figure 4.2 shows a graphical representation of a Lorentzian line shape. Although the Lorentzian line width is denoted by $\Delta\nu_L$ If the line is broadened by natural riadiative decay then it is often denoted by $\Delta\nu_R$. If the lower level is not the ground level but is instead also a radiating level with a lifetime τ_l then the natural radiative line shape of the transition is still a Lorentzian line shape but the line width is given by

$$\Delta\nu_L = \frac{1}{2\pi}\left(\frac{1}{\tau_u} + \frac{1}{\tau_l}\right). \tag{4.6}$$

One can understand Equation 4.6 by thinking of the radiation as being emitted from an excited level u with a width in energy of $\Delta W_u = h\Delta\nu_u = h/2\pi\,\tau_u$ and into a lower level with a width in energy of $\Delta W_l = h\Delta\nu_l = h/2\pi\,\tau_l$ When viewed this way the natural width of a transition is seen to be a result of the broadening introduced by the uncertainty principle because of the natural radiative lifetimes of the upper and lower levels. For a Lorentzian line shape one finds $g(0) = 2/\pi\Delta\nu_L$ rather than the approximate value $g(0) \approx 1/\Delta\nu_L$ used in Chapter 3. As previously mentioned the natural or radiative broadening results in a homogeneous line shape since all atoms in the radiating ensemble have the same line shape and the same center frequency.

4.1.2 *Collisionally Broadened Line Shapes*

An atomic (or molecular) transition can be broadened by collisions as well as by the natural radiative lifetime broadening. During a collision the frequency emitted or absorbed by an atom is changed. The reason for this is shown in Figure 4.3 where two interatomic potentials are presented as

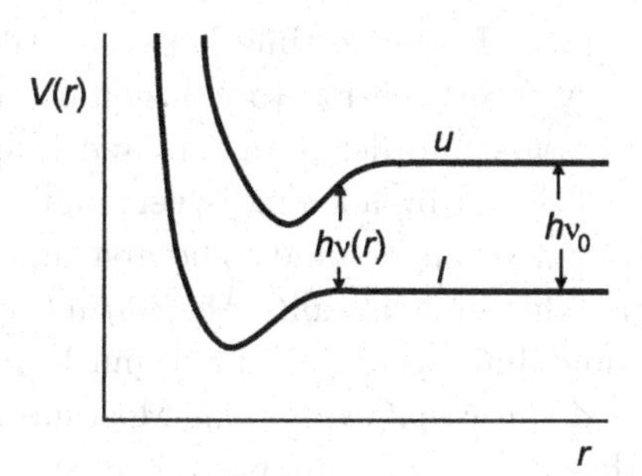

Fig. 4.3 The potential curves for an atom colliding with another atom as a function of the internuclear separation.

a function of the internuclear separation for the molecule. A molecule is formed for a short time when the atom, whose transition is being studied makes a collision with another atom. A molecule composed of an excited atom and a ground level atom in the separated atom limit (i.e. in the limit as r becomes very large) has an interatomic potential with a different functional dependence on the internuclear separation than a molecule composed of two ground level atoms in the separated atom limit so that the energy difference and hence the frequency of the emitted radiation is altered during the collision. Therefore the electric field of the radiation emitted by the atom looks like a decaying sine wave but with the frequency of the wave modulated during each collision. The magnitude of the electric field emitted by an atom when making a transition from an upper level to a lower level as observed at a given point in space is shown in Figure 4.4.

The average time between collisions is called τ_c, and the average

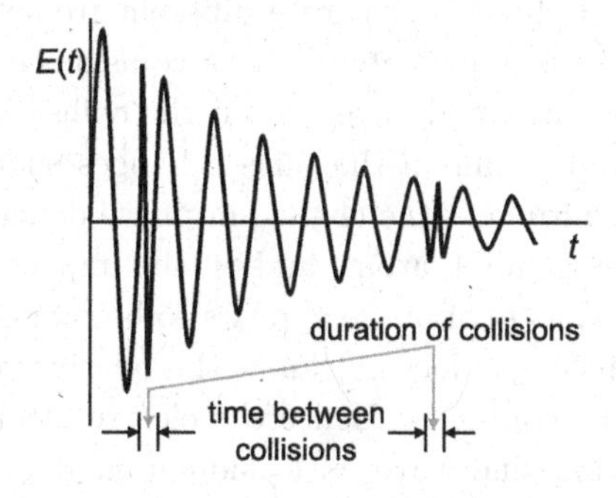

Fig. 4.4 The electric field emitted by an atom making collisions, which change the emitted frequency as a function of the time.

duration of a collision τ_d. Both the time between collisions and the duration of collisions vary from collision to collision so that averaging over these quantities is necessary in order to understand collisional broadening. In the general case collisional broadening is very difficult to treat, but in two limits it is fairly easy to understand. The first limit occurs when τ_c is much larger than τ_d. This limit is called the *impact approximation*. The other easy to understand limit occurs when τ_c is much smaller than τ_d. This limit is called the *quasi-static approximation*. Most gas lasers operate with gas densities such that the impact approximation is applicable for their analysis. Collisional broadening in the impact approximation is discussed in the following section of the text, and the quasi-static approximation is discussed in section 4.1.6. Collisional broadening results in a homogeneous line shape.

4.1.3 *Impact Approximation*

In the impact approximation where the time between collisions, τ_c, is much larger than length of the collision, τ_d, the collisions are well isolated in time, and the magnitude of the electric field of a photon emitted by an excited atom can be approximated by an exponentially decaying sine wave between collisions. If the magnitude of $\omega - \omega_0$ is much less than $1/\tau_d$ then the frequency distribution in the wave is determined almost entirely by the decaying sine wave between collisions and is little affected by the emission during the collision. The Fourier spectrum of the emission during a collision will be spread over the very large range of frequencies from $\omega - \omega_0 = -1/\tau_d$ to $1/\tau_d$ and will have a very small amplitude compared to the Fourier spectrum of the radiation between collisions. During a collision the phase of the field changes at a rate different from the rate of change between collisions. Thus immediately after a collision the phase of the field is changed from what it would have been if the collision had not occured. Since τ_c is much larger than τ_d the phase changes suddenly on the time scale of τ_c. The magnitude of the change in phase depends on the relative velocity between the colliding atoms and on the impact parameter for the collision so that the change of phase appears to be random for an ensemble of colliding atoms since one does not know the relative velocities or impact parameters for all the collisions. Hence the electric field appears to be an exponentially decaying sine wave with sudden random changes of phase occurring at random times, but with an average time between the changes of phase equal to τ_c as is shown in Figure 4.5.

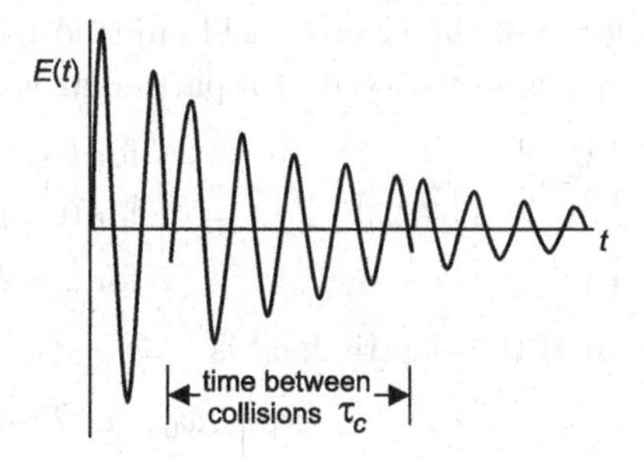

Fig. 4.5 The electric field emitted by an atom making collisions as a function of the time in the impact approximation.

The line shape that results from collisional broadening is obtained by calculating the distribution of frequencies present in an exponentially decaying sine wave that exists for a time between collisions, T, and then averaging over the distribution of times between collisions. Again the simple case where the upper level decays to the ground level is treated. As mentioned earlier the time between collisions is not a constant. In order to average over the distribution of collision times one must know the probability that an atom has made no collisions. The number of atoms per unit volume, Δn, that make an initial collision in a time Δt is given by

$$\Delta n = \frac{n(t)\,\Delta t}{\tau_c} \tag{4.7}$$

where $n(t)$ is the number of atoms per unit volume that have not made a collision. Thus

$$\frac{dn}{dt} = -\frac{n}{\tau_c} \tag{4.8}$$

so that

$$n(t) = n_0 \exp(-t/\tau_c)\,. \tag{4.9}$$

Therefore the probability that an atom has made no collisions after a time T is given by

$$P(T) = \frac{\exp(-T/\tau_c)}{\tau_c}\,, \tag{4.10}$$

where $P(T)$ is normalized so that the integral of $P(T)$ over all times T is equal to 1. The average time between collisions is given by τ_c as can be seen from

$$\overline{T} = \int_0^\infty T\,P(T)\,dT = \tau_c\,. \tag{4.11}$$

The time dependence of the electric field emitted by an atom between two successive collisions and observed at a particular location is given by

$$E(t) = \begin{cases} 0 & \text{for } t < 0, \\ E_0 \exp\left[i(\omega_0 t + \phi) - \frac{t}{2\tau_c}\right] & \text{for } 0 \leq t \leq T, \\ 0 & \text{for } t > T\,. \end{cases} \tag{4.12}$$

The Fourier transform of the electric field is

$$E(\omega) = \int_{-\infty}^{\infty} E(t)\, e^{-i\omega t}\, dt = E_0\, e^{i\phi}\, \frac{\exp\left[i(\omega_0 - \omega)T - \frac{T}{2\tau_u}\right] - 1}{i(\omega_0 - \omega) - \frac{1}{\tau_u}}\,. \tag{4.13}$$

It is now necessary to average $E(\omega)$ over the distribution of collision times, T. This yields

$$\begin{aligned} \overline{E(\omega)} &= \int_0^{\infty} E(\omega)\, P(T)\, dT \\ &= -E_0\, e^{i\phi}\, \frac{1}{i(\omega - \omega_0) - \left(\frac{1}{2\tau_u} + \frac{1}{\tau_c}\right)}\,. \end{aligned} \tag{4.14}$$

This leads to a fluence given by

$$F(\omega) \propto \overline{E_0(\omega)} \cdot \overline{E^*(\omega)} = \frac{E_0^2}{(\omega - \omega_0)^2 + \left(\frac{1}{2\tau_u} + \frac{1}{\tau_c}\right)^2} \tag{4.15}$$

The normalized line shape for a collisionally broadened line in the impact approximation is therefore a Lorentzian line shape

$$g(\nu - \nu_0) = \frac{\frac{\Delta\nu_L}{2\pi}}{(\nu - \nu_0)^2 + \left(\frac{\Delta\nu_L}{2}\right)^2} \tag{4.16}$$

where

$$\Delta\nu_L = \frac{1}{2\pi}\left(\frac{1}{\tau_u} + \frac{2}{\tau_c}\right)\,. \tag{4.17}$$

If both the upper and lower levels can radiate then the line shape is still given by Equation 4.16, but the line width is given by

$$\Delta\nu_L = \frac{1}{2\pi}\left(\frac{1}{\tau_u} + \frac{1}{\tau_l} + \frac{2}{\tau_c}\right)\,. \tag{4.18}$$

In the impact approximation the collisionally broadened line shape is Lorentzian, the same shape as for natural radiative broadening, but with a line width (full width at half maximum) that depends on both the lifetimes of the upper and lower levels *and* on the average time between collisions. It should be noted that collisions are twice as effective as the radiative decay in broadening the line. If collisional broadending is much greater than the natural radiative broadening the line width is often denoted by $\Delta\nu_C$. In Section 4.1.6 the relationship between the collision times that go into the expression for collisional broadening of a line and the collision cross section are analyzed. That section also discusses collisional broadening in the quasi-static approximation.

4.1.4 *Inhomogeneous Line Shapes*

An inhomogeneous line shape results from different atoms having different frequency distributions. An individual group of atoms will have a homogeneous line shape $g(\nu - \nu_i)$ with a center frequency ν_i while another group of atoms will have the same homogeneous line shape but a different center frequency ν_j. The inhomogeneous line shape is determined by the relative numbers of atoms having each center frequency. The probability of an atom having a center frequency ν_i is called the inhomogeneous distribution, $h(\nu_i - \nu_0)$. Since $h(\nu_i - \nu_0)$ is a probability, the integral of $h(\nu_i - \nu_0)$ over all frequencies, ν_i, is one (i.e. $h(\nu_i - \nu_0)$ is the normalized envelope for the inhomogeneous distribution) so that

$$\int_0^\infty h(\nu_i - \nu_0)\, d\nu_i = 1\,. \tag{4.19}$$

The envelope for the inhomogeneous distribution, $h(\nu_i - \nu_0)$, is written as though it has a center frequency ν_0 even though there is no requirement that an inhomogeneous distribution be symmetric about any frequency. This is, however, a useful notation because one of the most important inhomogeneous distributions, the Doppler distribution, is symmetric about the frequency corresponding to zero component of the velocity parallel to the direction of propagation of the light beam.

Figure 4.6 illustrates an inhomogeneous distribution, showing an individual homogeneous packet at the frequency ν_i beneath the inhomogeneous distribution. The full width at half maximum for the inhomogeneous envelope is denoted by $\Delta\nu_I$, and the full width at half maximum for a homogeneous packet of atoms is denoted by $\Delta\nu$. Although the distribution of frequencies, ν_i, is continuous it is common to say that an inhomogeneous line shape has $\Delta\nu_I/\Delta\nu$ homogeneous packets within the width of the inhomogeneous distribution.

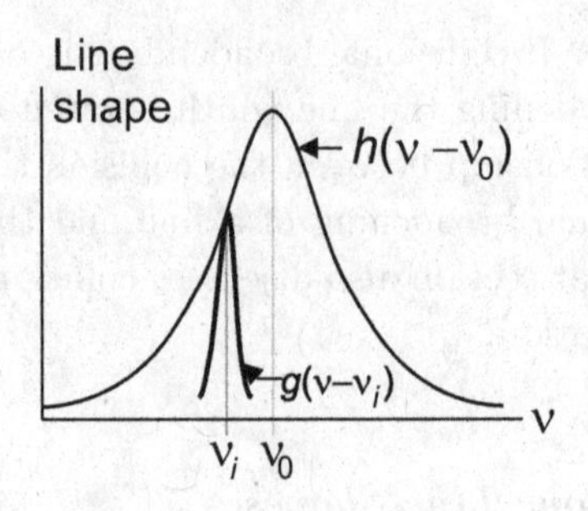

Fig. 4.6 An inhomogeneously broadened line shape.

For an inhomogeneous line the absorption cross section is given by

$$\begin{aligned}\sigma_{\mathrm{ABS}} &= \int_0^\infty \sigma_{lu}(\nu - \nu_i)\, h(\nu_i - \nu_0)\, d\nu_i \\ &= \int_0^\infty \frac{g_u}{g_l} \frac{\lambda_i^2}{8\pi} A_{ul}\, g(\nu - \nu_i)\, h(\nu_i - \nu_0)\, d\nu_i\,. \end{aligned} \tag{4.20}$$

The line shape of the inhomogeneous line is

$$\begin{aligned}G(\nu - \nu_0) &= \int_0^\infty g(\nu - \nu_i)\, h(\nu_i - \nu_0)\, d\nu_i \\ &= \int_0^\infty \frac{(\Delta\nu_L/2\pi)}{(\nu - \nu_i)^2 + (\Delta\nu_L/2)^2}\, h(\nu_i - \nu_0)\, d\nu_i \end{aligned} \tag{4.21}$$

where it is assumed that the line shape of the homogeneous packets is Lorentzian and where the full width at half maximum of the homogeneous packet is taken as $\Delta\nu = \Delta\nu_L$. If $\Delta\nu_I$ is much larger than $\Delta\nu_L$ then $h(\nu_i - \nu_0)$ is almost constant over a homogeneous packet so that the shape of the inhomogeneous envelope is approximately given by

$$\begin{aligned}G(\nu - \nu_0) &= \int_0^\infty g(\nu - \nu_i)\, h(\nu_i - \nu_0)\, d\nu_i \\ &\approx h(\nu - \nu_0) \int_0^\infty g(\nu - \nu_i)\, d\nu_i = h(\nu - \nu_0)\,. \end{aligned} \tag{4.22}$$

The above approximation is useful provided $\Delta\nu_I$ is more than about 10 times $\Delta\nu_L$. It is also important that $\nu - \nu_0$ be of the order of magnitude or less than a few times $\Delta\nu_I$ because the inhomogeneous line shape function may fall rapidly as $\nu - \nu_0$ increases so that the inhomogeneous line shape lacks appreciable wings. The homogeneous Lorentzian packets always contribute significantly to the line shape in the wings of the line where $\nu - \nu_0$ is large since the Lorentzian line shape only decreases like

$1/(\nu - \nu_0)^2$ for large values of $\nu - \nu_0$. In the next section of this Chapter it is shown that a Doppler line shape is inhomogeneous. The inhomogeneous distribution of center frequencies is Gaussian. The inhomogeneous distribution decreases very rapidly as $\nu - \nu_0$ increases. This leads to the result that for small values of $\nu - \nu_0$ the line shape is nearly Gaussian but for large enough values of $\nu - \nu_0$ the line shape is essentially Lorentzian. A line shape that is Gaussian for small values of $\nu - \nu_0$ and Lorentzian for large values of $\nu - \nu_0$ is called a Voigt line shape.

4.1.5 *The Doppler Line Shape*

The most common inhomogeneous line shape encountered is the Doppler line shape. The Doppler line shape results from the spread in absorption frequencies due to the thermal motion of atoms. The frequency, at which an atom absorbs light is given by

$$\nu = \nu_0 \left(1 + \frac{v_z}{c}\right) \tag{4.23}$$

where ν_0 is the absorption frequency for an atom at rest, v_z is the component of the atom's velocity in the direction anti-parallel to the direction of propagation of the light, and c is the speed of light. For a gas in thermal equilibrium the Maxwell-Boltzmann distribution predicts that the probability of an atom having a z component of velocity between v_z and $v_z + dv_z$ is given by

$$P(v_z)\, dv_z = \left(\frac{M}{2\pi\, kT}\right)^{1/2} \exp\left[-\frac{M\, v_z^2}{2kT}\right] dv_z \tag{4.24}$$

where M is the mass of the atom, k=1.381×10^{-23} JK^{-1} is Boltzmann's constant, and T is the Kelvin or absolute temperature. Solving Equation 4.23 for v_z and substituting into Equation 4.24 yields

$$\begin{aligned} P(\nu - \nu_0)\, d\nu &= P(v_z)\, \frac{dv_z}{d\nu}\, d\nu \\ &= \left(\frac{M}{2\pi\, kT}\right)^{1/2} \exp\left[-\frac{Mc^2\,(\nu - \nu_0)^2}{2\nu_0^2\, kT}\right] \frac{c}{\nu_0}\, d\nu\, . \end{aligned} \tag{4.25}$$

The inhomogeneous frequency distribution function $h(\nu - \nu_0)$ is equal to $P(\nu - \nu_0)$. With some rearrangement and relabeling one obtains

$$h(\nu - \nu_0) = \frac{2(\ln 2)^{1/2}}{\pi^{1/2}\, \Delta\nu_D}\, e^{-\frac{4(\ln 2)(\nu - \nu_0)^2}{\Delta\nu_D^2}} \tag{4.26}$$

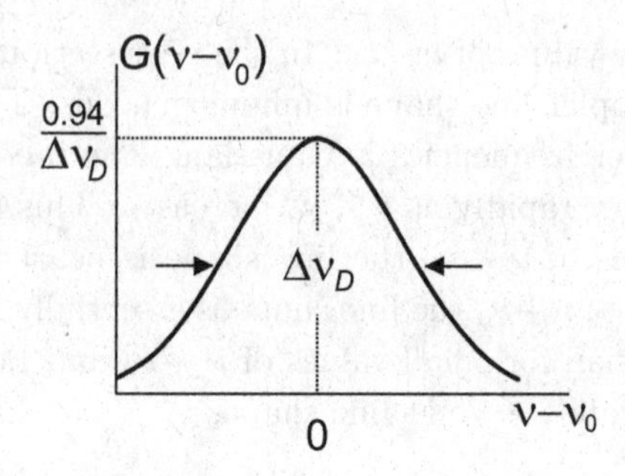

Fig. 4.7 A normalized Doppler line shape.

where

$$\Delta\nu_D = 2\nu_0 \left(\frac{2(\ln 2)kT}{Mc^2}\right)^{1/2} \tag{4.27}$$

and $\Delta\nu_D$ is the full width at half maximum of the Doppler distribution. This distribution is a Gaussian distribution.

The line shape that results from combining a Doppler distribution with a Lorentzian shape for the homogeneous packets is given by

$$G(\nu - \nu_0) = \int_0^\infty \frac{\frac{\Delta\nu_L}{2\pi}}{(\nu - \nu_i)^2 + \left(\frac{\Delta\nu_L}{2}\right)^2} \frac{2(\ln 2)^{1/2}}{\pi^{1/2}\,\Delta\nu_D} e^{-\frac{4(\ln 2)(\nu-\nu_0)^2}{\Delta\nu_D^2}} d\nu_i \,. \tag{4.28}$$

This called a *Voigt line shape.* The line shape is a Gaussian line shape only if $\Delta\nu_D$ is much larger than $\Delta\nu_L$ and if $\nu - \nu_0$ is comparable to or less than a few times $\Delta\nu_D$.

The value of a Gaussian line shape at line center is

$$G(0) = \frac{2(\ln 2)^{1/2}}{\pi^{1/2}\,\Delta\nu_D} = \frac{0.94}{\Delta\nu_D}. \tag{4.29}$$

For this case it is seen that $G(0)$ is not exactly $1/\Delta\nu_D$ but is nearly so. The full width at half maximum for the Doppler distribution can be written as

$$\Delta\nu_D = 2\nu_0 \left(\frac{2(\ln 2)\,kT}{Mc^2}\right)^{1/2} = 7.16\times 10^{-7}\,\nu_0 \left(\frac{T}{A}\right)^{1/2} \tag{4.30}$$

where A is the atomic mass in atomic mass units. Figure 4.7 shows a Doppler line shape as a function of the frequency.

4.1.6 *Collision Times and the Quasi-Static Approximation*

In this section material on the relationship of collision times to the collision cross sections is presented, and material on collisional broadening in the quasi-static approximation is discussed. The average time between collisions is determined by the cross section for the radiating atom to make a collision, σ_c. The average collision time is given by

$$\tau_c = \frac{1}{n\,\langle \sigma_c\, v\rangle} \tag{4.31}$$

where n is the number density of atoms, with which the radiating atoms are colliding, and where $\langle \sigma_c\, v\rangle$ is the average over the velocity distribution of the collision cross section times the relative velocity v of the radiating atom and the colliding atom. In this section a brief discussion of the phase resetting cross section is presented following the analysis of Weisskopf.

The collision cross section depends on the interatomic potentials as illustrated in Figure 4.1. Relatively distant collisions produce phase shifts large enough to result in randomizing the phase of the radiated electromagnetic wave so that the interatomic potentials at large distances determine the collision cross sections for resetting the phase of the radiating atom. At large distances the interatomic potentials are relatively simple and have the following form

$$V(R) \propto \frac{1}{R^p}\,, \tag{4.32}$$

where R is the interatomic separation and p is an integer that depends upon the details of the collision. Accounting for shifts in the interatomic potentials of both the upper (u) and lower (l) molecular levels:

$$\Delta V_u = \frac{C_s^u}{R^s}$$
$$\Delta V_l = \frac{C_m^l}{R^m} \tag{4.33}$$

where the C's are constants, and s and m are integers representing different values of p. In general s and m are not the same integer. The dominant term in the dephasing is due to the longest range potential i.e. the potential with the smaller value of s or m. The potentials are written as ΔV to indicate that only the change in the potentials from their value at $R = \infty$ is important. The integer s (or m) can only take the values 2, 3, 4, and 6 (see Table 4.1). The integer 2 corresponds to the interaction of a hydrogen atom (or hydrogen like ion) in an excited level colliding with an ion or electron where the linear Stark effect shifts the energy levels of the atom proportionally to the electric field produced at the atom by the ion or electron. The electric field from the ion or electron varies as $1/R^2$. The integer 4 corresponds to the interaction of all atoms other than hydrogen colliding

Table 4.1 Summary of collisional broadening interactions. The last column lists the expected temperature dependence of the line width calculated from Eq. 4.41.

R^{-p}	Interaction	Components	Temp.
p=2	linear Stark effect	H with electrons and ions	$T^{-0.5}$
p=3	resonant dipole-dipole	like atoms, one in resonant level	none
p=4	quadratic Stark effect	atoms with electrons and ions	$T^{0.17}$
p=6	induced dipole-dipole	any atom/molecule	$T^{0.3}$

with an ion or electron where the normal (or quadratic) Stark effect shifts the energy levels of the atom proportionally to the square of the electric field at the atom. The integer 3 corresponds to the resonant dipole-dipole interaction. When an atom in an excited level that is connected optically to the ground level collides with an identical atom in the ground level the correct quantum mechanical eigen-function for the system is a linear superposition of the wave function with atom-1 in the excited level and atom-2 in the ground level plus or minus atom-1 in the ground level and atom-2 in the excited level. This means that both atoms must simultaneously have an electric dipole moment. The atomic interaction that shifts the energy levels of the radiating atom is due to the electric field from the dipole moment of one atom interacting with the dipole moment of the other atom so that the energy shift varies as R^{-3} since the electric field of a dipole moment varies as $1/R^3$. The integer 6 corresponds to the Van der Waal's interaction. In this interaction non-resonant atoms interact via an induced dipole-dipole interaction. The electric field from one atom's dipole moment induces a dipole moment on the other atom so that the shift of the atomic energy levels, that is proportional to the product of the dipole moment of the radiating atom times the electric field produced at the radiating atom by the other atom, varies as R^{-6}.

The shift in the frequency radiated during a collisions if s=m is given by

$$\Delta\nu = \frac{\Delta V_u - \Delta V_l}{h} = \frac{C_s^u - C_s^l}{h\,R^s} = \frac{\alpha_s}{R^s}\,. \tag{4.34}$$

If s is not equal to m then the lowest integer produces the dominant interaction at large R so that the form of the frequency shift is the essentially same as when the s=m. The phase shift, Φ, during a collision is given by

$$\Phi = \int_{-\infty}^{\infty} 2\pi\,\Delta\nu\,dt = \int_{-\infty}^{\infty} \frac{2\pi\,\alpha_s}{R^s}\,dt$$
$$= \int_{-\infty}^{\infty} \frac{2\pi\alpha_s}{(b^2+v^2t^2)^{s/2}}\,dt = \frac{2\pi\alpha_s\,a_s}{v\,b^{s-1}} \quad (4.35)$$

where the integration is over the duration of the collision. In this equation for the phase shift v is the relative velocity of the colliding atoms and b is the impact parameter; It isassumed that the trajectory of the atoms during the collision is a straight line. It can be shown that the constant a_s is given by

$$a_s = \int_{-\pi/2}^{\pi/2} \cos^{s-2}\theta\,d\theta \quad (4.36)$$

and that a_2=π, a_3=2, $a_4 = \pi/2$, and a_6=$3\pi/8$.

In order to estimate the cross section for randomization of the phase it is arbitrarily assumed that a phase shift of 1 radian or larger during the collision is necessary for phase randomization. The impact parameter, b_c, that results in a phase shift of 1 radian is

$$b_c = \left(\frac{2\pi\,\alpha_s\,a_s}{v}\right)^{\frac{1}{s-1}} . \quad (4.37)$$

The corresponding collision cross section is given by

$$\sigma_c = \pi b_c^2 = \pi\left(\frac{2\pi\,\alpha_s\,a_s}{v}\right)^{\frac{2}{s-1}} . \quad (4.38)$$

The average collision time is therefore

$$\tau_c = \frac{1}{n\,\langle\sigma_c\,v\rangle} = \frac{1}{n\,\pi\,v_{\mathrm{RMS}}}\left(\frac{v_{\mathrm{RMS}}}{2\pi\,\alpha_s\,a_s}\right)^{\frac{2}{s-1}} \quad (4.39)$$

where in the last expression it is assumed that the average of $\sigma_c\,v$ over the velocity distribution can be approximated by the product of the cross section at the RMS speed times the RMS speed.

The line broadening due to collisions alone is

$$\Delta\nu_C = \frac{1}{2\pi}\frac{2}{\tau_c} = n\,v_{\mathrm{RMS}}\left(\frac{2\pi\,\alpha_s\,a_s}{v_{\mathrm{RMS}}}\right)^{\frac{2}{s-1}} . \quad (4.40)$$

For a gas with a temperature, T, the RMS speed, $v_{\mathrm{RMS}} = (3k\,T/m)^{1/2}$, is proportional to $T^{0.5}$ so that

$$\Delta\nu_C \propto T^{\left(\frac{1}{2}-\frac{1}{s-1}\right)} . \quad (4.41)$$

A common method for determining the type of collision that produces the broadening is to examine the temperature dependence of the collisional line width under conditions where the impact approximation is valid. The magnitude of the line width is used to determine the magnitude of α_s. It is noted that for s=3 the line width is independent of the temperature. Better theories of the impact approximation predict that the line center is also shifted, but the shift of the line center is usually much less than the line broadening.

The collisional line broadening in the situation where τ_d is much larger than τ_c is considered next. The impact approximation is a theory that is applicable at low densities, and the quasi-static approximation is a theory that is applicable at high densities. In this approximation the radiating atom is essentially always participating in a collision. The probability of an atom being in a shell between R and $R + dR$ of the radiating atom is

$$P(R)\, dR = \frac{4\pi\, R^2\, N}{V}\, dR \tag{4.42}$$

where V is the volume of the gas, and N is the total number of atoms in the volume V. An atom at a distance R from a radiating atom produces a frequency shift given by

$$\Delta\nu = \frac{\alpha_s}{R^s} \tag{4.43}$$

so that

$$R = \left(\frac{\alpha_s}{\Delta\nu}\right)^{1/s} . \tag{4.44}$$

Therefore the probability of the frequency of a radiating atom having its frequency shifted by an amount between $\Delta\nu$ and $\Delta\nu + d\Delta\nu$ is

$$\begin{aligned} P(\Delta\nu)\, d\Delta\nu &= \left(\frac{4\pi\, N}{V}\right)\left(\frac{\alpha_s}{\Delta\nu}\right)^{2/s} \frac{\alpha_s^{1/s}}{\Delta\nu^{(s+1)/s)}}\, d\Delta\nu \\ &= \left(\frac{4\pi}{V}\right) \frac{\alpha_s^{3/s}}{\Delta\nu^{(s+3)/s}}\, d\Delta\nu . \end{aligned} \tag{4.45}$$

The line shape is proportional to $P(\Delta\nu)$ so that

$$g(\nu - \nu_0) \propto \frac{4\pi\, N}{V} \frac{\alpha_s^{3/s}}{(\nu - \nu_0)^{(s+3)/s}} . \tag{4.46}$$

This expression for the line shape is valid only if the magnitude of $\Delta\nu = \nu - \nu_0$ is small enough to correspond to an R large enough that the expressions for $\Delta\nu$ at large distances are valid, and $\Delta\nu$ is large enough to be removed from line center. The line shape is treated as due to a single perturber in the shell of thickness dR. This expression breaks down near line center. The line shape does not diverge as $1/(\nu - \nu_0)^{(s+3)/s}$ near line center. In order to treat the line shape near line center one must consider the interactions from distant perturbers i.e. one must consider the perturbation for very large R. For large R the volume $dV = 4\pi R\,dR$ is large and the effects of multiple perturbers in the shell of thickness dR must be treated. This is beyond the scope of this book.

The line shapes obtained in the quasi-static approximation are homogeneous but are not in general Lorentzian. In the case of s=3 the wings of the line shape in the quasi-static approximation *are* the same as for a Lorentzian line shape.

For densities between those where the impact approximation and the quasi-static approximation are valid there are no simple theories for the line shape. In these cases experiment is the best guide to line shape and line width.

4.1.7 *Example: Sodium D_1 Line*

It may be useful at this point to determine the line shapes and line widths for a particular transition under different conditions. Before discussing a particular example it should be pointed out that many, even most, of the lines observed in absorption or emission are not the result of a transition from one level to another, but are instead are the result of transitions from one or several closely spaced levels to one or several closely spaced levels. If the separation of the transitions is smaller than the line widths of the individual transitions then the observed line is a blending of the several transitions. The shape of a blended line is obtained simply by taking the linear sum of the absorption cross sections for the individual transitions.

As an example consider the line shape for the $3^2S_{1/2} \rightarrow 3^2P_{1/2}$ absorption transition in sodium at 589.6 nm (also known as the D_1 line). The electronic angular momentum is 1/2 for both the upper and lower levels. The measured lifetime of the upper level is τ_u=16 ns. Since the lower level is the ground level, the natural radiative line width depends solely on the lifetime of the upper level, $\Delta\nu_R = 1/2\pi\,\tau_u = 1.0 \times 10^7$ Hz. If the Na vapor is maintained at a temperature of T=600 K the Doppler line width is much larger than the natural radiative line width being $\Delta\nu_D$=1.9× 10^9 Hz. In fact for the Doppler line width to be as small as the natural radiative line width the temperature would have to be T=1.7×10^{-2} K. While this

Table 4.2 Collisional broadening of the 589.6 nm $3^2S_{1/2} \rightarrow 3^2P_{1/2}$ D_1 line of sodium by various collision partners.

Perturber	(MHz cm^3)	Temperature	Source
Na	3.1×10^{-13}	-	Ref. a
He	6.1×10^{-16}	460-490 K	Ref. b
Ne	3.6×10^{-16}	460-490 K	Ref. b
Ar	8.8×10^{-16}	460-490 K	Ref. b
Xe	8.4×10^{-16}	460-490 K	Ref. b
e^-	1.0×10^{-13}	T_e=0.4 eV	Ref. c
	2.4×10^{-13}	T_e=3.4 eV	Ref. c

[a] J. Huennekens and A. Gallagher, Phys. Rev. A **27**, 1851 (1983).
[b] D. G. McCartan and J. M. Farr, J. Phys. B **9**, 985 (1976).
[c] Hans R. Griem, *Spectral line broadening by plasmas*, (1974).

seems exceptionally small, temperatures much lower than T=17 mK can be obtained by laser cooling of atoms.

The total homogeneous line width of the $3^2S_{1/2} \rightarrow 3^2P_{1/2}$ transition contains contributions from both the natural radiative line width and collisional broadening. Along the lines of Equation 4.40, the collisional width can be written as $\Delta\nu_C = \beta n$ where n is the number density of colliding atoms per cubic centimeter and the factor β depends on the collision partner and the type of collision process (Sec. 4.1.6). Depending upon the collision interaction, *beta* will have a temperature dependence $T^{1/2-1/(s-1)}$. For the sodium D_1 line, the upper level of the transition is a resonance level, so for Na-Na collisions the s=3 resonant dipole-dipole interaction term dominates and there is no temperature dependence. For collisions between sodium and a foreign gas, the s=6 induced dipole-dipole interaction dominates and β depends weakly on the gas temperature. Collisional line broadening has been widely studied for Na, with some sample values of β listed in Table 4.2. As an example, let us calculate the pressure of helium gas that needs to be added to a sodium cell heated to 600 K to make the collisional width equal to the Doppler width. At 600 K, the Doppler width of the D1 line is $\Delta\nu_D = 1.9\times10^9$ Hz, so $n = 3.1\times10^{18}$ cm^{-3} which corresponds to a pressure of 190 Torr. It has also been found that for Na in He gas the center frequency for the $3^2S_{1/2} \rightarrow 3^2P_{1/2}$ transition is shifted by $\delta\nu = (6\times10^{-17}$ Hz cm$^3)\, n$. This shift (which is much smaller than the broadening of the line) arises from the shape of the potential energy curves of the temporary NaHe molecule that is formed during collisions (see Fig. 4.3).

Low-pressure sodium lamps (as might be found in a laboratory) operate at temperatures around 525 K. The vapor pressure of sodium at this temperature is approximately 2 mTorr, which yields a number density of 5×10^{13} cm^{-3}. Based upon the values in Table 4.2, the collisions between sodium atoms broaden the D1 line by $\approx$ 15 MHz. In addition to the sodium, these lamps also contain about 10 Torr of argon that broadens the D_1 line by 160 MHz. So the collisional broadening in a low-pressure sodium lamp is dominated by the argon buffer gas, but is still much less than the Doppler broadening (1.7 GHz). In contrast, consider a high-pressure sodium lamp (such as those used as street lamps) that operates at a temperature around 920 K and is filled with 400 Torr of argon. Due to sodium's rapid rise in vapor pressure with temperature, the partial pressure of sodium is , 60 Torr (versus 2 mTorr in the low-pressure lamp), at this temperature. As a result, in high pressure lamp the broadening is dominated by the self-broadening with $\Delta\nu_L \approx 200$ GHz (which is also well in excess of the Doppler width). The light emission in both types of lamps is driven by electron-atom collisions, and collisions between excited Na(3^2P$_{1/2}$) atoms and slow-electrons will result in some additional Stark broadening, but this is negligible in comparison to the atom-atom broadening.

So far, the 3^2S$_{1/2} \rightarrow 3^2$P$_{1/2}$ line as a single transition has been considered, but in reality things are a bit more complicated. The nuclear spin of ^{23}Na is I=3/2, so that both the upper and lower levels have two sub-levels with total angular momenta (the sum of the electronic and nuclear angular momenta) of F=2 and 1. The F=2 and F=1 sub-levels are called hyperfine levels and the energy separation between the hyperfine levels of an given electronic level is called the hyperfine energy of the electronic level. The hyperfine levels in the ground 3^2S$_{1/2}$ electronic level are separated by 1772 MHz, and the hyperfine levels in the excited 3^2P$_{1/2}$ electronic level are separated by 192 MHz. Thus, in order to understand the 3^2S$_{1/2} \rightarrow 3^2$P$_{1/2}$ absorption one must take the linear sum of the absorption cross sections for all four hyperfine transitions corresponding to $F=2 \rightarrow F=2$, $F=2 \rightarrow F=1$, $F=1 \rightarrow F=2$, and $F=1 \rightarrow F=2$ transitions with the line shape and width appropriate to the temperature and pressure of the system. This is discussed further in Sec. 5.5.

4.2 Line Shapes in Molecular Gases

The line shapes of atomic gases are sufficient for understanding many of the laser systems that are discussed in Chapters 10-9. These include the He-Ne laser, the Ar$^+$ ion laser, and the copper vapor laser. Nevertheless, many other gas lasers use molecular gases such as N_2, CO_2, or HF. The line shapes of these molecules are much the same as those of atoms after accounting for the more complicated energy levels of molecules.

4.2.1 *Molecular Energy Levels*

First let us consider a simple diatomic molecule made up of two atoms. If one envisions one atom being brought up from an infinite distance towards the other atom then the two atoms will interact. At very short ranges, this interaction is highly repulsive as the electron clouds overlap. At long ranges, however, the net interaction can be either attractive or repulsive, typically depending upon the spin of the electrons and the occupancy of the various molecular orbitals. If the interaction is attractive at large distances, a potential well is formed by the interaction of the atoms as a function of the interatomic separation. In this situation the two atoms can be bound as a stable molecule, with a 'bond length' equal to the interatomic separation corresponding to the minimum of the potential well. If the interaction is repulsive at all separations, no potential well is formed and the atoms repel each other. For example when two ground level hydrogen atoms are brought together if the spin of the two electrons is a triplet (a symmetric spin state) then the spatial state must be antisymmetric, which means that the electron density between the two protons is low, and the resulting electrostatic repulsion of the nuclei results in a repulsive interatomic potential with no bound state. On the other hand, if the spin state is a singlet (an antisymmetric spin state) then the spatial state must be symmetric and the interatomic potential is attractive. A H_2 molecule is bound by this potential well. Another way the atoms can be different from each other is if one of the atoms is in an excited electronic level. This is the situation in the the KrF excimer laser previously discussed in Sec. 3.2. When both the Kr and F atoms are in their ground states, the potential is repulsive. In contrast, if the krypton atom is an electronically excited state, a bound molecule can be formed.

The different electronic energy levels of a diatomic molecule are typically designated with a combination of Roman letters (a,b,c,...,A,B,C,...) and Greek letters for the quantum numbers ${}^{(2S+1)}\Lambda$ where S is the spin angular momentum quantized along the internuclear axis and Λ is the component of the electronic orbital angular momentum along the internuclear axis. The values of Λ are denoted by Greek letters with Σ and Π indicating zero and one unit of electronic orbital angular momentum along the internuclear axis.

The symbol 'X' is used to designate the ground state. Higher energy levels (for both bound and repulsive curves) are typically labeled starting with 'A' or 'a', with the case of the letter depending upon the spin multiplicity. The designation can also contain a superscript '+' or '-' and a subscript 'g' or 'u' to denote various symmetries of the wave function. For a Σ^+ level, the superscript '+' denotes a level where the wave function is the same about a reflection through any plane passing through the internuclear axis, whereas it reverses sign for a Σ^- level. For homonuclear molecules (i.e., H_2 or N_2 but not HF or KrF) a subscript is added to denotes whether the wave function is symmetric ('g') or asymmetric ('u') under an interchange of the two nuclei. For example, the nitrogen laser operates on transitions between the $N_2(C\ ^3\Pi_u)$ and $N_2(B\ ^3\Pi_g)$ levels.

For a given electronic configuration, the potential energy of the molecule as a function of the internuclear separation can be described empirically using a Morse potential,

$$V(R) = D_e \left[1 - \mathrm{e}^{-\beta(R-R_0)}\right]^2, \tag{4.47}$$

with three parameters: the depth of the potential well D_e, the equilibrium atomic separation R_0 and an additional constant β. The bottom of the potential well is essentially a parabola, so near the minimum it can be approximated as a simple harmonic oscillator varying as $k(R-R_0)^2$ where r is the interatomic separation, and $k = D_e\beta^2$ is the spring constant for the potential well. As a result, in addition to electronically excited states as in atoms, molecules also have vibrational energy levels corresponding to oscillations of the nuclei. The solution of the Hamiltonian for a simple harmonic oscillator is covered in introductory quantum mechanics. The energy of these vibrational levels is approximately given by

$$E_v = \left(v + \tfrac{1}{2}\right)\, h\, f_v \tag{4.48}$$

where v is an integer (the vibrational quantum number), $f_v = (1/2\pi)(k/m)^{1/2}$ is the lowest vibrational frequency of the diatomic molecule and m is the reduced mass. This is only an approximation since the potential is not a pure simple harmonic well. Only the lower levels are more-or-less equally spaced, whereas the higher vibrational levels become more closely spaced as the potential energy curve deviates from a simple parabola. Of course the vibrational levels of the well can go no higher than the dissociation energy of the molecule i.e. up to an energy where the two atoms have an energy as large as the energy of the atoms when they are far apart.

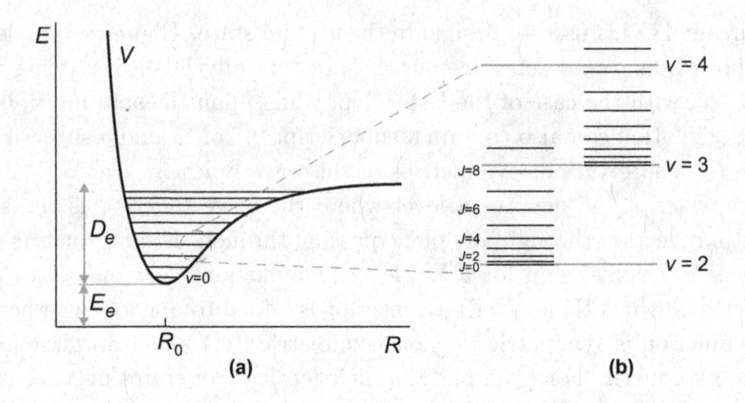

Fig. 4.8 (a) The vibrational levels of a diatomic molecule in an electronic potential well. Note that only the first few levels are equally spaced. (b) Sample rotational levels for a sample vibrational level. For clarity, only the lowest v=0 vibrational level is labeled in (a) and only the even values of J are labeled in (b). Compared to a typical spacing between vibrational levels, the rotational spacing has also been greatly ($\times$20) exaggerated.

A diatomic molecule can also rotate and has rotational energy levels. Assuming the distance between the nuclei is fixed, the rotational Hamiltonian of a diatomic molecule is $H = J^2/(2I)$ where J is the rotational angular momentum and I is the moment of inertia of the molecule. As a result, the rotational energy levels of a molecule are equal to

$$E_r = h^2\, J(J+1)/(8\pi^2 I) = J(J+1)\; h\, f_R\;, \tag{4.49}$$

where J is the rotational angular momentum quantum number of the level and $f_R = h/(8\pi^2 I)$. This too is only an approximation for a 'real' molecule, and the rotational energies deviate from this simple formula. For example, due to vibrations, the average distance between nuclei varies with the vibrational quantum number. Hence, the moment of inertia I and resulting f_R are functions of v. Figure 4.8 shows an interatomic potential well and the vibrational and rotational levels associated with it. The total energy of a molecular energy level is given by

$$E = E_e + \left(v + \tfrac{1}{2}\right)\, h\, f_v + J(J+1) h\, f_R\;, \tag{4.50}$$

where E_e is the electronic energy component, equal to the energy of the bottom of the electronic potential well. Note that the energy of the lowest possible level (with v=0 and J=0) is not quite at the bottom of the well, but $1/2\, h\, v_f$ above the bottom. It is stressed that the energy levels are given by Equation 4.50 only for levels with low v and J.

The separability of electronic, vibrational and rotational components of the molecular energy levels is possible because there is a natural hierarchy among these components. The electronic excitations are typically much larger than the vibrational excitations, and vibrational excitations are typically much larger than the rotational excitations. For example, let us consider the N_2 molecule. For the N_2 ground level, the spacing between first two vibrational energy levels is 2360 cm^{-1}, whereas the spacing between the first two rotational energy levels is only 4 cm^{-1}. These differences are in turn, dwarfed by the energy difference between the ground level and first excited electronic level that is greater than 50,000 cm^{-1}.

Since the difference between rotational energy levels is typically much less than room temperature (approximately 200 cm^{-1}), molecular gases are characterized by a rotational population distribution rather than all the molecules being in the lowest J=0 level.

4.2.2 *Molecular Spectra and Line Shapes*

There are three different types of molecular transitions. In an *electronic transition* the molecule decays from a particular vibrational-rotational (vibro-rotational) level in an electronically excited level to a particular vibro-rotational level in a lower electronic level. In a *vibrational transition* the molecule makes a transition from a particular vibrational-rotational level in a given electronic potential well to another vibrational-rotational level in the same potential well. In a purely *rotational transition* the molecule makes a transition from a particular rotational level in a given vibrational level to another rotational level in the same vibrational level. The electronic transitions have relatively large energies and typically emit photons in the near-infrared, visible, or ultraviolet regions of the spectrum. The vibrational transitions are typically in the infrared region of the spectrum, and the rotational transitions are typically in the far-infrared region of the spectrum. The N_2 laser (Sec. 11.4) operates on an electronic transition and lases at 337.1 nm in the near-ultraviolet, while the CO_2 (Sec. 11.1) and HF (Sec. 11.8) lasers both operate on vibrational transitions and lase in the infrared.

The selection rules for molecular quantum numbers depend upon the type of transition. For example, rotational and vibrational transitions are only allowed in a molecule with a permanent dipole moment. For most types of transitions, there is no firm restriction on Δv, but only transitions with $\Delta J = \pm 1$ are allowed (with $\Delta J = 0$ also allowed in some cases).

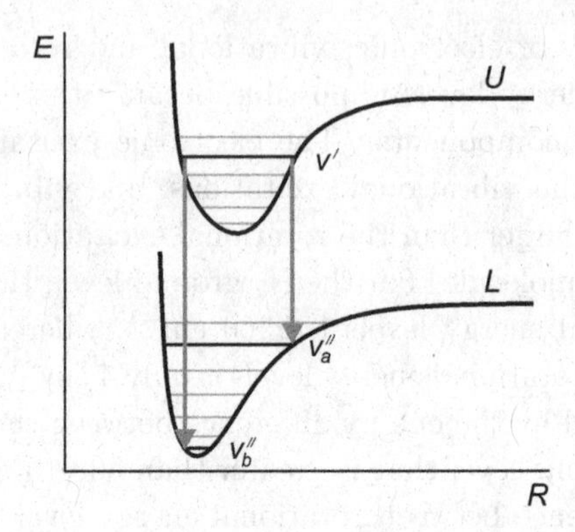

Fig. 4.9 The Frank-Condon principle states that the Einstein A coefficient out of the indicated v' vibrational level of the upper U electronic level is largest for the v_a'' and v_b'' vibrational levels of the lower electronic level L. Levels with intermediate v'' values, as well as those with $v'' > v_a''$ will tend to have smaller, but generally non-zero, A values.

There is, however, one major consideration for molecular radiation that merits discussion. The decay from an electronically excited level to a lower electronic level occurs very fast compared to the vibration times. Thus the atomic nuclei must have the same separation and momentum after the radiation is emitted as before, i.e. the radiation must go from a given interatomic separation R and momentum in the upper potential well to the same separation R and momentum in the lower potential well. This is described by saying the transition is vertical as it goes from the upper level to the lower level since R is unchanged. This is called the *Frank-Condon principle* and has a strong effect on the Einstein A coefficients for the radiation. For a classic harmonic oscillator, the kinetic energy and hence the momentum is zero only at the turning points at either side of the potential well. This means that a molecule spends much more time near the turning points than moving through the interior of the potential well. Hence the most probable electronic decays of a molecule from an excited electronic level occur when the molecule is near a classical turning point with the molecule radiating to a level in the lower electronic well that is also near a turning point that has the same interatomic separation and momentum. This results in relatively large Einstein A coefficients for these transitions and relatively small A coefficients for the transitions to other levels in the lower electronic well. A quantum analysis yields the same result due to the overlaps of the wave functions. This effect is illustrated in Fig. 4.9.

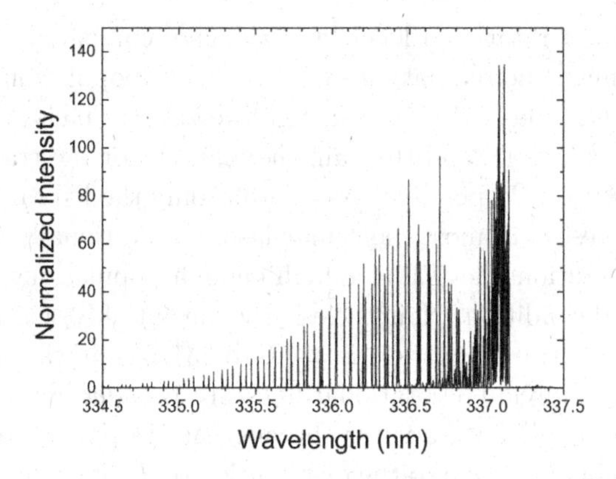

Fig. 4.10 Synthetic N_2 emission spectrum for the C $^3\Pi_u \rightarrow$ B $^3\Pi_g$ (0,0) band for a rotational (gas) temperature of 300 K.

Because of the large number of vibrational and rotational levels within a particular electronic level, molecular spectra are very much more complicated than atomic spectra. A *system* is composed of all emission lines due to transitions from one electronic level to another. A system is composed of a number of *bands*, where each band is the collection of lines from the vibrational level v' to the lower level v''. A band, in turn, is composed of a large number of individual emission lines correspond to all the possible J' to J'' transitions. For example, in nitrogen, the *Second Positive system* consists of all the emission lines from decays out the C $^3\Pi_u$ level into the B $^3\Pi_g$ level. The N_2 laser lases on the $v' = 0 \rightarrow v'' = 0$ band. Initially the molecules start out in the X $^1\Sigma_g^+$ ground level. At room temperature, only the $v = 0$ vibrational level is substantially populated, but well over twenty N_2 rotational levels are substantially populated. Electron collisions excite the molecules to the C $^3\Pi_u$ level. Electron collisions do not appreciably alter the rotational distribution of the molecules, so within the $v' = 0 \rightarrow v'' = 0$ band are many, many rotational lines as illustrated in Fig. 4.10. Interestingly, the spectrum for a laser is considerably simpler since only lines with gain will be present. A plain N_2 discharge, following the selection rules for J will exhibit transitions corresponding to both the P-branch ($\Delta J = J' - J'' = -1$), Q-branch ($\Delta J = 0$), and R-branch ($\Delta J = +1$). For an N_2 laser, the gain (and thus the intensity) of the individual transitions depends on the population inversion between the upper and lower laser levels. Consider the choices for an excited N_2 molecule in a particular J' upper rotational level. Since the rotational energy spacing between the lower rotational levels is very small compared to thermal energies, the primary factor that affects the relative populations of the lower rotational levels is the $(2J''+1)$ statistical weighting. Thus, the population inversion will be largest, and the gain the highest, for the transition to the level with the lowest J'' possible. As a result, only the P-branch transitions are typically present in most molecular lasers. Additionally, only a few of the upper J' rotational levels have high enough populations to satisfy the lasing threshold condition. Nonetheless, the output of N_2 laser still consists of over ten strong transitions in the 337.0 to 337.14 nm wavelength range.

Each of the individual rotational lines are broadened by the same effects as atomic lines, i.e. radiative broadening, Doppler broadening, and collisional broadening. Due to the distribution of J rotational values across the different molecules, the the rotational structure of a band is an extreme form of inhomogeneous broadening. When a molecular emission spectrum is viewed with a low resolution spectrometer, the lines of a molecular band appear to blend together to form a near continuum even though it is actually made up of many different lines arising from different transitions. Nonetheless, if one excites a particular electronically excited vibrational-rotational level by using a very narrow bandwidth laser then one can observe the decays from that particular level. True continuous molecular spectra occur when an electronically excited molecule decays to a repulsive lower electronic level such as in the KrF laser (page 46). The line shape for a continuous spectra of this type depends on width of the potential well of the upper level as it maps onto the repulsive lower level. Additional details on vibrational transitions will be covered in Chapter 9 when the HF and CO_2 lasers are discussed.

4.3 Line Shapes in Liquids

The only significant laser that uses a liquid medium is the tunable dye laser (Sec. 12.2). Dyes are complicated molecules that are dissolved in a liquid solvent such as ethyl alcohol. As such, they are very similar to molecular gases, with the significant difference that the mean collision time between the dye and solvent is much shorter than the radiative lifetime of excited levels. This has two primary effects on the line shape.

First, due to quenching by the solvent, the only vibrational-rotational levels with any significant population are those within about kT of the bottom of the potential well for both the ground level and electronically excited levels. Dye molecules in the ground level absorb high energy photons from an external light source (usually a laser), and initially populate the high vibrational-rotational levels of an excited electronic level. They are then rapidly, non-radiatively, quenched to the lowest vibrational-rotational levels of the electronically excited level. These levels can decay into almost all vibrational-rotational levels (or at least those satisfying the Frank-Condon criteria) of the ground level. The result of this sequence of events is that the absorption band for a dye solution is at shorter wavelengths than the emission band for dyes dissolved in a solvent. This reduces any problems with the dye solution absorbing its own radiation. Regardless of what vibrational-rotational level the excited dye molecule decays to, the dye is rapidly quenched back to a level within kT of the bottom of the potential well. Thus the lower level of the laser transition usually has a very low population. As a result, it is quite easy to achieve a population inversion between the upper and lower laser levels as solvent collisions cause excited dye molecules to pool in the upper laser level *and* rapidly drain out of the lower level.

The second effect of the solvent collisions is that the homogeneous packet that results from collisional broadening is larger than kT near room temperature. Combined with the close packing of vibrational-rotational levels in the dye molecule, the wavelength of a dye laser is continuously tunable over a wide wavelength range. Additional details on dye lasers are found in Chapter 12.

4.4 Line Shapes in Solids

Many important classes of lasers use a solid as the active gain medium. Broadly speaking, these lasers can be classed into two catagories. The first, use ions in crystal lattices as the lasing medium. The second type of solid state laser is semiconductor diode lasers. Today, solid state lasers are far more common than gas lasers, but the previously discussed line shapes for atomic and molecular gases serve as a framework for understanding the line shapes of these laser systems.

4.4.1 *Ions in Crystal Fields*

There are several lasers that use ions in crystal lattices as the lasing medium. For example, the first laser created was the ruby laser utilizing Cr^{3+} ions in a sapphire (Al_2O_3) lattice. The Cr^{3+} impurity ion substitutionally replaces an Al^{3+} ion in the lattice. Another laser that utilizes

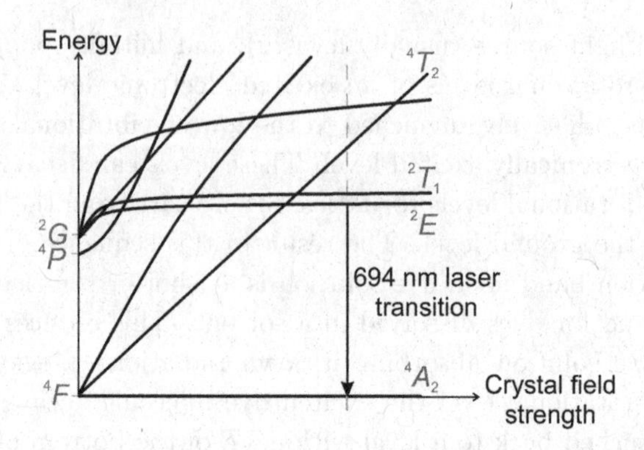

Fig. 4.11 Energy levels of the lowest levels of a d^3 configuration as a function of the octahedral crystal field strength. Unperturbed atomic levels (left) are labeled using *LS* notation, the perturbed levels are labeled using group theory notation (right). Dashed vertical line corresponds to the approximate field strength in the Al_2O_3:Cr^{3+} Ruby crystal. Laser emissions at 694 nm arise from the $^2E \rightarrow {}^4A_2$ transition. Adapted from Fig. 5 of L. G. Van Uitert in *Luminescence of Inorganic Solids* (1966).

ions in a crystal lattice is the Nd:YAG laser, with a Nd^{3+} ion in a yttrium aluminum garnet (YAG) crystal. The key common denominator in both lasers is the use of a transition metal ion. Let us first consider the case of Cr^{3+} in the ruby laser.

The ground state configuration of the Cr atom is $1s^2 2s^2 2p^6 3s^2 3p^6 4s^2 3d^4$. Cr^{3+} is formed by removing the two $4s$ electrons and one of the $3d$ electrons, leaving $3d^3$ as the open orbital. Consider the energy levels of an isolated Cr^{3+} ion. There are eight allowed *LS*-terms for three equivalent d electrons. From Hund's rule, the one with the highest multiplicty lies lowest, and of these the one with the largest L is the ground level. This is the ^4F level; progressively higher excited levels arise from the ^4P and ^2G terms. Since these higher levels arise from the same electronic configuration as the ground level, they all have the same parity and are dipole-forbidden to decay. The forbidden $^2G \rightarrow {}^4F$ transition, with a wavelength around 700 nm, has an A_{ul} value around 0.1 s^{-1} (a 10 s lifetime). From Eq. 4.6, the natural line width of this line would be very small. Now consider the effect of placing the ion in a crystal. The electric field of the surrounding lattice atoms perturbs the ion's energy levels. Normally, one would expect that this would smear out the energy levels into a band structure (as covered in Chapter 8), but in this case the effect of the crystal field on the ^2G term is relatively mild as indicated in Fig. 4.11. The ^2G energy levels vary only slightly with the size of the crystal field strength.

In the crystal, the levels are labeled using group theory notation, so the ^2G term becomes the ^2E. Vibrations of the Cr^{3+} ions within the Al_2O_3 lattice leads to some line broadening in a way similar to the collisional broadening discussed in Sec. 4.1.2. Due to the flatness of the ^2E curve, however, the vibration induced line broadening is minimal. Nevertheless, the line width still increases with temperature as the ions rattle around inside the lattice. Due to the regular nature of the crystal structure, the perturbation felt by every ion is the same, and the line is thus homogeneously broadened. At room temperature, $\Delta\nu \approx 11$ cm^{-1}. Other Cr^{3+} levels are broadened into wide bands, such as $^4F_1$ and $^4F_2$ bands (group theory notation) which absorb light from the blue and green regions of the spectrum. Similar to what happens to vibrationally excited molecules in dye lasers, ions pumped into these two bands are rapidly transfered into the ^2E level by non-radiative processes. The ruby laser operates on the emissions from one of the ^2E levels to the ground state with a wavelength of 694 nm.

The Nd:YAG laser also uses emissions from a transition metal ion in a crystal lattice. The Nd^{3+} ion has an open $4f^3$ shell. Three identical f electrons yield 17 allowed LS terms. The lowest two terms in order of increasing energy are ^4I and ^4F. In a crystal, the perturbation caused by the crystal field on the $4f$ electrons is reduced by the shielding provided by the outer $5s$ and $5p$ electrons, so that the ^4I and ^4F levels act more as discrete atomic levels rather than bands. Nevertheless, the field causes some mixing of the $4f$ orbital with other configurations which increases the strength of the otherwise forbidden $^4F \rightarrow {}^4I$ laser transition. As in the Ruby laser, ions are first optically pumped into a broad absorption band that decays non-radiatively into the upper $^4F_{3/2}$ laser level. Unlike the Ruby laser, the lower laser level is not the ground level, $^4I_{9/2}$, but the $^4I_{11/2}$ level approximately 2100 cm^{-1} above the ground level. At room temperature, this level is minimally populated, making it much easier to achieve a population inversion than in the Ruby laser as is addressed in Sec. 6.2.

At room temperature, the line width of the Nd:YAG laser operating at 1.06 μm is approximately 6 cm^{-1}. As with the Ruby laser, the regularity of the YAG crystal produces a homogeneous line. The $Nd^{3+}(^4F \rightarrow {}^4I)$ transition is also used in a number of other laser systems which use different host materials in replacement of the YAG crystal. Since the crystal field strength varies from one material to another, the host medium affects the wavelengths of both the absorption band and emission line, although the effect is typically much larger for the absorption bands. In addition to crystalline solids, ions can also be dissolved in glass or silica. Since glass is an amorphous material instead of a regular lattice, the spacing between each Nd^{3+} ion and its nearest neighbors is not fixed, so each ion feels a slightly different perturbation. As a result, the Nd:Glass laser has a inhomogeneous line width that is much greater than that of the Nd:YAG laser. Besides Nd, other rare earth elements dissolved in glass or silica fibers are also widely used as laser mediums, these are covered in Sec. 9.2.

A third solid state laser medium of great importance is the Titanium:sapphire laser (Sec. 9.4) which is widely used as a tunable laser. In this laser the Ti^{3+} ion replaces an Al^{3+} ion in the sapphire lattice. Unlike the Cr^{3+} and Nd^{3+} energy levels in the Ruby and Nd:YAG lasers, the energy levels of the Ti^{3+} ions in the sapphire crystal are more molecular-like than atomic-like. The Ti^{3+} ion vibrates about its equilibrium position. The three dimensional vibration levels are very closely spaced, forming a near continuum set of energy levels above both the ground level and above the lowest excited electronic level of the Ti ion. If a Ti^{3+} ion is in a highly excited vibrational level it rapidly decays by non-radiative processes to the lowest vibrational level of the electronic level (or more properly to one within kT of that level). If a Ti ion in the ground electronic level ($^2T_2$) absorbs a photon and is excited into a highly excited vibrational level of the first electronic level (^2E), it then rapidly decays non-radiatively to the lowest vibration levels of this excited electronic level. The excited ion then radiates to any of the excited vibration levels of the ground electronic level after which it decays non-radiative to the lowest vibration levels of the ground electronic level (see Fig. 9.6 on p. 226). Thus the absorption spectrum of the Ti^{3+} ion is at shorter wavelengths than the emission spectrum. In this sense the Ti:sapphire laser crystal is similar to the liquid dye laser. The homogeneous packet of the Ti:sapphire crystal is much wider than kT so that the line is effectively homogeneous.

4.4.2 *Line Shapes for Electrons in Semiconductors*

Semiconductors constitute another type of solid state laser medium (Chapter 8). Semiconductors like all crystalline solids have energy bands. An energy band consists of a large number of electron energy levels that are so closely spaced that they are essentially continuous. Energy bands are separated by forbidden gaps where there are no allowed electron energy levels. The highest nearly filled band in a semiconductor is called the valence band, and the lowest band that is nearly empty is called the conduction band.

It is possible to produce a situation where the valence band has a number of empty levels within kT of the top of the valence band, and where there are electrons in the conduction band. Interactions with the phonons in the semiconductor rapidly cause any electrons in the conduction band to decay into levels within about kT of the lowest levels in the band. The electrons in the conduction band can emit a photon and decay to the valence band. It is this radiation that is stimulated in a semiconductor laser. To be a little more precise in most situations the semiconductor is doped with impurities and the laser light results from the transitions between levels produced by these impurities. This subject is very complicated and is discussed in detail in the chapter on semiconductor lasers. Since electrons in levels within kT of the bottom of the conduction band are decaying into levels within kT of the top of the valence band, the width of the radiative transition is of the order of kT. Because an electron in the conduction band can radiate to any empty level in the valence band and because the phonons rapidly move the electrons in a band from one level to another the transition is essentially homogeneous.

Summary of Key Ideas

- Radiatively and collisionally broadened lines (in the impact approximation) have homogeneous Lorentzian line shapes. The normalized Lorentzian line shape is given by

$$g(\nu - \nu_0) = \frac{\frac{\Delta\nu_L}{2\pi}}{(\nu - \nu_0)^2 + \left(\frac{\Delta\nu_L}{2}\right)^2} \tag{4.5}$$

where the Lorentzian line width $\Delta\nu_L$ is equal to

$$\Delta\nu_L = \frac{1}{2\pi}\left(\frac{1}{\tau_u} + \frac{1}{\tau_l} + \frac{2}{\tau_c}\right) \tag{4.18}$$

- Doppler broadened lines have inhomogeneous Gaussian line shapes given by

$$h(\nu - \nu_0) = \frac{2(\ln 2)^{1/2}}{\pi^{1/2}\,\Delta\nu_D}\, e^{-\frac{4(\ln 2)(\nu-\nu_0)^2}{\Delta\nu_D^2}} \tag{4.26}$$

with a Doppler width of

$$\Delta\nu_D = 2\nu_0\left(\frac{2(\ln 2)\,kT}{Mc^2}\right)^{1/2} = 7.16\times 10^{-7}\,\nu_0\left(\frac{T}{A}\right)^{1/2} \tag{4.30}$$

- Combining a Doppler distribution with a Lorentzian shape results in a Voigt line shape (Eq. 4.28).

- In the *impact approximation* (Sec. 4.1.3), the time between collisions (τ_c) is much larger than the duration of the collisions (τ_d). The opposite limit when τ_c is much smaller than τ_d is called the *quasi-static approximation* (Sec. 4.1.6).
- Line shapes in molecules are similar to those in atoms, but there are many more transitions. Transitions between individual vibrational-rotational levels of molecules typically have both a natural (radiaitve) line width component and Doppler component. But each transition between different vibrational levels is broken into a large number of transitions depending upon the rotational quantum numbers (Sec. 4.2).
- The probability of a transition (for either absorption or emission) between two vibrational levels of different electronic states can be understood in terms of the Frank-Condon principle (p. 80).
- Line shapes for liquids and ions in crystal fields can be described using the same tools as used for atomic line shapes. The Ruby laser and the Nd:YAG laser, both have narrow homogeneous line shapes. The Nd:Glass laser has a broad inhomogeneous line shape. The Titanium:sapphire laser is a very useful laser with a tunable wavelength with a homogeneous line shape. Despite arising from transitions between bands, the line shape of semiconductor lasers typically have homogeneous line widths on the order of kT.

Suggested Additional Reading

Anne P. Thorne, *Spectrophysics 2nd Ed.*, Chapman and Hall (1988).

William T. Silfvast, *Laser Fundamentals*, Cambridge University Press (1996).

J. O. Hirschfelder, C. F. Curtis, and R. B. Bird, *Molecular Theory of Gases and Liquids*, John Wiley and Sons (1964).

A. C. G. Mitchell and M. W. Zemansky, *Resonance Radiation and Excited Atoms*, Cambridge University Press (1961).

A. E. Siegman, *An Introduction to Lasers and Masers*, McGraw Hill (1971).

A. E. Siegman, *Lasers*, University Science Books (1986).

V. Weisskopf, "Die Breite der Spektrallinien in Gasen [The width of the spectral lines in gases]" *Phys. Z.* **34**, 1 (1933).

V. Weisskopf and E. Wigner, "Berechnung der natürlichen Linienbreite auf Grund der Diracschen Lichttheorie [Calculation of the natural line width due to the Dirac light theory]" *Z. Phys.* **63**, 54 (1930).

R. G. Breene, "Chapter 1: Line Width", in *Handbuch Der Physik Vol. XXVII Spektroskopie I [Encyclopedia of Physics Vol. 27, Spectroscopy I]*, Springer-Verlag (1964).

D. C. Sinclair and W. E. Bell, *Gas Laser Technology*, Holt, Rinehart, and Winston (1969).

Gerhard Herzberg, *Molecular Spectra and Molecular Structure, I. Spectra of Diatomic Molecules*, Kriger (1989).

L. G. Van Uitert, "Chapter 9: Luminescence of Insulating Solids for Optical Masers" in *Luminescence of Inorganic Solids*, Academic Press (1966).

Problems

1. The lifetime of the $2p$ level in atomic hydrogen is 1.6 ns. What is the radiative width of the $2p \rightarrow 1s$ transition in both frequency and in wavelength units?

2. If an atomic hydrogen gas with a temperature of 1500 K is radiating on the $2p \rightarrow 1s$ transition what is the Doppler line width? How many homogeneous packets lie under the inhomogeneous line width?

3. At what temperature is the Doppler width of the $2p \rightarrow 1s$ transition in atomic hydrogen equal to the natural radiative width?

4. For atomic Na in Ne gas the $3^2\mathrm{P}_{1/2} \rightarrow 3^2\mathrm{S}_{1/2}$ transition at 589.6 nm is collisionally broadened by $(4.5\times10^{-10})n$ Hz where n is the Ne number density. The oscillator strength of the line is 1/3. (a) What is the line width at a number density that corresponds to a pressure of 3 atmospheres at T=273 K? (b) At what pressure is the collisional broadening equal to the Doppler width at T=273 K? (c) At what pressure is the collisional broadening equal to the natural radiative line width?

5. Using the information on the collisional broadening by Ne gas of the $3^2\mathrm{P}_{1/2} \rightarrow 3^2\mathrm{S}_{1/2}$ transition in Na provided in Problem 4, calculate the cross section for a phase resetting collision. To what impact parameter does the cross section correspond?

6. Using the results of Problem 5 calculate the average time between collisions for Na in one atmosphere of Ne gas.

7. The radiative lifetimes of the $4f$ and $3d$ levels in atomic hydrogen are respectively 73 ns and 15.6 ns. What is the natural radiative line width for the $4f \rightarrow 3d$ transition in atomic hydrogen?

8. For a transition where the collisional width of the homogeneous packets is 1×10^9 Hz and the Doppler width is 2×10^9 Hz estimate how far from line center the frequency must be in order that the Lorentzian wings of the packets are larger in magnitude than the wings of the inhomogeneous envelope.

9. For a Boltzmann distribution, the population in the Jth rotational

level is

$$N(J) = (2J+1)\exp\left[-B\,J(J+1)\,\frac{hc}{kT}\right],$$

where T is the rotational temperature and B is the rotational constant (units of cm^{-1}). (a) Take the derivative of $N(J)$ to find an expression for J_{max} as a function of T and B. (b) What is J_{max} for a N_2 laser with T=325 K and B=1.9 cm^{-1}? (c) Which rotational transition ($J'' \rightarrow J'$) is expected to be the strongest in an N_2 laser?

10. Compute the line widths in GHz for (a) the Ne(632.8 nm) line in a He-Ne laser (400 K), (b) the 488 nm line in a Ar^+ laser (3000 K), (c) the 694 nm Ruby laser line (300 K), and (d) the 1.06 μm Nd:YAG laser line (300 K).

11. Suppose you created a gas discharge Nd^{3+} laser operating at 1.06 μm. The vapor pressure of Nd metal heated to 1500 K is approximately 2 mTorr. By passing an electrial current through the vapor, perhaps 1% of these atoms can be ionized to form Nd^{3+} ions. (a) Compare the Doppler width of the 1.06 μm transition for a gas phase 1500 K Nd^{3+} ion with the linewidth of a solid state Nd:YAG laser at room temperature. (b) The number density of Nd^{3+} ions in a 1% doped Nd:YAG crystal is approximately 1×10^{20} cm^{-3}. Assuming the discharge tube and crystal have the same cross sectional area, how long would the discharge tube need to be to have the same number of Nd^{3+} ions as a 1 cm long crystal?

Chapter 5

Saturation

5

Saturation of the gain or absorption for an optical transition is of the utmost importance in understanding laser action in an active medium or in understanding the interaction of a laser beam with a passive medium. Saturation occurs because the laser beam interacts with the medium to reduce the magnitude of the population difference between the upper and lower levels thereby reducing the magnitude of the gain or absorption coefficient. The saturation of an optical transition is different depending upon whether the line is homogeneously or inhomogeneously broadened. This chapter treats first saturation of a homogeneously broadened line (Secs. 5.1 and 5.3) and then saturation of an inhomogeneously broadened line (Sec. 5.4). In this chapter the saturation is treated with the assumption that the intensity of the laser beam is not affected by its interaction with the medium, through which it is passing. The change in the intensity of the laser beam as it passes through the medium is treated in the next chapter.

The saturation of a homogeneously broadened transition depends upon the number of possible levels the upper level can decay to. When the upper level of the saturated transition can decay only to the lower level of that transition the system is called a *closed two level system*, and when the upper level can decay to several lower levels the system is called an *open multi-level system*. These two situations are treated separately for clarity.

5.1 Homogeneously Broadened Line for a Closed Two Level System

Saturation of a closed two level homogeneously broadened transition is straight forward to understand. Figure 5.1 shows the energy levels of a closed two level system. The system is such that the upper level u with

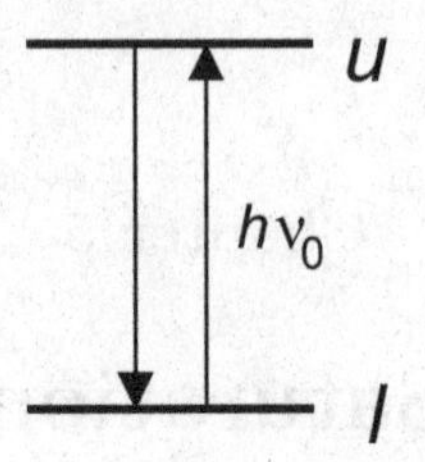

Fig. 5.1 The energy levels for a closed two level system.

statistical weight g_u has a population number density n_u and the lower level l with statistical weight g_l has a population number density n_l. It is assumed that a single mode laser beam is passing through the system without any alteration in its intensity. The rate equations for the population densities of the upper and lower levels are

$$\begin{aligned}\frac{dn_u}{dt} &= -\left(\sigma_{ul}\, n_u - \sigma_{lu}\, n_l\right) n_{\mathrm{ph}}\, c - \frac{n_u}{\tau_u} \\ &= -\sigma_{ul}\left(n_u - \frac{g_u}{g_l}\, n_l\right) n_{\mathrm{ph}}\, c - \frac{n_u}{\tau_u} \qquad (5.1)\end{aligned}$$

and

$$\begin{aligned}\frac{dn_l}{dt} &= \left(\sigma_{ul}\, n_u - \sigma_{lu}\, n_l\right) n_{\mathrm{ph}}\, c + \frac{n_u}{\tau_u} \\ &= \sigma_{ul}\left(n_u - \frac{g_u}{g_l}\, n_l\right) n_{\mathrm{ph}}\, c + \frac{n_u}{\tau_u} \qquad (5.2)\end{aligned}$$

where it is assumed that the lower level is the ground level and therefore does not decay and where use is made of the relationship between the stimulated emission and absorption cross sections, $\sigma_{lu} = (g_u/g_l)\,\sigma_{ul}$. In Equations 5.1 and 5.2 the velocity of light is taken as c, the velocity of light in a vacuum, and the index of refraction of the medium is assumed to be one. In Equations 5.1 and 5.2 the density of photons in the single mode laser beam, n_{ph}, is treated as a constant. The reader is reminded that photons in a single mode have a unique frequency, phase, and wave vector. It should be noted that τ_u^{-1} is the total decay rate of the upper level from all processes and is not just the radiative decay rate. It should also be noted that since the system is a closed two level system $dn_u/dt = -dn_l/dt$. Because no atoms make a transition to any level other than the two levels u and l the total population of these levels is constant. This is expressed as

$$n_u + n_l = n_T \qquad (5.3)$$

where n_T is the total atomic number density in both the upper and lower levels. The rate equations depend on the frequency of the laser beam, ν, because the cross section for stimulated emission depends on the laser frequency, being given by [see Eqs. 2.52 and 2.54]

$$\sigma_{ul} = \frac{\lambda_0^2}{8\pi} A_{ul}\, g(\nu - \nu_0) = \frac{g_l}{g_u} \sigma_{lu} \,. \tag{5.4}$$

In the steady state

$$\frac{dn_u}{dt} = -\frac{dn_l}{dt} = -\sigma_{ul}\left(n_u - \frac{g_u}{g_l} n_l\right) n_{\rm ph}\, c - \frac{n_u}{\tau_u} = 0\,. \tag{5.5}$$

Combining Equations 5.3 and 5.5 yields

$$\frac{n_u}{n_T} = \frac{\sigma_{lu} X}{1 + (\sigma_{lu} + \sigma_{ul}) X} \tag{5.6}$$

where $X = n_{\rm ph}\, c\, \tau_u = I_\nu\, \tau_u / h\nu$. In a similar manner one can solve for n_l/n_T with the result

$$\frac{n_l}{n_T} = \frac{1 + \sigma_{ul} X}{1 + (\sigma_{lu} + \sigma_{ul}) X}\,. \tag{5.7}$$

Combining Equations 5.6 and 5.7 one obtains

$$\frac{\alpha}{\alpha_0} = \frac{\sigma_{lu}\, n_l - \sigma_{ul}\, n_u}{\sigma_{lu}\, n_T} = \frac{1}{1 + (\sigma_{lu} + \sigma_{ul}) X}\,, \tag{5.8}$$

where α_0 is absorption coefficient in the absence of any saturation, defined as $\alpha_0 = \sigma_{lu}\, n_T$. This leads to

$$\frac{\alpha}{\alpha_0} = \frac{1}{1 + \frac{I_\nu}{I_S}} \tag{5.9}$$

where $I_\nu = n_{\rm ph}\, c\, h\nu$ is the intensity of the single mode laser beam and I_S is the *saturation intensity* for the closed two level system, which is equal to the intensity needed to reduce the absorption coefficient by a factor of two. The value of the saturation intensity can be calculated from

$$I_S = \frac{h\nu}{(\sigma_{lu} + \sigma_{ul})\, \tau_u} = \frac{h\nu}{\left(\frac{g_u}{g_l} + 1\right) \frac{\lambda^2}{8\pi} A_{ul}\, g(\nu - \nu_0)\, \tau_u}\,. \tag{5.10}$$

One often sees the saturation intensity written as

$$I_S = \frac{h\nu}{2\sigma_{lu}\, \tau_u} = \frac{h\nu}{2\, \frac{\lambda^2}{8\pi} A_{ul}\, g(\nu - \nu_0)\, \tau_u}\,, \quad \text{when } g_u = g_l\,. \tag{5.11}$$

Since the saturation intensity varies as the inverse of the line shape function, $g(\nu - \nu_0)$, which depends on the difference between the frequency of

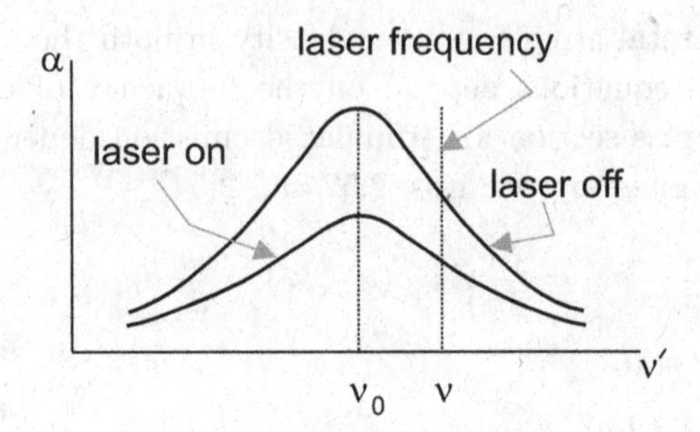

Fig. 5.2 The absorption coefficient of a homogeneously broadened line with a fixed frequency saturating laser on and off. The absorption coefficient is measured with a broad bandwidth light source. Note that $\Delta\nu$ is unchanged (i.e., the FWHM remains unchanged despite the reduction in maximum value).

the saturating laser and the frequency corresponding to the line center for the transition, the saturation intensity depends on the difference between the frequency of the saturating laser and the line center frequency. The saturation intensity is least when the frequency of the saturating laser is at line center and increases when that frequency is further from line center.

It is important to understand what is meant by saturation. In order to understand saturation in a homogeneously broadened line let us consider the following experiment. A laser at the frequency ν is used to saturate a homogeneously broadened line with a line center frequency of ν_0. A broad bandwidth light source acts as a probe across the entire line shape (Fig. 5.2). The probe has an intensity much less than the saturation intensity for the transition. The ratio of the absorption coefficient with the saturating laser on to the absorption coefficient without the saturating laser, α/α_0, is measured as a function of the probe frequency. It is seen that the absorption coefficient is reduced when the saturating laser is on by the factor $1/(1 + I_\nu/I_S)$ from its value when the saturating laser is off. This reduction in the absorption coefficient is independent of the probe frequency (i.e., it is the same for all frequencies of the broadband light probe).

One might wonder why a broadband light source is used as the probe instead of using a second low-power laser that can be swept over the frequency ν'. While conceptually equivalent, the coherent light from the probe laser interacting with the atoms and with the saturating laser can cause many surprising effects, particularly for 'real' atoms where the idea of a closed two-level system is only an approximation. If the problem is with two lasers, then why not simplify the problem even more and use one laser for both the 'pump' (i.e., the saturating laser) and the 'probe' (i.e., the one used to measure α)? It turns out that yields a very different result that is the topic of the following Section 5.2.

It is stressed here that, although the saturation of the absorption coefficient has been calculated, the gain coefficient will saturate in exactly the same fashion since absorption and gain are essentially the same processes.

When an intense laser beam is incident on an absorbing material the absorption coefficient may be very small since the population difference between the upper and lower levels is reduced due to saturation. This means that the absorption coefficient for an intense laser beam is much less than the absorption coefficient for a weak laser beam of the same frequency incident on the same material. This results in an effect that is referred to as the laser burning its way through the material. An intense laser can burn its way through the material with the laser beam intensity decreasing only slowly (almost linearly). Nevertheless as the laser intensity decreases the absorption coefficient increases. This can lead to a situation where an intense laser beam burns its way through a substantial length of material and then is rapidly absorbed in a relatively short distance.

5.2 Power Broadening of a Line

As was discussed in the previous section when an experiment to measure the absorption coefficient is carried out using a weak broad bandwidth probe it is found that the absorption coefficient is decreased at all frequencies (ν') when a saturating laser (at frequency ν) is 'on' by the same factor, $1/(1 + I_\nu/I_S)$ across the entire line. In this situation there is no broadening of the line since the line shape is not altered. The only effect of the saturating laser is to decrease the absorption coefficient by the same factor at all frequencies. There is, however, another experiment one can envision carrying out. The saturating laser could be swept across the line shape while maintaining the intensity of the saturating laser constant and measuring the absorption coefficient at the frequency of the saturating laser. In this situation the line is more saturated at line center than away from line center so that the line shape will be altered causing the line appears to be *power broadened.*

The concept of power broadening of an optical line shape often seems difficult to understand. This concept is not actually difficult to grasp, but it is necessary for the reader to carefully study the ideas to entirely understand them. In this section the basic ideas related to power broadening of a homogeneously broadened line shape is presented.

The absorption coefficient at the frequency ν for a line subject to the influence of a saturating laser operating at the frequency ν is given by

$$\frac{\alpha(\nu-\nu_0)}{\alpha_0(0)} = \frac{\alpha(\nu-\nu_0)\,\alpha_0(\nu-\nu_0)}{\alpha_0(\nu-\nu_0)\,\alpha_0(0)} = \frac{\frac{g(\nu-\nu_0)}{g(0)}}{1+\left[I_\nu/\frac{I_S(0)\,g(0)}{g(\nu-\nu_0)}\right]}$$

$$= \left(\frac{g(0)}{g(\nu-\nu_0)} + \frac{I_\nu}{I_S(0)}\right)^{-1} \quad (5.12)$$

where $\alpha_0(0) = \sigma_{lu}(0)\,n_T$ is the absorption coefficient at line center in the absence of the saturating laser so that all the atoms are in the ground level and $I_S(0) = h\nu/[2\sigma_{lu}(0)\,\tau_u]$ is the saturation intensity at line center. The above equation can be derived from $\alpha(\nu-\nu_0)/\alpha_0(\nu-\nu_0) = 1/[1+I_\nu/I_S(\nu-\nu_0)]$, $I_S(\nu-\nu_0) = I_S(0)g(0)/g(\nu-\nu_0)$, and $\alpha_0(\nu-\nu_0)/\alpha_0(0) = g(\nu-\nu_0)/g(0)$. It should be clear that the absorption coefficient observed when the frequency of the saturating laser is swept across the line shape is not identical to the absorption line shape observed when the saturating laser is maintained with a constant frequency and the line is examined with a weak broad bandwidth probe. The line shape observed when the saturating laser is swept across the line is broader than the line observed with a weak probe and this effect is called power broadening. For a Lorentzian line the absorption coefficient becomes

$$\alpha(\nu-\nu_0) = \alpha_0(0)\,\frac{\left(\frac{\Delta\nu_L}{2}\right)^2}{(\nu-\nu_0)^2 + \left(\frac{\Delta\nu_L}{2}\right)^2\left[1+\frac{I_\nu}{I_S(0)}\right]} \quad (5.13)$$

where $\Delta\nu_L$ is the FWHM of the Lorentzian line width in the absence of power broadening. It is seen that for a Lorentzian line the line shape observed when the saturating laser's frequency is swept across the line shape is still Lorentzian but with a width that depends on the ratio of the intensity of the saturating laser to the saturation intensity at line center. With power broadening the Lorentzian FWHM is increased from $\Delta\nu_L$ to $\Delta\nu_L[1+I_\nu/I_S]^{1/2}$. More generally, for lines that are not Lorentzian the line shape as observed by sweeping the frequency of the saturating laser is altered into a broader shape.

A line is power broadened due to the fact that the absorption coefficient is squashed down more at line center than in the wings of the line. It is not broadened by maintaining the absorption coefficient at line center and simply increasing the line width. In fact the absorption coefficient of the line is everywhere reduced. It is reduced more at line center than in the wings of the line, which results in the line appearing broadened as illustrated in Figure 5.3.

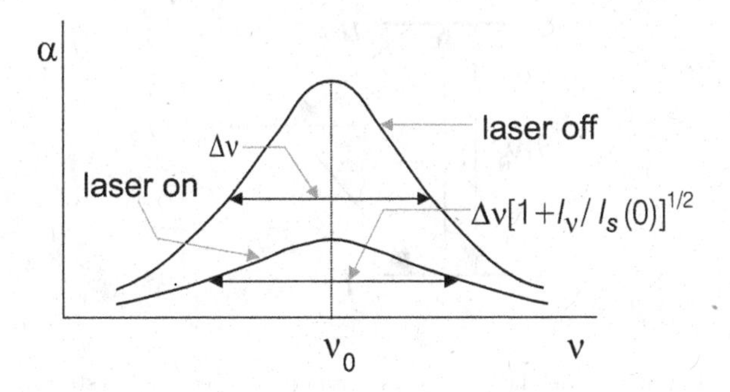

Fig. 5.3 The absorption coefficient of an homogeneously broadened line as measured by sweeping the saturating laser at high and at low power. The absorption coefficient is measured at the frequency of the saturating laser. Note the power broadening.

5.3 Homogeneously Broadened Line for an Open System with Optical Pumping

When there are more than two levels that can be populated the saturation is somewhat different from the saturation for a closed two level system. Optical pumping is the transfer or "pumping" of atoms from one level (often the ground level) to another level that is very long lived. Figure 5.4 shows a three level system where the upper level is labeled u, the lower level is labeled l, and the third and intermediate energy level is labeled i. The upper level can decay spontaneously to either of two lower levels, l or i. It is assumed that both the upper and intermediate levels decay radiatively. For this system we assume for simplicity that $g_l = g_u = g_i$ so that $\sigma_{lu} = \sigma_{ul}$. The Einstein A coefficients for the upper level are given by $A_{ul} = \Gamma_{ul} A_u$ and $A_{ui} = \Gamma_{ui} A_u$ where $A_u = \tau_u^{-1}$ is equal to the inverse of the lifetime of the upper level and where Γ_{ul} and Γ_{ui} are the branching ratios to levels l and i respectively for the radiative decay from the upper level. Note that $\Gamma_{ul} + \Gamma_{ui} = 1$.

The rate equations for the three levels are given by

$$\frac{dn_u}{dt} = -\sigma_{ul}\,(n_u - n_l)\,n_{\text{ph}}\,c - A_u\,n_u\;, \tag{5.14}$$

$$\frac{dn_l}{dt} = \sigma_{ul}\,(n_u - n_l)\,n_{\text{ph}}\,c + \Gamma_{ul} A_u\,n_u\; + A_i\,n_i, \tag{5.15}$$

and

$$\frac{dn_i}{dt} = \Gamma_{ui}\,A_u\,n_u - A_i\,n_i\;, \tag{5.16}$$

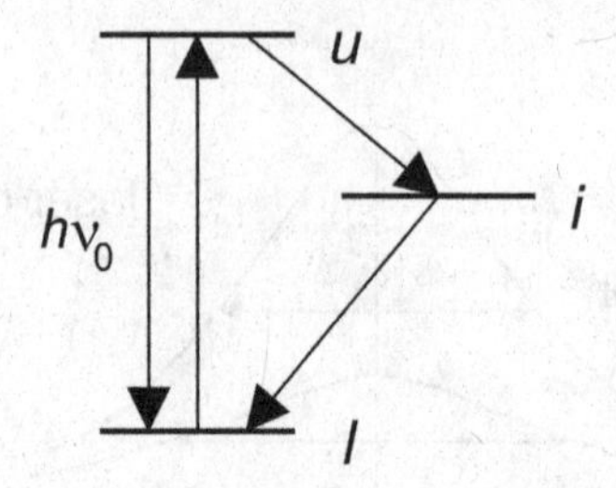

Fig. 5.4 The energy levels for an open three level system.

The total number density of atoms in the system is $n_T. = n_u + n_l + n_i$. It is assumed that n_T is constant i.e. that atoms can not make transitions to levels other than the three levels l, u, and i. In the above equations the speed of light is taken as equal to the speed of light in a vacuum. Equations 5.14, 5.15, and 5.16 are solved in the steady state where the rates of change for the atomic number densities is taken as zero. The result is

$$\frac{n_i}{n_u} = \frac{\Gamma_{ui} A_u}{A_i}, \tag{5.17}$$

$$\frac{n_u}{n_l} = \frac{\sigma_{ul} X}{1 + \sigma_{ul} X}, \tag{5.18}$$

and

$$\frac{n_i}{n_l} = \frac{\Gamma_{ul} A_u}{A_i} \frac{\sigma_{ul} X}{1 + \sigma_{ul} X} \tag{5.19}$$

where $X = n_{\text{ph}}\, c\, A_u^{-1} = n_{\text{ph}}\, c\, \tau_u$. If A_i is much greater than $\Gamma_{ui} A_u$ then the intermediate level decays rapidly to the lower level and the saturation is similar to the saturation of a closed two level system, but if A_i is much less than $\Gamma_{ui} A_u$ then n_u is much less than n_i (or n_l) and atoms accumulate in level i (i.e., they are pumped into i). This is the condition for optically pumping atoms into the intermediate energy level. For an open three level system with optical pumping it will be found that saturation occurs at relatively low laser powers. In this situation the product $\sigma_{ul} X$ is much less than one. In this situation n_T is nearly the same as $n_l + n_i$. Under this condition one finds that ratio of the absorption coefficient with the laser on to the absorption coefficient with the laser off is given by

$$\begin{aligned} \frac{\alpha}{\alpha_0} &= \frac{(n_l - n_u)\sigma_{lu}}{n_T\, \sigma_{lu}} \cong \frac{n_l}{n_T} = \left[1 + \frac{\Gamma_{ui} A_u}{A_i} \sigma_{ul} X\right]^{-1} \\ &= \frac{1}{1 + \Gamma_{ul}\, \sigma_u l\, n_{\text{ph}}\, c\, A_i^{-1}} = \frac{1}{1 + (I_\nu / I_S^*)} \end{aligned} \tag{5.20}$$

where the saturation intensity for a situation where optical pumping occurs is indicated by I_S^*, which is given by

$$I_S^* = \frac{h\nu\, A_i}{\Gamma_{ul}\,\sigma_{ul}} = \left(\frac{2A_i}{\Gamma_{ui}\, A_u}\right) I_S \tag{5.21}$$

where I_S is the saturation intensity for a closed two level system when $g_u = g_l$. Since $\Gamma_{ui}A_u$ is much larger than A_i the saturation intensity I_S^* with optical pumping is much less than the saturation intensity for a closed two level system.

The saturation and power broadening of an open system with optical pumping are the same as for the closed two level system except that the saturation intensity is given by Equation 5.21 rather than by Equation 5.11. Saturation with more than three levels is similar to the three level system.

5.4 Inhomogeneously Broadened Line

The saturation of an inhomogeneously broadened line is quite different from the saturation of a homogeneous line. As discussed in Chapter 3 an inhomogeneous line shape occurs when there is a distribution of center frequencies for the homogeneous packets that make up the line. If one has a saturating laser operating with a frequency ν the homogeneous packets centered near the frequency ν will be most strongly saturated and the homogeneous packets centered away from the frequency of the saturating laser will be less saturated. The absorption coefficient at a given frequency is made up of the sum of the absorption coefficients of each of the packets with their different saturation factors. The absorption coefficient at the frequency ν' is given by

$$\alpha(\nu') = \int_0^\infty \sigma_{lu}(\nu' - \nu_\xi)\, n_T \left[1 + \frac{I_\nu}{I_S(\nu - \nu_\xi)}\right]^{-1} h(\nu_\xi - \nu_0)\, d\nu_\xi \tag{5.22}$$

where the center frequency of a homogeneous packet is ν_ξ, and where the inhomogeneous distribution has a center frequency ν_0. Using a Lorentzian shape for the homogeneously broadened packets one obtains after much algebra

$$\alpha(\nu') = \int_0^\infty \frac{\sigma_{lu}(0)\, n_T \left(\frac{\Delta\nu}{2}\right)^2 h(\nu_\xi - \nu_0) \left[(\nu - \nu_\xi)^2 + \left(\frac{\Delta\nu}{2}\right)^2\right] d\nu_\xi}{\left[(\nu' - \nu_\xi)^2 + \left(\frac{\Delta\nu}{2}\right)^2\right] \left[(\nu - \nu_\xi)^2 + \left(\frac{\Delta\nu}{2}\right)^2 \left(1 + \frac{I_\nu}{I_S(0)}\right)\right]}\,. \tag{5.23}$$

Needless to say, this expression for $\alpha(\nu')$ is very complex. In order to understand the saturation of an inhomogeneously broadened line, consider

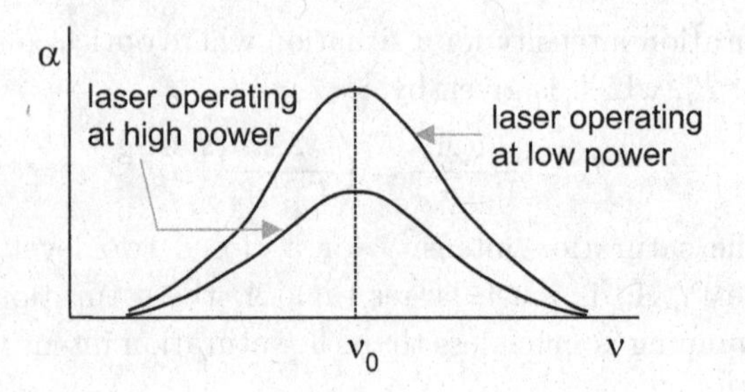

Fig. 5.5 The absorption coefficient of an inhomogeneously broadened line as measured by sweeping a saturating laser at high and at low power. The absorption coefficient is measured at the frequency of the saturating laser. Note the absence of power broadening as compared to Fig. 5.3 on page 99.

the absorption coefficient at the frequency of the saturating laser i.e. when $\nu = \nu'$. One finds that the absorption coefficient is given by

$$\alpha = \sigma_{lu}(0)\, n_T \int_0^\infty \frac{\left(\frac{\Delta\nu}{2}\right)^2 h(\nu_\xi - \nu_0)\, d\nu_\xi}{(\nu - \nu_\xi)^2 + \left(\frac{\Delta\nu}{2}\right)^2 \left[1 + \frac{I_\nu}{I_S(0)}\right]} \,. \tag{5.24}$$

If the inhomogeneous distribution is very broad compared to the width of the homogeneous packets, one can remove the inhomogeneous distribution from inside the integral treating it as a constant. Since ν_ξ is large it is possible to extend the lower limit for the integration from 0 to $-\infty$ without altering the value of the integral significantly. Using these approximations and carrying out the integration one obtains the result that

$$\alpha = \frac{\sigma_{lu}(0)\, n_T\, h(\nu - \nu_0)\, \pi \frac{\Delta\nu}{2}}{\sqrt{1 + \frac{I_\nu}{I_S(0)}}} = \frac{\alpha_0}{\sqrt{1 + \frac{I_\nu}{I'_S}}} \tag{5.25}$$

where the saturation intensity for an inhomogeneously broadened line, I'_S, is given by $I_S(0)$ the saturation intensity for a homogeneous packet at the center frequency of the packet.

An inhomogeneously broadened line saturates as $1/(1+I_\nu/I'_S)^{1/2}$ rather than as $1/(1 + I_\nu/I_S)$, which occurs for a homogeneously broadened line. The saturation intensity for an inhomogeneously broadened line is the intensity that reduces the absorption coefficient by $2^{-1/2}$. If the saturating laser is swept across the inhomogeneously broadened line the absorption coefficient is decreased by the same factor $1/(1 + I_\nu/I'_S)^{1/2}$ at all frequencies since the saturation intensity for the inhomogeneously broadened line, $I'_S = I_S(0)$ is independent of the frequency. There is no power broadening of an inhomogeneously broadened line.

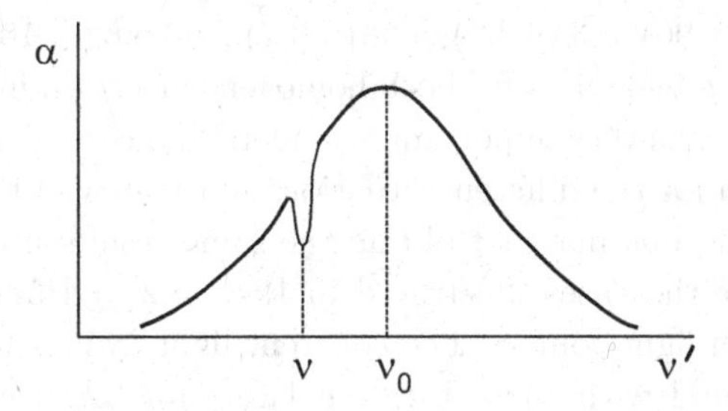

Fig. 5.6 The absorption coefficient of an inhomogeneously broadened line with a fixed frequency saturating laser. The absorption coefficient is measured with a broad bandwidth light source. Note the effects of hole burning.

If one has a saturating laser operating at a constant frequency ν and uses a weak broad bandwidth probe to measure the absorption coefficient as a function of the frequency across an inhomogeneously broadened line then using Equation 5.23 one finds that for probe frequencies far from the saturating laser frequency the absorption coefficient is the same as if the saturating laser were off, but for probe frequencies near the frequency of the saturating laser the absorption coefficient is reduced from the value with the saturating laser off since the homogeneous packets centered at frequencies near the frequency of the saturating laser are saturated. The population difference for those packets centered near the frequency of the saturating laser are reduced from the population difference when the saturating laser is off. This effect is called burning a hole in the inhomogeneous distribution and is illustrated in Figure 5.6. Hole burning is very important in many applications.

An interesting result is that the hole burned in an inhomogeneous distribution is centered at the frequency of the saturating laser and that the hole has a width in frequency given by

$$\Delta\nu_{\text{hole}} = \Delta\nu\sqrt{1 + \frac{I_\nu}{I'_S}} . \tag{5.26}$$

The frequency width of the hole varies as $(1 + I_\nu/I'_S)^{1/2}$ because as I_ν increases atoms with center frequencies further from the laser frequency are saturated. It is again noted that the saturation intensity for an inhomogeneously broadened line, I'_S, is equal to the saturation intensity for the homogeneously broadened packets at packet center, $I_S(0)$. The saturation intensity for the homogeneously broadened packets can be either the saturation intensity for a closed or an open system depending on the energy levels of the atom under study.

Figures 5.2 (p. 96), 5.3 (p. 99), 5.5 (p. 102), and 5.6 (p. 103) illustrate the various saturation behaviors for both homogeneously and inhomogeneously broadened lines. It is very important to understand carefully the differences in the saturation for the different situations illustrated in these figures. As mentioned earlier, one may not obtain the same results if one uses a laser as the probe for the cases illustrated in Figs. 5.2 and 5.6 rather than a broad-bandwidth light source. The coherent light from a laser interacting with the atom and with the saturating laser can cause many surprising effects that are very complicated to explain and are beyond the scope of this textbook.

5.5 Laser Saturation Spectroscopy

Saturation spectroscopy is a technique for measuring with high accuracy the line center for an atomic or molecular transition. Although saturation spectroscopy is an application of lasers we include a brief discussion of it because it illustrates many of the ideas related to line shapes from Chapter 4 and the ideas of laser saturation discussed in this chapter. Typically, the measured line width of an atomic transition in a gas is dominated by the Doppler width, and is typically on the order of a few GHz wide. This is much greater than the MHz width of the homogeneous natural line width. Using saturated absorption spectroscopy it is possible to perform Doppler-free measurements and easily resolve individual hyperfine transitions with line widths on the order of the natural line width.

Although the technique of saturation spectroscopy had been previously used with fixed frequency gas lasers, Hänsch, Shahin and Schawlow introduced the use of the technique using tunable lasers that made possible the accurate wavelength measurements for a much wider class of atoms and molecules. The basic idea of their experiment is illustrated in Fig 5.7. A tunable laser beam is split and reflected in almost opposite directions through a Na vapor cell. The weak beam going to the photo detector is called the *probe beam.* The beam going in the opposite direction is strong enough to saturate the Na vapor and is called the *saturating beam.* An atom in the Na vapor cell moving with an axial velocity v sees the two beams as having frequencies approximately given by $\nu(1+v/c)$ and $\nu(1-v/c)$ where ν is the frequency of the tunable laser. The Na atoms have a Doppler line shape. In the beam frequencies seen by the Na atom it is assumed that the two beams are both going in opposite directions along the axis of the

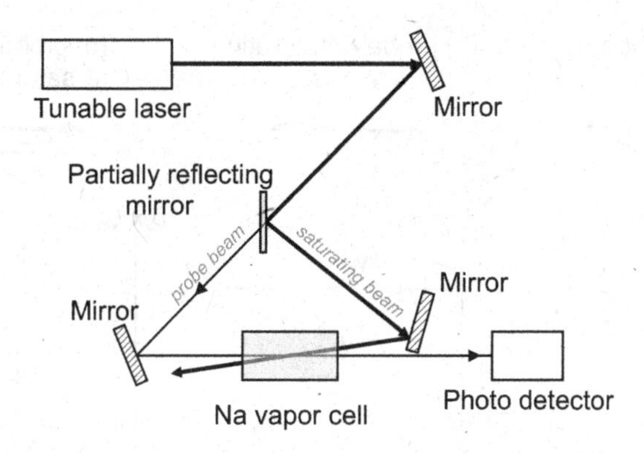

Fig. 5.7 Saturated absorption spectroscopy experiment of Hänsch, Shahin and Schawlow.

Na vapor cell. In Fig. 5.7 the two beams are shown at a significant angle for clarity. In an actual experimental set up the beams are going in nearly opposite directions along the axis of the cell. For Na at a temperature of 600 K the Doppler line width is 1.9 GHz. If the saturating beam were not present then the probe light reaching the photo detector would record a very wide absorption dip corresponding to the Doppler line shape of the transition as the frequency of the tunable laser is swept across the transition as illustrated in Fig. 5.8a.

When the saturating beam is present the situation is different. While the probe beam samples a segment of the Boltzmann velocity distribution that is near the velocity $+v$, the saturating beam burns a hole in the segment of the velocity distribution near the velocity $-v$. The width of the velocity distribution sampled by the two beams is determined primarily by the frequency width of the laser beam. As long as the probe and saturating beams sample different segments of the velocity distributions then the probe beam reaching the photo detector is not affected by the saturating beam. When, however, the two beams are each interacting with the segment of the velocity distribution near v=0 then the saturating beam decreases the population difference between the lower and the upper level of the transition and the probe beam absorption by the Na vapor is decreased and the intensity reaching the photo detector is increased. The increase in the probe beam intensity reaching the photo detector corresponds to Doppler free line shape as illustrated in Fig. 5.8b. In order to measure a nearly Doppler free line shape it is necessary that the laser bandwidth be very small.

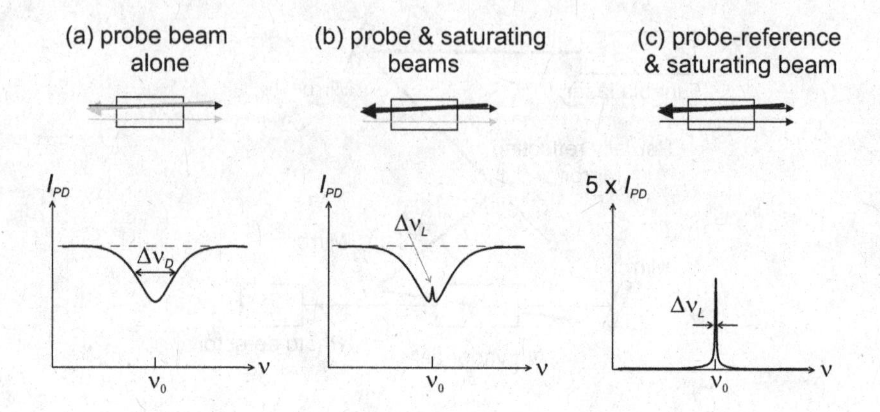

Fig. 5.8 Signals recorded in an idealized saturated absorption spectroscopy experiment. (a) Photo detector signal for probe beam recorded with saturating beam blocked. (b) Doppler-free signal recorded with saturating beam. (c) Use of a reference beam to subtract off Doppler profile (and changes in laser intensity).

The explanation provided in the previous paragraphs has treated Na as a closed two-level system, but including the hyperfine structure, there are actually four closely spaced hyperfine transitions contained within the Doppler line width. Sodium has only one stable isotope, ^{23}Na, which has nuclear spin of 3/2. The ground level of Na is $3^2S_{1/2}$ so that the electronic angular momentum is due entirely to the electron spin and is equal to 1/2 since the orbital angular momentum is L=0. Therefore the electronic spin and the nuclear spin can add up to a total angular momentum of F=2 or 1. The first excited electronic level of Na is $3^2P_{1/2}$. In this level the total electronic angular momentum J is 1/2 and the total electronic angular momentum and the nuclear spin can add up a total angular momentum of F'=2 or 1. The allowed absorption transitions from the ground level to the first excited level must satisfy ΔF=1 or 0. This leads to four allowed transitions corresponding to $F = 1 \rightarrow F' = 1$, $F = 1 \rightarrow F' = 2$, $F = 2 \rightarrow F' = 1$, and $F = 2 \rightarrow F' = 2$. These transitions are separated only by small differences in frequency due to the hyperfine separations in the ground (1772 MHz) and first excited levels (192 MHz).

In the experiment of Hänsch, Shahin, and Shawlow a pulsed N_2 laser was used to pump a dye laser that was the tunable laser. The pulsed dye laser bandwidth was narrowed using an external etalon. When the saturating beam is blocked, then the absorption profile of the probe beam consists of a very wide absorption dip caused by the Doppler line shapes of the four

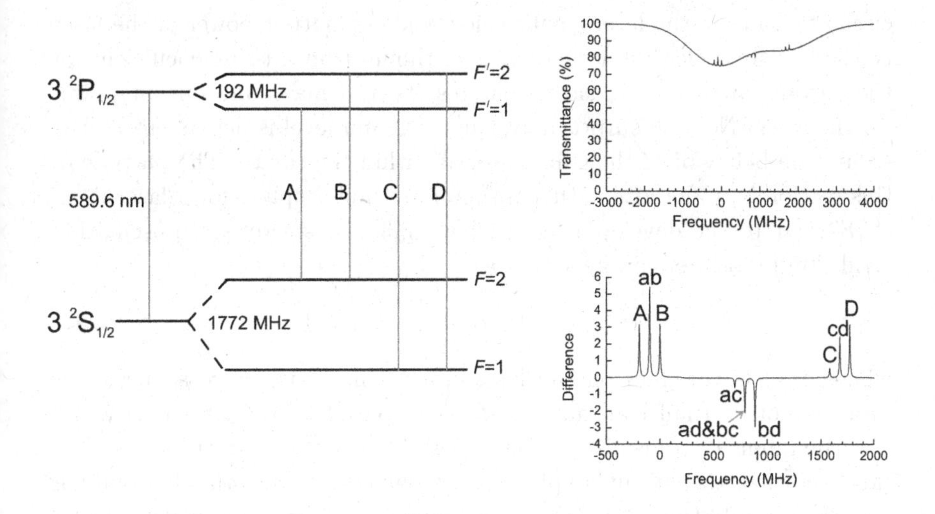

Fig. 5.9 Doppler-free saturated absorption profile for the Na D_1 line including hyperfine transitions (capital letters) and crossover resonances (pairs of lowercase letters). The zero of the frequency scale was set to that of the $F' = 2 \rightarrow F = 2$ transition.

Na hyperfine transitions. With the saturating beam present, however, the four hyperfine transitions emerge clearly (Fig. 5.9). Additional *crossover* peaks occur in the spectrum when the hole burned by the saturating beam for one Doppler-shifted hyperfine transition corresponds to the Doppler-shifted frequency of a different hyperfine transition for the probe beam since one of the levels sampled by the probe is altered by the pump. Optical pumping from one hyperfine level to the other by the saturating beam can also increase the absorption of the probe beam, leading to crossover dips. Some of these effects in the sodium saturated absorption spectrum of Na are discussed by C. J. Anderson *et al* (see additional readings). Saturation spectroscopy has been used to measure accurately the transitions in a wide variety of atoms and molecules.

5.6 Optimum Output Coupling

The optimum output coupling for a laser is an important problem. If the laser mirrors are perfectly reflecting then the intensity of the laser beam circulating inside the laser cavity is large, but the output power of the laser is zero. If one of the laser mirrors has a low reflection coefficient and a correspondingly high transmission coefficient then the intensity of beam circulating inside the laser cavity is low and the output power of the laser is also low. It is clear that there is an optimum transmission coefficient for the output mirror of the laser, that results in a maximum output power for the laser. Not too surprisingly, this optimum level is dependent on the saturation behavior of the gain material within the cavity. The purpose of this section is to determine this optimum output coupling for a laser.

Similar to the development in Chapter 3, the steady state oscillation condition for a laser can be stated as

$$\exp\{ [\alpha(\nu) - \alpha'] \, 2L - T \} = 1 \tag{5.27}$$

where $\alpha(\nu)$ is the gain on the laser transition, α' is the loss due to all processes other than transmission, L is the length of the laser cavity and T is the intensity transmission coefficient of the output mirror for the laser (also known as the output coupler). It is assumed that the other laser mirror is perfectly reflecting and that neither mirror has significant absorption or diffraction losses around the edges of the mirrors. The term in the exponent is equal to zero so that

$$\alpha(\nu) = \frac{\alpha_0}{1 + (I_\nu / I_S)} = \alpha' + \frac{T}{2L} \tag{5.28}$$

where the laser has been assumed to be operating at line center on a transition that is homogeneously broadened and with the saturation appropriate to that condition. If one solves for I_ν and equates the output power to $P = T\, I_\nu\, A$, where I_ν is the circulating intensity of the laser inside the laser cavity and A is the cross sectional area of the output laser beam, one finds that

$$P = T\, I_\nu\, A = T\, A \left(\frac{\alpha_0\, I_S}{\alpha' + (T/2L)} - I_S \right) . \tag{5.29}$$

If one differentiates P with respect to T and sets the derivative equal to zero one finds that the optimum output transmission is equal to

$$T_{\text{opt}} = \left[(\alpha_0\, \alpha')^{1/2} - \alpha' \right] 2L \, . \tag{5.30}$$

The output power when using the optimum output coupling is given by

$$P(T_{\text{opt}}) = 2 I_S\, A \left(\sqrt{\alpha_0} - \sqrt{\alpha'} \right) . \tag{5.31}$$

Thus the optimum output power of a laser depends on the saturation intensity of the laser, on the unsaturated gain, and on the losses.

Summary of Key Ideas

- A laser passing through a medium reduces the population difference between the upper and lower levels so that gain or absorption of the medium, α, is reduced by an amount that depends on the laser's intensity and the line shape of the medium.
- For a closed two level system with homogeneously broadened lines,

$$\frac{\alpha}{\alpha_0} = \frac{1}{1 + I_\nu / I_S} \tag{5.9}$$

where the saturation intensity is given by

$$I_S = \frac{h\nu}{\left(\frac{g_u}{g_l} + 1\right) \frac{\lambda^2}{8\pi} A_{ul}\, g(\nu - \nu_0)\, \tau_u} \tag{5.10}$$

- For a homogeneously broadened open system with optical pumping, α/α_0 is the same as above (Eq. 5.9), but with I_S replaced by I_S^* given by Eq. 5.21. The saturation intensity for an open system can be much smaller than for a closed level system.
- For saturation of an inhomogeneously broadened line a hole is burned in the inhomogeneous line shape. The gain at frequency of the saturating laser is

$$\frac{\alpha}{\alpha_0} = \frac{1}{\sqrt{1 + I_\nu / I'_S}} \tag{5.25}$$

where $I'_S = I_S(0)$. The width of the hole is equal to

$$\Delta\nu_{\text{hole}} = \Delta\nu \sqrt{1 + \frac{I_\nu}{I'_S}} \tag{5.26}$$

- When a saturating laser is swept across a homogeneously broadened line, the line width increases from $\Delta\nu$ to $\Delta\nu\,(1 + I_\nu/I_S)^{1/2}$ an effect known as power broadening (Sec. 5.2). There is no power broadening for inhomogeneously broadened lines.
- Saturation reduces the gain of the active medium as the laser power is increased. The optimum output power can be obtained by adjusting the reflectivity of the output coupler using Eq. 5.30. The optimum power is a function of the saturation intensity, unsaturated gain of the medium and the losses in the optical cavity (Eq. 5.31).

Suggested Additional Reading

William T. Silfvast, *Laser Fundamentals*, Cambridge University Press (1996).

A. E. Siegman, *Lasers*, University Science Books (1986).

B. E. A. Saleh and M. C. Teich, *Fundamentals of Photonics*, John Wiley and Sons (1991).

A. Yariv, *Quantum Electronics 3rd Ed.*, John Wiley and Sons (1989).

Donald C. O'Shea, W. Russell Callen and William T. Rhodes, *An Introduction to Lasers and their Applications*, Addison Wesley Publishing Co. (1977).

M. S. Feld, M. M. Burns, T. U. Kuhl, P. G. Pappas, and D. E. Murnick, "Laser-saturation spectroscopy with optical pumping" *Opt. Lett.* **5**, 79 (1980).

M. Sargent III, M. O. Scully, and W. E. Lamb, *Laser Physics*, Addison Wesley Publishing Co. (1974).

T. W. Hänsch, I. S. Shahin, and A. L. Schawlow, "High-Resolution Saturation Spectroscopy of the Sodium D Lines with a Pulsed Tunable Dye Laser" *Phys. Rev. Lett.* **27**, 707 (1971).

C. J. Anderson, J. E. Lawler, L. W. Anderson, T. K. Holley, and A. R. Filippelli "Radiative-decay-induced four-level crossover signals in saturation spectroscopy" *Phys. Rev. A* **17**, 2099 (1978).

A. Corney, *Atomic and Laser Spectroscopy*, Oxford University Press (1977).

C. J. Foot, *Atomic Physics*, Oxford University Press (2005).

Problems

1. Calculate the saturation intensity for the $3^2S_{1/2} \rightarrow 3^2P_{1/2}$ resonance transition in sodium. Assume the idealized situation where one ignores the nuclear spin of the sodium nucleus. The temperature of the sodium vapor is 600 K, and the gas density is low enough that collisional broadening is negligible. The lifetime of the upper level is 1.6×10^{-8} s and the wavelength of the transition is 589.6 nm. The oscillator strength of the line is 1/3.

2. For the sodium vapor in Problem 1 what intensity of the saturating laser is necessary to reduce the absorption coefficient at line center, ν_0, by a factor of 10 if the saturating laser has a frequency ν_0?

3. For the saturating laser of Problem 2 how wide is the hole burned in the inhomogeneous line shape as observed using a weak broad bandwidth probe?

4. If the frequency of the saturating laser in Problem 2 is changed to be one Doppler frequency above ν_0 without changing the intensity of the laser by what factor is the absorption coefficient measured using a weak broad bandwidth probe light source reduced at the saturating laser frequency? at ν_0?

5. Consider the Na $3^2S_{1/2} \rightarrow 3^2P_{1/2}$ resonance transition described in Problem 1, but now with the sodium in a helium buffer gas with a density such that the line width of the Na resonance transition is 10^{11} Hz. What is the saturation intensity at line center? You may assume that Na is a closed two level system.

6. For the conditions of Problem 5 what saturating laser intensity at line center is needed to reduce the absorption coefficient, as measured using a weak broad bandwidth probe, at line center by a factor of 10?

7. For the same intensity and the frequency of the saturating laser as in Problem 6 by what factor is the absorption coefficient reduced at a frequency 10^{11} Hz below the line center as measured by a broad bandwidth probe?

8. Explain carefully why a laser's optimum output power depends on the saturation intensity.

9. For a homogeneously broadened three level system the upper level

has a total decay rate of 10^8 s^{-1} and the intermediate level has a total decay rate of 10^6 s^{-1}. If the upper level has a branching ratio for decay to the lower level of 2/3 what is the ratio of the saturation intensity with optical pumping to what it would be if there were no optical pumping?

10. Repeat Problem 9 for the same situation except that the line is inhomogeneously broadened.

6

Chapter 6

Laser Photon Densities

The coupled rate equations for both the atomic level number densities and for the photon densities determine the behavior of a laser. In this chapter these equations are analyzed and the behavior of a laser under various conditions is exhibited. In the previous chapter the saturation of an atomic transition was studied with the assumption that the laser intensity is unchanged as it passes through the medium. This could be done using rate equations that did not include a separate rate equation for the photon density in the laser beam. Obviously this approximation is valid in only a few situations of restricted interest. This chapter analyzes the coupled rate equations that include the change in the photon density in the laser beam as the beam passes through an active medium.

6.1 Photon Densities Both Above and Below Threshold

The population difference between the upper and lower laser levels and the photon density in a given mode inside a laser cavity are important properties of a laser. Both the population difference and the photon density in the laser's oscillating mode are remarkable functions of the input excitation power. For cw laser operation below threshold for laser action the population difference between the upper and lower laser levels increases linearly as a function of the input excitation power into the active laser medium, and the photon density in the oscillating mode is very low. Above threshold for laser action the photon density in the oscillating mode increases linearly as a function of the input excitation power while the population difference remains constant. These remarkable properties occur because of the nonlinear coupling between the photon density and the population difference between the upper and lower laser levels. This section discusses how the

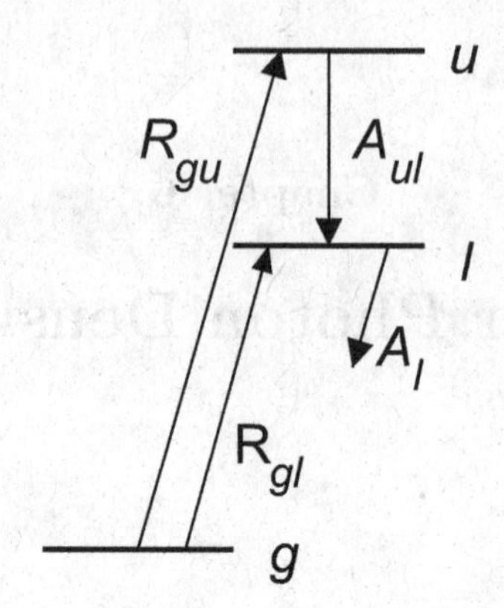

Fig. 6.1 The energy levels for a laser excited by a gas discharge. The excitation rates per unit volume into levels u and l are R_{gu} and R_{gl}.

non-linear coupling produces these properties and also some of the results of these properties.

In order to understand the photon density in a laser, consider the operation of a cw laser using the level scheme shown in Figure 6.1. The population inversion needed for laser action is assumed to be produced by electron excitation collisions in a gas discharge. In the system illustrated most of the atoms or molecules are in the ground level which is indicated by g. Some of the atoms are excited from the ground level to the upper laser level u by electron impact excitation at a rate per unit volume of

$$R_{gu} = n_g \, n_e \, \langle \sigma_{gu} \, v_{eg} \rangle \tag{6.1}$$

where n_g is the density of ground level atoms, n_e is the electron density, σ_{gu} is the electron impact excitation cross section from the ground level into the upper laser level, v_{eg} is the relative velocity between the electron and the ground level atom, and the angular brackets indicate an average over the electron velocity distribution in the gas discharge. In a similar manner atoms can be excited from the ground level into the lower laser level l at a rate per unit volume of

$$R_{gl} = n_g \, n_e \, \langle \sigma_{gl} \, v_{eg} \rangle \tag{6.2}$$

where σ_{gl} is the electron impact excitation cross section from the ground level into the lower laser level l. In order to simplify the problem it is assumed that the upper level cannot decay radiatively to the ground level. Electron impact excitation can excite atoms or molecules to levels that are not connected to the ground level by an allowed radiative transition. It is also assumed that the upper laser level can only decay radiatively to the lower laser level so that $A_{ul} = A_l$, and that the lower laser level decays to still lower levels very rapidly (i.e. A_l is very large). For simplicity it is asumed that the upper and lower levels have the same statistical weights ($g_u{=}g_l$), and that only a single laser mode is oscillating. It is further assumed that at any given time only a small fraction of the atoms are excited into the laser levels so that most of the atoms are in the ground level, and hence n_g is nearly constant. The rate equations for the densities of the upper and lower laser levels as well as the photon density are given by a set of coupled differential equations

$$\frac{dn_u}{dt} = R_{gu} - A_u\, n_u - (n_u - n_l)\, \sigma_{ul}\, n_{\text{ph}} c\,, \tag{6.3}$$

$$\frac{dn_l}{dt} = R_{gl} + A_u\, n_u - A_l\, n_l + (n_u - n_l)\, \sigma_{ul}\, n_{\text{ph}} c\,, \tag{6.4}$$

and

$$\frac{dn_{\text{ph}}}{dt} = (n_u - n_l)\, \sigma_{ul}\, n_{\text{ph}} c + \gamma\, A_u\, n_u - \frac{n_{\text{ph}}}{\tau_{\text{ph}}} \tag{6.5}$$

where n_{ph} is the density of photons in the single oscillating mode and the speed of light is taken as c (i.e., the index of refraction of the active medium is taken as one). The constant γ requires a brief discussion. The spontaneous radiative decay rate per atom in the upper level is A_u. The photons emitted in the spontaneous decay go intothe $p = 8\pi\nu^2\, \Delta\nu\, V/c^3$ (Eq. 2.9) modes within the line width $\Delta\nu$ of the transition. The constant γ is the fraction of the photons that go into the single oscillating mode. Thus $\gamma = 1/p$. It should be noted that every stimulated emission results in the addition of a photon into the oscillating mode and every absorption removes a photon from the oscillating mode so that in the photon rate equation the term $(n_u - n_l)\, \sigma_{ul}\, n_{\text{ph}} c$ does not have a constant γ multiplying it.

The rate equations appear to be very simple, but he non-linear term $(n_u - n_l)\, \sigma_{ul}\, n_{\text{ph}} c$ leads to interesting and complex results that did not show up in the linear rate equations used to discuss saturation in Chapter 5 where it was assumed that the photon density was constant. The solutions to Equations 6.3, 6.4, and 6.5 are discussed in the steady state where all the rate equations are set equal to zero.

In the steady state the density of atoms in the lower laser level is given by

$$n_l = A_l^{-1}\, \frac{R_{gl} + (A_u + \sigma_{ul}\, n_{\text{ph}}\, c)\, n_u}{1 + \sigma_{ul}\, n_{\text{ph}}\, c\, A_l^{-1}}\,. \tag{6.6}$$

If $R_{gl}\, A_l^{-1}$ is much smaller than $R_{gu} A_u^{-1}$ and if $A_l^{-1}\, A_u$ and $A_l^{-1}(\sigma_{ul}\, n_{\text{ph}}\, c)$ are much less than one then n_l is very small compared to n_u and can be taken as zero. A large value of A_l is necessary for steady state laser action, and it is assumed that A_l is large enough that n_l is nearly zero. With the assumption n_l=0 the steady state rate equations (Equations 6.3 and 6.5) become

$$\frac{dn_u}{dt} = R_{gu} - A_u\, n_u - n_u\, \sigma_{ul}\, n_{\text{ph}}\, c = 0 \tag{6.7}$$

and

$$\frac{dn_{\text{ph}}}{dt} = n_u\, \sigma_{ul}\, n_{\text{ph}}\, c + \gamma\, A_u\, n_u - \frac{n_{\text{ph}}}{\tau_{\text{ph}}} = 0\,. \tag{6.8}$$

Even in this simplified form the rate equations show the important effects of the non-linear coupling between the atomic and photon densities.

Solving Equation 6.7 for n_u in the steady state one obtains

$$n_u = \frac{A_u^{-1}\, R_{gu}}{1 + \sigma_{ul}\, n_{\text{ph}}\, c\, A_u^{-1}}\,. \tag{6.9}$$

This equation can be rewritten as

$$n_u = \frac{A_u^{-1}\, R_{gu}}{1 + (n_{\text{ph}}/n_S)}\,, \tag{6.10}$$

where n_S is the photon saturation density,

$$n_S = \frac{A_u}{\sigma_{ul}\, c}\,. \tag{6.11}$$

For the beam in a standing wave laser cavity the photon density is made up of photons going in both directions so the intensity going in one direction is given by $n_{\text{ph}}\, h\nu\, c/2$. The intensity corresponding to the saturation photon density is equal to

$$I_S = \frac{1}{2} n_S\, h\nu\, c = \frac{h\nu}{2\sigma_{ul}\, A_u^{-1}} \tag{6.12}$$

which is the same as the saturation intensity calculated in the previous chapter. If the threshold excitation rate for the upper level is defined as $R_{\text{th}} = \Delta n_{\text{th}}\, A_u$ where Δn_{th} is the threshold population difference between the upper and lower laser levels and $r = R_{gu}/R_{\text{th}}$, then Equation 6.10 can be written as

$$n_u = \frac{r\, \Delta n_{\text{th}}}{1 + (n_{\text{ph}}/n_S)}\,. \tag{6.13}$$

It is interesting to determine the saturation photon density for laser operation at line center. At line center

$$\sigma_{ul} = \frac{\lambda^2}{8\pi} A_{ul}\, g(0) \cong \frac{\lambda^2}{8\pi} A_{ul} \frac{1}{\Delta\nu} . \tag{6.14}$$

Since the upper laser level decays only to the lower laser level $A_{ul} = A_u$. Combining Equations 6.12 and 6.14 one finds that at line center

$$n_S = \frac{8\pi\, \nu^2\, \Delta\nu}{c^3} = \frac{p}{V} . \tag{6.15}$$

Thus the saturation photon density at line center is equal to the number of modes per unit volume within the line width of the laser transition. Equation 6.13 can be rewritten at line center as

$$n_u = \frac{r\, \Delta n_{\rm th}}{1 + (n_{\rm ph}\, V/p)} = \frac{r\, \Delta n_{\rm th}}{1 + (N_\nu/p)} \tag{6.16}$$

where $N_\nu = n_{\rm ph} V$ is the total number of laser photons within the laser volume V in the single oscillating mode. It is seen that the total number of laser photons within the laser volume at the saturation intensity is p.

Equation 6.8 can be solved for $n_{\rm ph}$ with the result that

$$n_{\rm ph} = \frac{N_\nu}{V} = \frac{\gamma\, A_u\, n_u\, \tau_{\rm ph}}{1 - n_u\, \sigma_{ul}\, c\, \tau_{\rm ph}} . \tag{6.17}$$

It is interesting to examine this expression for $n_{\rm ph}$ at line center. At line center the expression $\sigma_{ul}\, c\, \tau_{\rm ph}$ is approximately given by

$$\sigma_{ul}\, c\, \tau_{\rm ph} \cong \frac{\lambda^2}{8\pi} A_u \frac{1}{\Delta\nu} c\, \tau_{\rm ph} = \frac{c^3}{8\pi\, \nu^2\, \Delta\nu} \tau_{\rm ph}\, A_u = \frac{p}{V} \tau_{\rm ph}\, A_u = \frac{1}{\Delta n_{\rm th}} . \tag{6.18}$$

Since $\Delta n_{\rm th} = p/(V\, \tau_{\rm ph}\, A_u)$ one finds that $\tau_{\rm ph}\, A_u = p/(V\, \Delta n_{\rm th})$. Combining Equations 6.17 and 6.18 with the expression for $\tau_{\rm ph}\, A_u$ one obtains

$$N_\nu = \frac{\gamma\, p\, (n_u/\Delta n_{\rm th})}{1 - (n_u/\Delta n_{\rm th})} = \frac{(n_u/\Delta n_{\rm th})}{1 - (n_u/\Delta n_{\rm th})} \tag{6.19}$$

where use is made of $\gamma\, p$=1. If X is defined as $X = n_u/\Delta n_{\rm th}$ then one can express Equation 6.16 as

$$N_\nu = p\left(\frac{r}{X} - 1\right) = p\left(\frac{r - X}{X}\right) \tag{6.20}$$

and one can express Equation 6.19 as

$$N_\nu = \frac{X}{1 - X} . \tag{6.21}$$

If one solves Equations 6.20 and 6.21 for X one obtains

$$X = \frac{n_u}{\Delta n_{\text{th}}} = \frac{(r+1) \pm \left[(r-1)^2 + \frac{4r}{p}\right]^{1/2}}{2\left(1 - \frac{1}{p}\right)}$$

$$= \frac{(r+1) \pm (r-1)\left[1 + \frac{4r}{p\,(r-1)^2}\right]^{1/2}}{2\left(1 - \frac{1}{p}\right)} . \tag{6.22}$$

One must appreciate that the number of modes within the line shape is typically very large. For a laser operating at a wavelength of 600 nm and with a line width of 2×10^9 Hz the number of modes per unit volume within the line width is $p/V = 8\times10^{12}$ cm^{-3}. For any reasonable laser volume p is an exceedingly large number. As a first approximation one might ignore the terms containing p^{-1} in Equation 6.22. For this approximation one finds that $X = r$ for the positive root of the equation and $X = 1$ for the negative root of the equation. The positive root of the equation corresponds to $n_u = r\,\Delta n_{\text{th}}$ which is the expression for n_u for a small excitation rate and with the population difference below threshold. The negative root of the equation corresponds to $n_u = \Delta n_{\text{th}}$ which is the expression for n_u for a large excitation rate and with a population difference equal to the threshold population difference so that laser action is occurring. Thus for a cw laser the population difference increases linearly for small excitation rates and the population difference clamps at Δn_{th} for large excitation rates.

One cannot ignore the terms containing p^{-1} when r is near one because near r=1 the term $[1 + 4r/p(r-1)^2]^{1/2}$ becomes significant. One can estimate how near to one, the approximations used in the previous paragraph, are useful by expanding the square root in the solution for X in a Taylor's series and keeping only the first term in p^{-1} the result is

$$X = \frac{(r+1) \pm (r-1)\left[1 + \frac{2r}{p\,(r-1)^2} + \cdots\right]}{2\left(1 - \frac{1}{p}\right)} . \tag{6.23}$$

The plus sign corresponds to the solution for small values of r and the minus sign corresponds to the solution for large r. The solution corresponding to the plus sign is given by

$$X = \frac{p}{p-1}\,r + \frac{r}{(p-1)(r-1)} + \cdots \tag{6.24}$$

where the first term is nearly equal to r, and the second term is negligible unless $r/(r-1)$ is comparable to $p-1$. Thus the second term is negligible

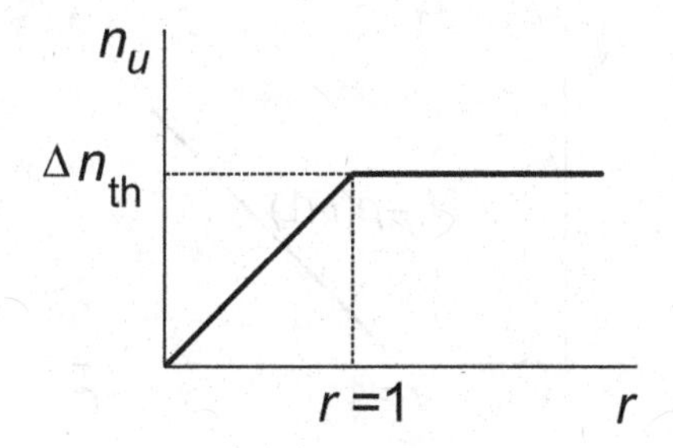

Fig. 6.2 The variation of n_u as a function of r. Since n_l is zero $n_u = \Delta n$.

unless $r-1$ is comparable to $1/(p-1)$. Thus $n_u = r\,\Delta n_{\mathrm{th}}$ from r=0 to r=$1-1/(p-1)$. Since p is exceedingly large the solution holds essentially from r=0 to r=1.

As was stated the negative sign corresponds to the solution for large values of r, and that solution is given by

$$X = \left(\frac{p}{p-1}\right)\left[1 - \frac{r}{p\,(r-1)} + \cdots\right] . \tag{6.25}$$

The first term is nearly equal to 1 and the second term is negligible unless $r/(r-1)$ is comparable to p. Thus the solution $n_u = \Delta n_{\mathrm{th}}$ is valid for values of r from r=$1+1/p$ to r=∞. Hence the simple solutions for X are valid essentially for all value of r. The variation of n_u as a function of r is shown in Figure 6.2. If one solves Equations 6.20 and 6.21 for N_ν rather than X one finds that

$$N_\nu = \frac{p}{2}\left[(r-1) + \left((r-1)^2 + \frac{4r}{p}\right)^{1/2}\right] . \tag{6.26}$$

In the expression for N_ν the plus sign for the square root has been selected in order to assure that N_ν is positive. The term $(r-1)$ in the square root must be positive so that for r less than one Equation 6.26 can be written as

$$\frac{N_\nu}{p} = \frac{1}{2}\left[(r-1) + (1-r)\left(1 + \frac{4r}{p\,(1-r)^2}\right)^{1/2}\right] \tag{6.27}$$

and for r greater than one Equation 6.26 can be written as

$$\frac{N_\nu}{p} = \frac{1}{2}\left[(r-1) + (r-1)\left(1 + \frac{4r}{p\,(r-1)^2}\right)^{1/2}\right] . \tag{6.28}$$

Examination shows that for r less than one N_ν/p is very small and for r

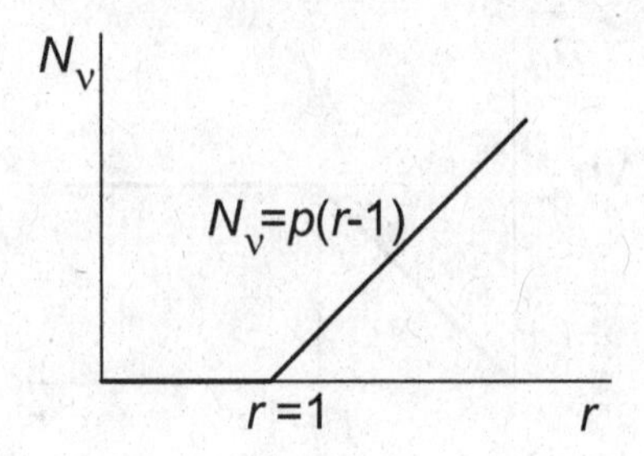

Fig. 6.3 The variation of N_ν as a function of r.

greater than one $N_\nu/p = r - 1$. These solutions are accurate except for values of r very near one.

Figure 6.3 shows N_ν/p as a function of r. Although Figure 6.3 seems to show N_ν as having a discontinuous derivative at r=1 in fact N_ν as a function of r is continuous and has a continuous derivative for all values of r. The derivative appears discontinuous at r=1 merely because N_ν is changing so rapidly at r=1. The actual value of the derivative at r=1 can be calculated by differentiating N_ν as expressed Equation 6.26 with respect to r and evaluating the expression at r=1. The result is

$$\left(\frac{dN_\nu}{dr}\right)_{r=1} = \frac{p}{2}\left[1 + \frac{1}{\sqrt{p}}\right] \cong \frac{p}{2}\,. \tag{6.29}$$

The slope of N_ν as a function of r is finite at r=1 but the slope is extremely large. In order to understand how fast N_ν is changing at r=1 the reader should study Figure 6.4, which shows N_ν as a function of r on a log-log plot. The value of p is arbitrarily taken as 10^{12}. Near threshold measurements of N_ν are very difficult because of the rapid variation of N_ν as a function of r near the threshold value of r=1.

It is interesting to examine both the fluorescence power, P_F, emitted from the active medium into all modes and the laser power, P_L, emitted from the active medium into the single laser mode as functions of r. The fluorescence power emitted from the active medium below threshold (i.e. for r less than one) is given by

$$P_F = h\nu\, A_u\, n_u\, V = h\nu\, A_u\, r\, \Delta n_{\text{th}}\, V = \frac{r\, p\, h\nu}{\tau_{\text{ph}}} \tag{6.30}$$

and above threshold (i.e. for r greater than one) is given by

$$P_F = h\nu\, A_u\, n_u\, V = h\nu\, A_u\, n_{\text{th}}\, V = \frac{p\, h\nu}{\tau_{\text{ph}}}\,. \tag{6.31}$$

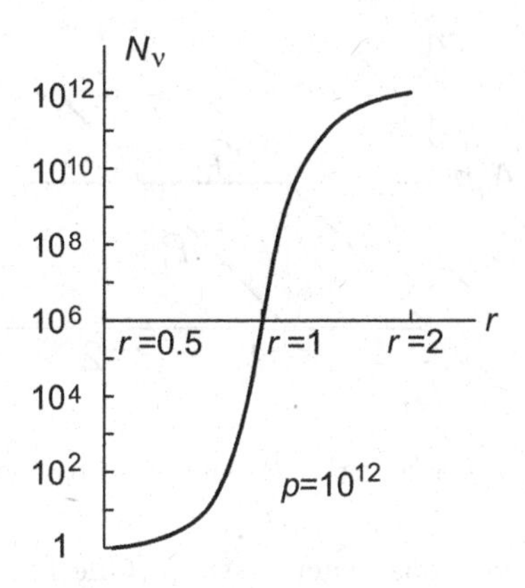

Fig. 6.4 The variation of N_ν as a function of r both below and above threshold.

The power emitted into the laser mode is approximately $P_L=0$ below threshold. Above threshold (i.e. for r greater than one) P_L is given by

$$P_L = \frac{(r-1)\,h\nu\,p}{\tau_{\rm ph}} \,. \tag{6.32}$$

Below threshold the fluorescent power increases linearly with r with a slope equal to $p\,h\nu/\tau_{\rm ph}$ and above threshold the fluorescent power is constant independent of the normalized excitation rate r. The power into the laser mode is nearly zero below threshold and increases linearly with r above threshold with a slope identical to the slope of the fluorescent power below threshold. Thus below threshold almost all the excitation goes into fluorescence, and above threshold the fluorescence is nearly constant and all the excitation above that required to reach threshold goes into the laser mode. The variation of P_F and P_L with r is shown in Figure 6.5. These results are quite general for cw lasers and are not dependent on the particular level scheme or excitation mechanism. A particular level system and excitation mechanism was selected merely to make the discussion concrete.

The relationship $N_\nu = X/(1-X)$ can be solved for X yielding $X = n_u/\Delta n_{\rm th} = N_\nu/(N_\nu+1)$, from which it is obvious that the population difference between the upper and lower laser levels is always less than $\Delta n_{\rm th}$ for a cw laser. This occurs because the photon loss from the laser mode as described by $\tau_{\rm ph}$ is balanced primarily by the stimulated emission into the mode but a small amount of the loss from the mode is balanced by spontaneous emission into the mode. In the earlier analysis the spontaneous emission into the single laser mode was ignored with the result that

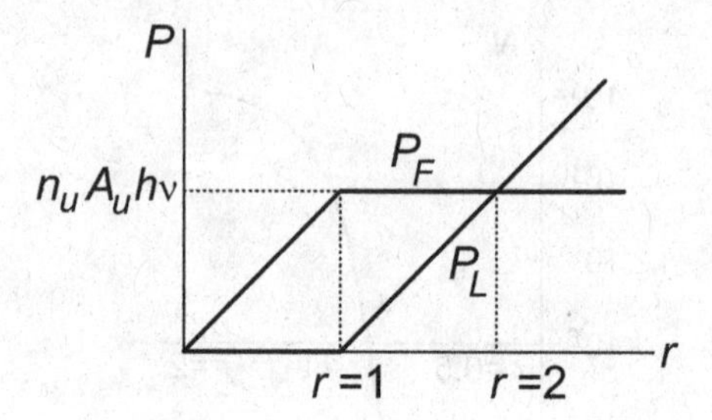

Fig. 6.5 The variation of P_F and P_L as functions of r.

the population difference was exactly $\Delta n_{\rm th}$. One further point should be stressed. The result that the population difference between the upper laser level and the lower laser level clamps at the threshold population difference occurs only for a cw laser. With an intense fast pulsed excitation rate a laser can be driven fast enough that that the population difference exceeds the threshold population difference by a large factor before laser oscillation begins. When the laser oscillation begins the excited level population is rapidly depleted by the resulting large pulse of laser light.

6.2 Lasers with Different Level Schemes

There are important differences in the characteristics of lasers with different level schemes. This section discusses some of these differences. The discussion for cw lasers is carried out in a simplified manner. The rate equations are written ignoring the non-linear term $[n_u - (g_u/g_l)\, n_l]\, \sigma_{ul}\, c\, n_{\rm ph}$. As discussed in the previous section of this chapter for a cw laser the solution obtained from the simplified equations for the population difference $n_u - (g_/g_l)\, n_l$ is correct for all excitation rates below the threshold excitation rate (except very near threshold) and above the threshold excitation the population difference is clamped at the threshold population difference.

A major difference in lasers is related to whether or not a laser operates with the ground level as the lower laser level. If the lower laser level is the ground level then it is necessary that more than half the total population be raised into the upper laser level in order to obtain a population inversion that is large enough for laser action to occur. This usually requires that the population be pumped into an excited level that decays rapidly and non-radiatively into the upper laser level. The upper laser level must have a decay rate to the ground level that is slow enough that a population inversion can be obtained. A laser of this type is called a three level laser. If the lower laser level is not the ground level then it is possible to have a large enough population inversion for laser action with only a small fraction of the total population raised into the upper level. In this situation typically some of the atoms are excited out of the ground level into an excited level that decays rapidly and non-radiatively to the upper laser level. A population inversion occurs if the lifetime of the lower laser level is very short compared to the lifetime of the upper laser level. A laser of this type is called a four level laser. Since it is necessary with a three level laser to excite more than half the total population into the upper laser level to obtain laser action whereas it is only necessary to excite a small fraction of the total population with a four level laser it is usually much easier to obtain laser action with a four level laser than with a three level laser. This section analyzes both three level lasers and four level lasers.

6.2.1 *Three Level Laser*

Figure 6.6 shows a three level laser. It employs the ground level of the atom as the lower level in the laser. For the purposes of this discussion, we will use the ruby laser (Section 9.1) as the prototypical three level laser system. In a ruby laser Cr^{3+} ions are the active medium. For the laser illustrated in Figure 6.6 the Cr^{3+} ions are excited using optical absorption from the ground level denoted as l to a higher level denoted as h. The light source used to drive the l to h absorption can be a broad bandwidth discharge lamp or a laser. Ions in the level h decay very rapidly by a non-radiative mechanism involving phonon emission to the upper laser level denoted by u. For simplicity we take $g_u = g_l$. The rate equations for the levels denoted

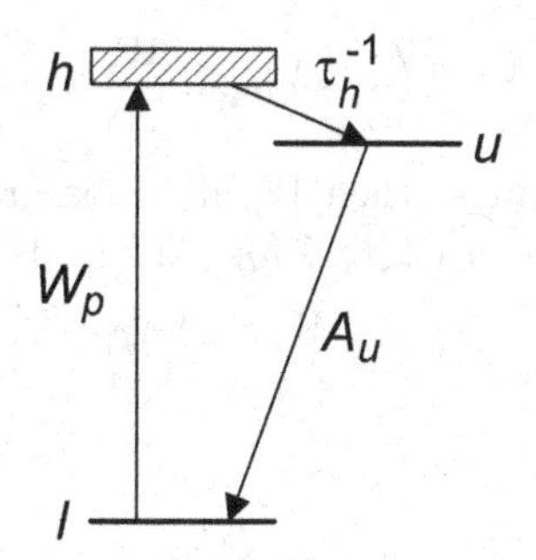

Fig. 6.6 The energy levels for a three level laser.

by h and u are given by

$$\frac{dn_h}{dt} = W_p\,(n_l - n_h) - \gamma_h\,n_h \tag{6.33}$$

and

$$\frac{dn_u}{dt} = \gamma_h\,n_h - A_u\,n_u \tag{6.34}$$

where n_h and n_u are number densities in levels h and u, W_p is the absorption rate per ion for absorption from level l to level h, and γ_h is the decay rate of the level h into the level u. Note that γ_h is the decay rate from the broad level h and is not related to $\gamma = p/V$. Obviously the constant W_p depends on the intensity of the light source used to generate the light for the l to h absorption. Equation 6.34 assumes that level h decays almost entirely into level u and does not decay back to the ground level l at a significant rate i.e. that the non-radiative decay of level h into level u is very rapid compared to the radiative decay of level h back to the ground level.

Since there are only three levels involved and since the ions must be in one of the three levels the ground level population can be obtained from $n_h + n_u + n_l = n_T$ where n_T is the total number density of ions. The number densities of ions in each level are calculated in the steady state. The steady state number density of ions in level h is

$$n_h = \frac{W_p\,n_l}{W_p + \gamma_h}\,. \tag{6.35}$$

If $\gamma_h = \tau_h^{-1}$ is very large so that W_p/γ_h is much less than one then

$$n_h \cong \frac{W_p\,n_l}{\gamma_h} \ll n_l \leq n_T\,. \tag{6.36}$$

In the steady state n_u is equal to

$$n_u = \left(\frac{\gamma_h}{A_u}\right) n_h = \frac{W_p}{A_u}\,n_l\,. \tag{6.37}$$

If n_u is to be larger than n_l then W_p must be greater than A_u. In this situation $n_T = n_h + n_u + n_l \cong n_u + n_l$. This leads to the result that

$$n_u = \frac{W_p\,A_u^{-1}\,n_T}{1 + W_p\,A_u^{-1}} \tag{6.38}$$

and

$$n_l = \frac{n_T}{1 + W_p\,A_u^{-1}}\,. \tag{6.39}$$

The population difference between the upper and the lower laser levels is

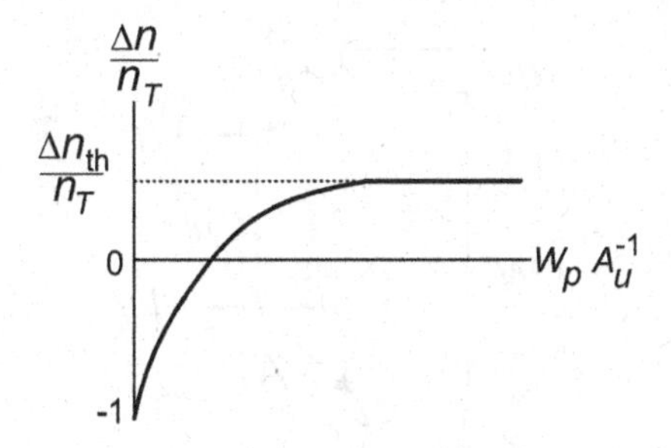

Fig. 6.7 The population difference for a three level laser as a function of W_p/A_u.

given by

$$\frac{\Delta n}{n_T} = \frac{n_u - n_l}{n_T} = \frac{W_p\,A_u^{-1} - 1}{W_p\,A_u^{-1} + 1}\,. \tag{6.40}$$

The expression for the population difference as a function of W_p/A_u is shown graphically in Figure 6.7. Since the calculations did not include the non-linear term that leads to laser action the calculated population difference increases as the excitation rate increases until almost all the ions are in the upper level u. If the non-linear term had been included then the population difference would clamp at the threshold population difference. The population difference is shown clamping at $\Delta n = \Delta n_{\text{th}}$, and the solution for Δn given by Equation 6.39 is ignored for values of Δn greater than Δn_{th} where Δn is simply equated to Δn_{th}.

6.2.2 *Four Level Laser*

Now consider a four level system, for which the lower laser level is not the ground level. The energy levels of such a system are shown in Figure 6.8. A Nd:YAG laser (Section 9.2) is a laser that operates with a four level scheme similar to that shown in Figure 6.8. In a Nd:YAG laser Nd^{3+} ions serve as the laser medium. The YAG portion of the name refers to yttrium aluminum garnet, which is the host crystal for the Nd ions. The ions are excited from the ground level denoted by g to a broad higher level denoted by h. Typically, the excitation is accomplished by absorption of radiation from a diode laser (or an array of diode lasers). The higher level h decays rapidly through a non-radiative mechanism to the upper laser level denoted by u. The lower laser level is denoted by l and is very short lived decaying rapidly to lower levels. For simplicity it is assumed that the statistical

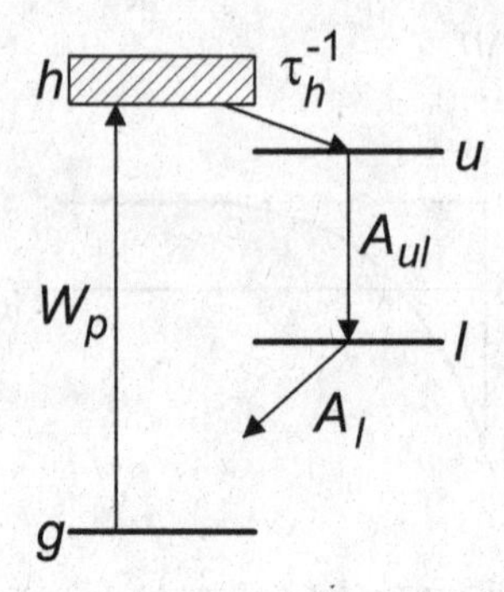

Fig. 6.8 The energy levels for a four level laser.

weight of all the levels is the same. The rate equation determining the population of the highest level h is

$$\frac{dn_h}{dt} = W_P\,(n_g - n_h) - \gamma_h\, n_h\ . \tag{6.41}$$

In the steady state the population of the level h is given by

$$n_h = \frac{W_p\, n_g}{\gamma_h + W_p} \cong \frac{W_p}{\gamma_h}\, n_g \tag{6.42}$$

where it is assumed that γ_h is large enough that W_p/γ_h is much less than one. The rate equations for the upper and lower laser levels are

$$\frac{dn_u}{dt} = \gamma_h\, n_h - A_u\, n_u \tag{6.43}$$

and

$$\frac{dn_l}{dt} = A_{ul}\, n_u - A_l\, n_l\ . \tag{6.44}$$

In the steady state one finds that

$$n_l = \left(\frac{A_{ul}}{A_l}\right) n_u\ . \tag{6.45}$$

If A_{ul}/A_l is much less than one then the population of the lower laser level is always very small compared to the population of the upper laser level. In this situation $n_T = n_g + n_h + n_u + n_l \cong n_g + n_u$, and the steady state population of the upper laser level is given by

$$\frac{n_u}{n_T} = \frac{\Delta n}{n_T} = \frac{W_p\, A_u^{-1}}{1 + W_p\, A_u^{-1}}\ . \tag{6.46}$$

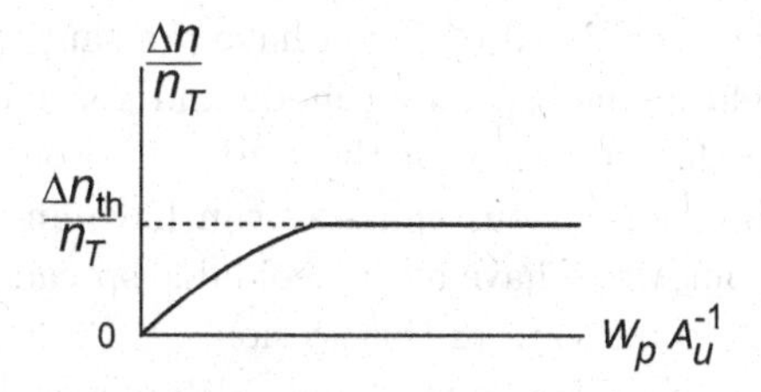

Fig. 6.9 The population difference for a four level laser as a function of W_p/A_u.

Since the population of the lower laser level is very small the population difference between the upper and lower laser levels is equal to $\Delta n = n_u - n_l \cong n_u$. Figure 6.9 shows the population difference between the upper and lower laser levels as a function of W_p/A_u. Again note that Δn clamps at Δn_{th}.

From Figures 6.7 and 6.9 it is clear that it is much easier to achieve a threshold population difference for the four level laser where the lower laser level is not the ground level than for the three level laser where the lower laser level is the ground level. The value of W_p/A_u at which the threshold population difference is achieved is much lower for the four level laser than for the three level laser. This is the result of the fact that to achieve laser action for the three level laser it is necessary to excite $(n_T + \Delta n_{\text{th}})/2$ ions into the upper level leaving only $(n_T - \Delta n_{\text{th}})/2$ ions in the ground level thereby producing a population difference of Δn_{th}. Whereas for the four level laser it is only necessary to excite Δn_{th} ions into the upper laser level. For the three level laser it is necessary to substantially depopulate the ground level to reach threshold, whereas for the four level laser it is only necessary to excite a small fraction of the ions out of the ground level.

6.3 The Minimum Bandwidth of a CW Laser

The minimum bandwidth of a cw laser is determined by the competing effects of stimulated emission and spontaneous emission into the laser mode and hence by the effects discussed in Section 6.1 of this chapter. The bandwidth of the laser is equal to the spread of frequencies present in the laser light. The minimum bandwidth of a cw laser is typically much narrower than the pass band of the laser cavity. The minimum bandwidth of a laser arises because of spontaneous emission into the oscillating mode.

Photons emitted by stimulated emission have the same phase as the total electromagnetic field in the oscillating mode and hence do not change the phase or therefore the frequency of the laser. Photons emitted into the oscillating mode by spontaneous emission can have any phase and hence do cause the laser output to have an unavoidable spread in the frequencies present. This effect is analyzed in this section.

In the discussion of the minimum bandwidth of a cw laser it is assumed that the laser operates on only one mode. The population difference between the upper and the lower laser levels is clamped at the threshold population difference and the number of photons in the oscillating mode is very large. If there are N_ν photons in the oscillating mode then the energy of the photons in the oscillating mode is $N_\nu \, h\nu$. The energy in the electromagnetic field of the laser beam can be written as

$$N_\nu \, h\nu = \frac{1}{2} \int \left(\varepsilon \boldsymbol{E}^2_{\mathrm{RMS}} + \mu \boldsymbol{H}^2_{\mathrm{RMS}}\right) dV = \varepsilon \overline{\boldsymbol{E}^2_{\mathrm{RMS}}} \, V \qquad (6.47)$$

where RMS indicates the root mean square and where the overbar indicates the average over the volume of the laser cavity. In Equation 6.47 use has been made of the fact that the energy in the magnetic field of a wave is equal to the energy in the electric field, so that $\varepsilon E^2/2 = \mu H^2/2$. Solving Equation 6.47 for the average value of the RMS electric field yields

$$\overline{E_{\mathrm{RMS}}} = \left(\frac{N_\nu \, h\nu}{\varepsilon V}\right)^{1/2} . \qquad (6.48)$$

The electric field has both a magnitude (given by Equation 6.48), and a phase. The initial phase of the electric field is arbitrarily taken as zero at t=0 so that the electric field initially varies as $\cos \omega t$ where $\omega = 2\pi \, \nu$. For a cw laser operating in the steady state the number of photons in the oscillating mode is constant so that the number of photons emitted into the oscillating mode per second by stimulated emission plus the number of photons emitted into the oscillating mode per second by spontaneous emission must be equal to the number of photons lost from the oscillating mode per second. The photons emitted into the oscillating mode by stimulated emission have a phase that is identical to the phase of the electromagnetic field in the cavity. On the other hand the photons emitted into the oscillating mode by spontaneous emission have a phase that is random with respect to the electromagnetic field in the cavity. This randomness results in the minimum bandwidth of a cw laser.

A full treatment of the bandwidth must include the randomness in the phase of the spontaneous emission. This is a difficult subject. Here, a

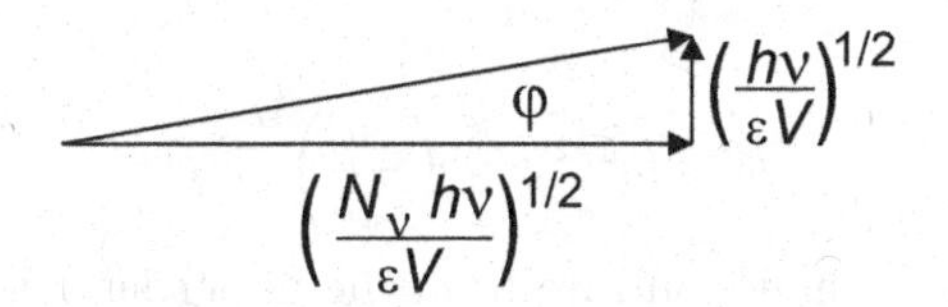

Fig. 6.10 The change in phase when a single spontaneous photon is emitted into a laser mode.

simple model is used to approximate the randomness by taking the phase of the spontaneous photons as being $\pm\pi/2$ with respect to the phase of the oscillating mode of the electromagnetic field in the cavity. The volume average of the electric field, E_{RMS}, of a single spontaneous photon emitted is given by

$$\overline{E_{\text{RMS}}} = \left(\frac{h\nu}{\varepsilon V}\right) . \tag{6.49}$$

The electric field of the spontaneous photon and the electric field due to all the other photons in the mode must be added together as phasors in order to treat the effect of their having different phases. Figure 6.10 shows the phasor field due to all the other photons in the mode added vectorially to the phasor field due to a single spontaneous photon. In Figure 6.10 the spontaneous photon leads the field in the cavity by $\pi/2$, but the phase could with equal probability have lagged the field in the cavity by $-\pi/2$ since it was assumed that the change in phase would be $\pm\pi/2$. The change in the phase of the total electromagnetic field in the cavity that results from the spontaneous emission of a single photon into the lasing mode is given by plus or minus $\phi = N_\nu^{-1/2}$ so that if the phase before the emission of the spontaneous photon is ωt then the phase after the emission of a spontaneous photon is $\omega t \pm N_\nu^{-1/2}$. The electromagnetic field in the cavity now induces photons to be emitted with this new phase. With the spontaneous emission of more photons into the oscillating mode the phase of the electromagnetic field of the oscillating mode changes by $\pm N_\nu^{-1/2}$ for each spontaneous photon that is emitted into the oscillating mode. Because the spontaneous photons are emitted with a random phase, the phase of the electromagnetic field in the oscillating mode changes in a random fashion, sometimes increasing and sometimes decreasing as spontaneous photons are emitted into the oscillating mode. In the simplified model where the phase of the spontaneous photon is either $\pm\pi/2$ the phase change due to the random walk after the emission of m spontaneous photons into the oscillating mode is

$$\Phi = m^{1/2}\,\phi = \left(\frac{m}{N_\nu}\right)^{1/2} . \tag{6.50}$$

The spontaneous emission rate from the upper laser level to the lower laser level is $A_{ul}\,n_u\,V$. This spontaneous emission goes with equal probability into one of the $p = (8\pi\,\nu^2\,\Delta\nu/c^3)\,V$ modes within the linewidth. The rate of emission into a single mode is therefore $A_{ul}\,n_u\,V/p$. In a time t there are

$$m = \frac{n_u\,A_u\,V\,t}{p} \tag{6.51}$$

spontaneous photons emitted into the single oscillating mode. It is arbitrarily assumed that when the electromagnetic wave in the cavity changes by Φ=1 radian then the wave is dephased from the original wave. The time required for the phase to change by one radian is taken as the coherence time of the laser t_{coh}. This leads to the result that

$$t_{\text{coh}} = \frac{p\,N_\nu}{A_u\,n_u\,V} . \tag{6.52}$$

The bandwidth of the laser is given by

$$\Delta\omega_{\text{laser}} = \frac{1}{t_{\text{coh}}} = \frac{A_u\,n_u\,V}{p\,N_\nu} = \frac{A_u\,\Delta n_{\text{th}}}{p\,N_\nu}\left(\frac{n_u}{\Delta n_{\text{th}}}\right)V . \tag{6.53}$$

The threshold population difference is given by

$$\Delta n_{\text{th}} = \left(\frac{8\pi\,\nu^2\,\Delta\nu}{c^3}\right)\left(\frac{1}{A_{ul}\,\tau_{\text{ph}}}\right) = \frac{(p/V)}{A_{ul}\,\tau_{\text{ph}}} . \tag{6.54}$$

Using Equations 6.53 and 6.54 and noting that $A_u \approx A_{ul}$ one obtains

$$\Delta\omega_{\text{laser}} = \frac{1}{N_\nu\,\tau_{\text{ph}}}\left(\frac{n_u}{\Delta n_{\text{th}}}\right) = \frac{h\nu}{(N_{\nu u}\,h\nu/\tau_{\text{ph}})\,\tau_{\text{ph}}^2}\left(\frac{n_u}{\Delta n_{\text{th}}}\right) . \tag{6.55}$$

The expression $N_\nu\,h\nu/\tau_{\text{ph}}$ is approximately equal to the external power, P_e, emitted by the laser provided the primary loss mechanism for photons from the cavity is from transmission through the end mirror into the external laser beam. The quantity $\delta\omega_c = 1/\tau_{\text{ph}}$ is the width in angular frequency of the individual cavity modes. Hence one obtains

$$\Delta\omega_{\text{laser}} = \frac{h\nu}{P_e}\,(\delta\omega_c)^2\left(\frac{n_u}{\Delta n_{\text{th}}}\right) \tag{6.56}$$

which can be rewritten as

$$\Delta\nu_{\text{laser}} = \frac{2\pi\,h\nu}{P_e}\,(\delta\nu_c)^2\left(\frac{n_u}{\Delta n_{\text{th}}}\right) \tag{6.57}$$

Equation 6.57 is identical to the expression given by Schawlow and Townes in their original paper proposing the laser.

The laser bandwidth predicted by Equation 6.57 is very narrow. For example the laser bandwidth for a 1 mW He-Ne laser 30 cm in length with a loss of 1% per pass due to transmission out of the laser cavity has a fractional bandwidth of about (see problem 2 of this chapter)

$$\left|\frac{\Delta\nu_{\text{laser}}}{\nu}\right| = \left|\frac{\Delta\lambda_{\text{laser}}}{\lambda}\right| = 2.6 \times 10^{-19}\,. \tag{6.58}$$

Summary of Key Ideas

- Below threshold population density almost all the input power produces photons that are emitted into fluorescence in all modes within the line width. Above the threshold population density almost all the power above that needed to reach threshold goes into producing photons in the single lasing mode and the population density clamps just infintesimally below the threshold population density (see Fig. 6.5).
- For a three level laser (Fig. 6.6) more than half the atoms or molecules must be excited out of the ground level to reach threshold laser action.
- For a four level laser (Fig. 6.8) the threshold can be reached with a much smaller fractional excitation out of the ground level.
- The minimum possible bandwidth of a laser is due to spontaneous emission into the lasing mode (Sec. 6.3). It varies as the inverse of the laser power and is extremely small for almost any cw laser.

Suggested Additional Reading

William T. Silfvast, *Laser Fundamentals*, Cambridge University Press (1996).

A. E. Siegman, *Lasers*, University Science Books (1986).

A. Yariv, *Quantum Electronics 3rd Ed.*, John Wiley and Sons (1989).

B. E. A. Saleh and M. C. Teich, *Fundamentals of Photonics*, John Wiley and Sons (1991).

M. Sargent III, M. O. Scully, and W. E. Lamb, *Laser Physics*, Addison Wesley Publishing Co. (1974).

A. L. Schawlow and C. H. Townes, "Infrared and Optical Masers" *Phys. Rev.* **112**, 1940 (1958).

L. W. Anderson and J. E. Lawler, "A simple derivation of the bandwidth of a laser oscillator" *Am. J. Phys.* **46**, 162 (1978).

Problems

1. A He-Ne laser operates with an output power of 10 mW. The optical cavity is 30 cm in length. If the transmission of the output mirror is 1% and if all other losses are negligible what is the photon lifetime in the cavity? What is the power of the laser beam inside the optical cavity?

2. For the laser described in Problem 1 what are the minimum laser bandwidth and the minimum fractional bandwidth?

3. For the He-Ne laser of Problem 1 what is the number of photons in the laser cavity summed over all modes, $\sum N_\nu$?

4. If the lifetime of the upper laser level for the He-Ne laser in Problem 1 is 2×10^{-8} s and if the operating temperature is $T = 400K$ what is the threshold population density difference for the laser? If the volume of the active medium is 3×10^{-2} cc and if the population of the lower laser level is very small how many excited atoms are present in the laser at threshold?

5. A cw laser operates on a transition for which there are p=10^{14} modes within the linewidth. What is the number of photons in a single oscillating mode of the laser cavity for r=0.5? for r=1.0? for r=2? for r=5?

6. For the laser of Problem 6 what is the ratio $n_u/\Delta n_{\text{th}}$ for r=0.5? for r=1.0? for r=2? for r=5?

7. Describe in your own words why it is much easier to obtain laser action for a laser when the lower laser level is not the ground level than when the lower level is the ground level.

8. For the laser of Problem 1 estimate the phase change for a single spontaneous photon emitted into the oscillating mode.

Chapter 7

Laser Optical Cavities

The number of optical modes interacting with an atom in the visible is usually very large. The number of modes within the line shape of a transition is equal to $p = 8\pi\nu^2\,\Delta\nu\,V/c^3$ (Eq. 2.9) where $\Delta\nu$ is the line width and where V is the volume of the optical cavity. Thus the number of modes interacting with an atom or molecule in the visible is 10^{12} or even larger. Schawlow and Townes proposed the use of a two mirror optical cavity that is otherwise open in their original paper proposing the concept of the laser. The two mirror cavity provides low losses for a small number of modes, perhaps even a single mode, along the optical axis and within the line width of the transition The result is a cavity with a long photon lifetime for a small number of modes within the line width of the lasing transition.

The first laser cavities were constructed with plane mirrors and were essentially Fabry Perot etalons with the active medium between the two mirrors. A cavity constructed with plane parallel mirrors is, as will be shown, rather lossy and operates at the edge of stability. It was therefore a difficult cavity to use. Much better cavities that used spherical mirrors were soon developed. Cavities using spherical mirrors are now widely used and are well understood.

In recent years "Super cavities" have been developed with extraordinary photon lifetimes. The extremely long photon lifetimes are achieved using spherical mirrors with extremely low fractional losses per round trip in the optical cavity due to scattering and absorption. The fractional losses can be as low as 10^{-6} or less.

At the other extreme, with a very high gain amplifying medium, a very lossy cavity with a short photon lifetime, that results from high transmission out of the cavity, is not only tolerable but is usually desirable. These cavities are called unstable resonators. Unstable resonators are very important for use with a high gain medium because they produce a near diffraction limited output laser beam. If one attempts to use a high gain medium in a conventional cavity the medium may lase in both on axis and off axis modes producing a poorly collimated beam with much of the light going in directions where it is not useable in a particular experiment or application.

The highest gain media such as the recombining plasmas used for X-ray lasers do not require any cavity at all. A long narrow cylindrical amplifying medium will lase if a photon spontaneously emitted along the axis at one end of the cylindrical medium is amplified to the saturation intensity before it reaches the other end of the medium. If the high gain medium has a diameter D and a length L the angular divergence of the laser beam will be the greater of the geometrical limit $2D/L$ or the diffraction limit λ/D.

This chapter contains several sections. Section 7.1 discusses the stability criterion for a laser cavity from the view points of both geometrical optics using a ray description (Sec. 7.1.1) and using wave optics based on Gaussian beams (Sec. 7.1.3). The influence of the optical cavity on the frequency distribution of the emitted laser light is discussed in Section 7.2. Section 7.3 covers cavity losses, and the remaining sections contain short discussions of closely related material.

7.1 Cavity Stability

In the following sub-sections the basic ideas of cavity stability are analyzed using both geometrical and Gaussian wave optics. Although many differences exist between the results for geometrical and wave optics it is interesting that the criteria for stability are the same for the two approaches.

7.1.1 *Geometrical Optics Stability Criterion*

The stability criterion for a laser cavity can be analyzed using geometrical optics. For this analysis the discussion is restricted to the stability of paraxial rays. A paraxial ray is one that makes a small angle with respect to the axis of the optical system so that one can approximate the angle, the tangent of the angle, and the sine of the angle as being identical. A ray at a position s along the direction of propagation can be characterized by two quantities $r(s)$ and $r'(s) = dr/ds$ where $r(s)$ is the transverse distance of the ray from the axis at the position s and $r'(s)$ is the slope of the ray at the position s . It is convenient to describe the ray as a vector with two

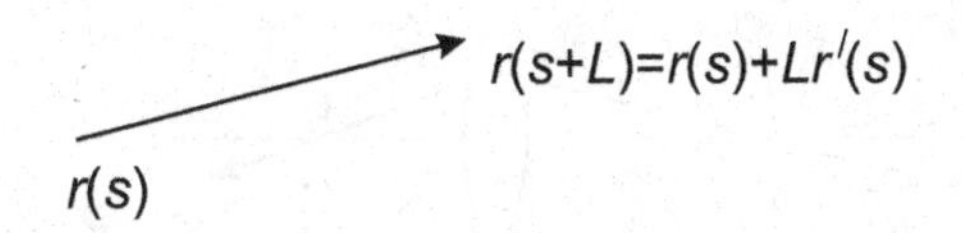

Fig. 7.1 The propagation of an optical ray in free space.

components

$$\begin{pmatrix} r(s) \\ r'(s) \end{pmatrix} . \tag{7.1}$$

Optical elements are represented as two by two matrices that act on the vector representing the ray. For example the propagation of a ray in a straight line through free space is represented by the matrix

$$\begin{vmatrix} 1 & L \\ 0 & 1 \end{vmatrix} \tag{7.2}$$

so that $r(s+L)$ and $r'(s+L)$ are given by

$$\begin{pmatrix} r(s+L) \\ r'(s+L) \end{pmatrix} = \begin{vmatrix} 1 & L \\ 0 & 1 \end{vmatrix} \begin{pmatrix} r(s) \\ r'(s) \end{pmatrix} . \tag{7.3}$$

The matrix changes r but not r' as the ray propagates through free space since the expressions for $r(s+L)$ and for $r'(s+L)$ from Equation 7.3 can be rewritten as

$$r(s+L) = r(s) + r'(s)\, L \tag{7.4}$$

and

$$r'(s+L) = r'(s) . \tag{7.5}$$

Figure 7.1 shows the propagation of the ray in a straight line through free space.

A thin lens of focal length f is represented by the matrix

$$\begin{vmatrix} 1 & 0 \\ -\frac{1}{f} & 1 \end{vmatrix} \tag{7.6}$$

so that the expressions for r and r' after passing through a lens are given by

$$\begin{pmatrix} r(s)_i \\ r'(s)_i \end{pmatrix} = \begin{vmatrix} 1 & 0 \\ -\frac{1}{f} & 1 \end{vmatrix} \begin{pmatrix} r(s)_o \\ r'(s)_o \end{pmatrix} \tag{7.7}$$

where the subscript o indicates the that the values of r and r' are evaluated in the object space just before the lens and the subscript i indicates that

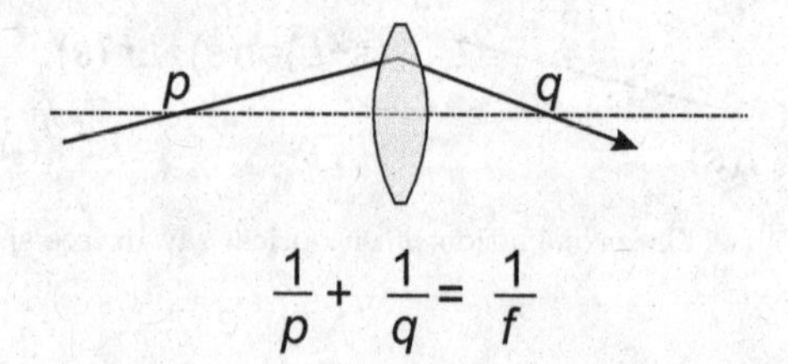

Fig. 7.2 The propagation of an optical ray through a thin lens.

the values of r and r' are evaluated in the image space just after the lens and where s is the position of the thin lens. Since the position of the ray off the axis r does not change as the ray passes through a thin lens the value of r is the same both before and after the ray passes through the lens so that $r(s)_o = r(s)_i$. The value of r' after passage through the thin lens is given by

$$r'(s)_i = -\frac{r(s)_o}{f} + r'(s)_o \,. \tag{7.8}$$

The passage of the ray through the thin lens is shown in Figure 7.2. If we call $p = s - s_o$ and $q = s_i - s$ so that $r'_o = r/p$ and $r'_i = -r/q$ where $r = r(s)_o = r(s)_i$ where s_0 and s_i are the locations of the object and the image. Equation 7.8 can be rewritten as

$$\frac{1}{p} + \frac{1}{q} = \frac{1}{f} \tag{7.9}$$

which is the usual expression for the thin lens equation.

The change in a ray at a dielectric interface is represented by the matrix

$$\begin{vmatrix} 1 & 0 \\ 0 & \frac{n_a}{n_b} \end{vmatrix} \tag{7.10}$$

here n_a and n_b are the indices of refraction on each side of the interface as shown in Figure 7.3. The displacement of the ray and the slope of the ray after it has crossed the interface are given by

$$\begin{pmatrix} r_b \\ r'_b \end{pmatrix} = \begin{vmatrix} 1 & 0 \\ 0 & \frac{n_a}{n_b} \end{vmatrix} \begin{pmatrix} r_a \\ r'_a \end{pmatrix} . \tag{7.11}$$

Clearly the interface matrix leaves r unchanged so that $r_b = r_a$. The second row of the matrix expresses Snell's law for paraxial rays

$$n_a \sin\theta_a = n_b \sin\theta_b \cong n_a\,\theta_a \cong n_b\,\theta_b \,. \tag{7.12}$$

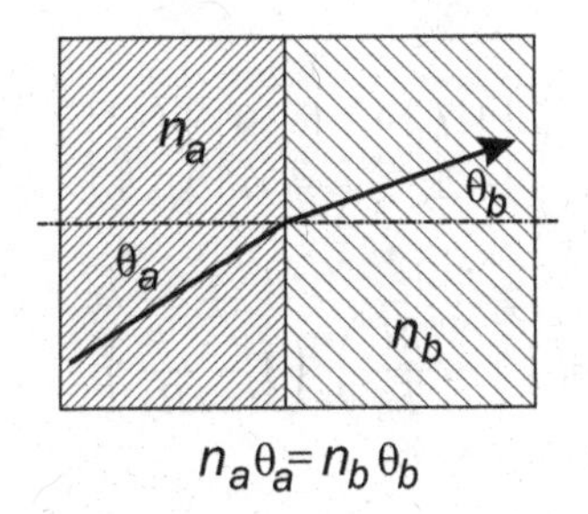

Fig. 7.3 The propagation of an optical ray across a dielectric boundary.

The effect of a plane mirror reflecting a ray at normal incidence is to reverse the direction of propagation of the ray. Our definition of r' is $r' = dr/ds$ where s is measured along the direction of propagation. Since our derivative is measured along the direction of propagation a plane mirror has no effect on either r or r'. The concept that a plane mirror is represented by the identity matrix is important since it leads us to recognize that we can unfold a two mirror cavity into an equivalent biperiodic sequence of lenses.

A concave spherical mirror of radius R is equivalent in our unfolded picture to a lens of focal length $f = R/2$. The matrix representing such a mirror is therefore

$$\begin{vmatrix} 1 & 0 \\ -\frac{2}{R} & 1 \end{vmatrix} . \tag{7.13}$$

Other optical components can also be represented by ray transfer matrices.

With the use of the preceding ray transfer matrices one can analyze the stability of an arbitrary two mirror cavity. Figure 7.4 shows an arbitrary two mirror cavity. One mirror has a radius of curvature of R_1 and the other has a radius of curvature R_2. The two mirrors are separated by a distance L. A round trip through the cavity is represented by the product of four

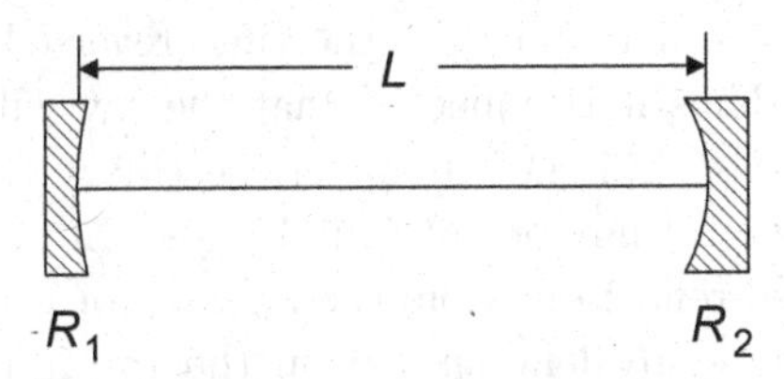

Fig. 7.4 A laser cavity with length L and mirrors with radii of curvature R_1 and R_2.

matrices

$$\begin{vmatrix} A & B \\ C & D \end{vmatrix} = \begin{vmatrix} 1 & L \\ 0 & 1 \end{vmatrix} \begin{vmatrix} 1 & 0 \\ -\frac{2}{R_1} & 1 \end{vmatrix} \begin{vmatrix} 1 & L \\ 0 & 1 \end{vmatrix} \begin{vmatrix} 1 & 0 \\ -\frac{2}{R_2} & 1 \end{vmatrix} \tag{7.14}$$

where

$$A = \left(1 - \frac{2L}{R_1}\right)\left(1 - \frac{2}{R_2}\right) - \frac{2L}{R_2} \tag{7.15}$$

$$B = L\left(1 - \frac{2L}{R_1}\right) + L \tag{7.16}$$

$$C = -\frac{2}{R_1}\left(1 - \frac{2L}{R_2}\right) - \frac{2L}{R_2} \tag{7.17}$$

and

$$D = 1 - \frac{2L}{R_1}\,. \tag{7.18}$$

It should be noted that $AD - BC{=}1$. The matrix for the round trip in the cavity leads to the following expression for the displacement and slope of a ray after a round trip in terms of the same quantities before the round trip

$$\begin{pmatrix} r_{N+1} \\ r'_{N+1} \end{pmatrix} = \begin{vmatrix} A & B \\ C & D \end{vmatrix} \begin{pmatrix} r_N \\ r'_N \end{pmatrix} \tag{7.19}$$

where r_{N+1} and r'_{N+1} are the displacement and slope of the ray at some point in the cavity after $N+1$ round trips and r_N and r'_N are the same quantities after N round trips. The equations can be written out explicitly yielding

$$r_{N+1} = A\,r_N + B\,r'_N \tag{7.20}$$

and

$$r'_{N+1} = C\,r_N + D\,r'_N\,. \tag{7.21}$$

The cavity is considered stable if after a very large number of round trips the ray remains close to the axis. The ray can not collapse to being along the axis because if it did then the time reversed ray would always be along the axis. Thus it is expected that the ray will oscillate in some fashion about the axis. It is not, however, necessary that the ray retrace itself after some integral number of round trips. In order to analyze the effect of the repeated reflections from the mirrors (or the effect of repeated passage through the equivalent lenses) on the ray it is advantageous to utilize a complex notation. A trial solution to the ray vector is taken as

$r_N = r_0\, e^{iN\theta}$ and $r'_N = r'_0\, e^{iN\theta}$ where both r_0 and r'_0 can be complex. The real parts of the complex r_N and r'_N are the actual values for r_N and r'_N. This leads to the equation

$$\begin{pmatrix} r_{N+1} \\ r'_{N+1} \end{pmatrix} = \begin{vmatrix} A & B \\ C & D \end{vmatrix} \begin{pmatrix} r_N \\ r'_N \end{pmatrix} = e^{i\theta} \begin{pmatrix} r_N \\ r'_N \end{pmatrix} . \tag{7.22}$$

Using Equation 7.22 one finds that

$$\begin{vmatrix} A - e^{i\,\theta} & B \\ C & D - e^{i\,\theta} \end{vmatrix} \begin{pmatrix} r_N \\ r'_N \end{pmatrix} = 0 . \tag{7.23}$$

Equation 7.23 has non-zero solutions if and only if

$$\left(A - e^{i\,\theta}\right)\left(D - e^{i\,\theta}\right) - BC = 0 \tag{7.24}$$

which leads to the result that

$$e^{i\theta} = \cos\theta + i\sin\theta = \frac{A+D}{2} \pm i\left[1 - \left(\frac{A+D}{2}\right)^2\right]^{1/2} \tag{7.25}$$

where use has been made of the fact that $AD - BC{=}1$. The constant θ is a real number if and only if

$$-1 \le \frac{A+D}{2} \le 1 . \tag{7.26}$$

Equation 7.26 is the condition for stability of a cavity. The trial solution works if Equation 7.26 is satisified. The stability criterion can be written in terms of the properties of the cavity by substituting the expressions for A and D and rearranging. The result is

$$0 \le \left(1 - \frac{L}{R_1}\right)\left(1 - \frac{L}{R_2}\right) \le 1 . \tag{7.27}$$

This is a common way that one sees the stability criterion written, but it is also commonly expressed using two new quantities $g_1 = 1 - L/R_1$ and $g_2 = 1 - L/R_2$. In terms of g_1 and g_2 the stability criterion becomes $0 \le g_1\, g_2 \le 1$. Figure 7.5 shows graphically the region of stability. Figure 7.6 shows a ray diagram for a confocal cavity. It is clear from Figure 7.6 that a ray always stays near the axis for a confocal cavity. For a cavity that is outside the region of stability a ray will diverge from the axis.

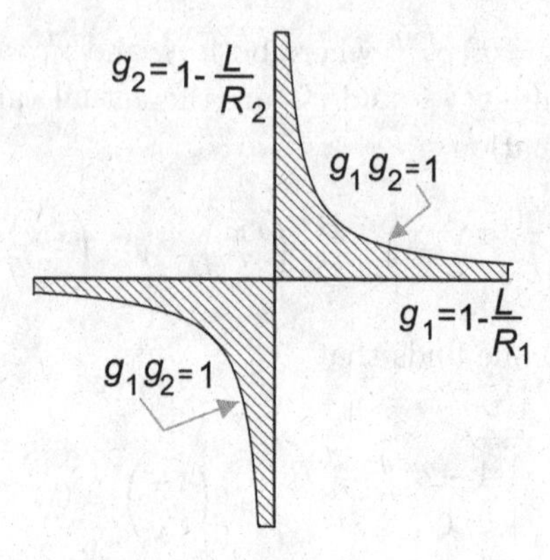

Fig. 7.5 The region of stability for a laser cavity.

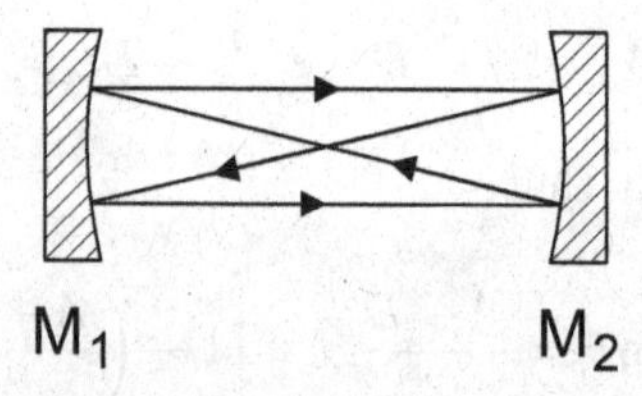

Fig. 7.6 A confocal cavity with a set of geometrical rays indicated. It is clear that the rays do not diverge from the axis.

7.1.2 *Gaussian Beams*

In the preceding section the criterion for stability for an optical laser cavity was derived using geometrical ray optics. It is also important to understand the stability of a cavity from the viewpoint of wave optics. In order to do this it is necessary to understand the properties of the electromagnetic modes of the cavity using wave optics. The lowest loss modes of a cavity are very well confined and always lie near the axis of the cavity. Modes that have a substantial intensity off axis spill over the edge of the mirror to some extent and have higher diffraction losses than the low loss modes that lie near the axis.

Although a formal proof is not presented, it is true that the best confined modes and hence the lowest loss modes have a Gaussian distribution for the intensity of the beam transverse to the beam axis. It will be shown that if a laser beam has a Gaussian distribution at one position along the beam then it will have a Gaussian distribution at all locations along the beam. This can be proven by using the fact that the Fourier transform of a Gaussian is a Gaussian and remembering that the Fraunhofer diffraction

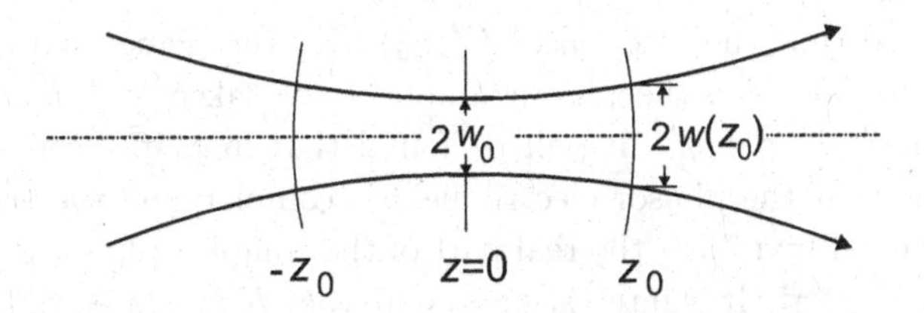

Fig. 7.7 A schematic diagram of a Gaussian beam.

of a plane wave is described by the Fourier transform of the plane wave. A Gaussian beam is "optimally" confined along the axis in both the near and the far fields since the beam is Gaussian in both regions. The properties of Gaussian beams are investigated in the intermediate field region between the near and far field regions in order to understand the stability of a cavity in the wave optics picture.

Fresnel diffraction theory is used to derive the properties of a Gaussian beam at all distances from the beam waist. The existence of a beam waist where the beam has a Gaussian intensity distribution and with a constant phase in a plane so that the phase front is flat is hypothesized. In fact there are an infinite number of modes that have a Gaussian intensity. These modes have different variations in the intensity as a function of the angle in a plane perpendicular to the direction of propagation of the beam. In this section only the fundamental Gaussian mode (also known as the TEM_{00} mode) is considered, higher order modes are discussed in Section 7.6. The fundamental Gaussian mode has an intensity that is independent of the angle in a plane perpendicular to the direction of propagation. In some situations the beam waist can be a virtual waist outside the cavity. Figure 7.7 shows a beam with the waist located at z=0. Fresnel diffraction is a scalar theory and does not permit the consideration of the vector character of the electromagnetic field. The only quantity calculated in Fresnel diffraction is an amplitude, and the intensity of the wave is taken as the absolute square of the amplitude. The amplitude is taken to be the electric field, E, even though the amplitude is a scalar and not a true vector electric field. The phasor electric field of the fundamental Gaussian mode (the complex amplitude divided by $e^{-i\omega t}$) in the plane of the waist is given by

$$E(\boldsymbol{r}_0) = E(x_0, y_0) = K \left(\frac{2}{\pi}\right)^{1/2} \frac{1}{w_0} \exp\left[-\frac{x_0^2 + y_0^2}{w_0^2}\right] \tag{7.28}$$

where x_0 and y_0 are the x and y coordinates in the plane z=0, and w_0 is the $1/e$ radius of the beam at the waist. The amplitude of the field is normalized so that the integral of $E^2(\boldsymbol{r}_0)$ over the plane z=0 is given by K^2. In the z=0 plane the phasor amplitude is taken as a real Gaussian function of $r_0^2 = x_0^2 + y_0^2$. It will be found that in planes other than the z = 0 plane that the phasor electric field is complex and the true electric field is obtained by taking the real part of the complex phasor electric field multiplied by $e^{-i\omega t}$. It is interesting to note that K is related to the electric field at the center of the waist (i.e., at $x = y = z = 0$). If E_0 is the field at the center of the waist then $E_0^2 = K^2(2/\pi\omega_0^2)$ and the integral of $E^2(\boldsymbol{r}_0)$ over the plane z=0 is given by $K^2 = (E_0^2/2)(\pi w_0^2)$.

With the assumption that the amplitude of the field in the plane z=0 has the form given in Equation 7.28 it is possible to calculate the field in different z planes using Fresnel diffraction. Huygens' principle gives a mathematical expression to Fresnel diffraction. The Fresnel diffraction expression for the electric field amplitude at a position $\boldsymbol{r}$ in terms of the electric field amplitude in the plane z=0 is

$$E(\boldsymbol{r}) = \frac{i}{\lambda} \int\int E(\boldsymbol{r}_0) \left(\frac{1+\cos\alpha}{2}\right) \frac{\exp[ik(\boldsymbol{r}-\boldsymbol{r}_0)]}{|\boldsymbol{r}-\boldsymbol{r}_0|}\, dx_0\, dy_0 \qquad (7.29)$$

where $(1+\cos\alpha)/2$ is the obliquity factor and where the integration is carried out over the plane z=0. Equation 7.29 is a mathematical statement that the amplitude at a given location is simply the sum at the position $\boldsymbol{r}$ of the Huygen's wavelets generated at each location $\boldsymbol{r}_0$ in the z=0 plane. In the treatment of Gaussian waves several simplifying assumptions are made. First the beams are restricted to those near the axis so that $\alpha \approx 0$ and the obliquity factor is nearly equal to one. Second, since only waves near the axis are considered the term $|\boldsymbol{r}-\boldsymbol{r}_0|$ in the denominator is taken as identical to z. Third, in the exponent $k\,|\boldsymbol{r}-\boldsymbol{r}_0|$ is approximared by

$$\begin{aligned} k\,|\boldsymbol{r}-\boldsymbol{r}_0| &= k\left[(x-x_0)^2+(y-y_0)^2+z^2\right]^{1/2} \\ &\approx k\left[z+\frac{(x-x_0)^2}{2z}+\frac{(y-y_0)^2}{2z}\right]. \end{aligned} \qquad (7.30)$$

This leads to the result that

$$E(\boldsymbol{r}) = \frac{i\,e^{-ikz}}{\lambda\,z} \int\int E(\boldsymbol{r}_0) \exp\left(-i\frac{k}{2z}\left[(x-x_0)^2+(y-y_0)^2\right]\right) dx_0\, dy_0\,. \qquad (7.31)$$

If one substitutes $E(\boldsymbol{r}_0)$ from Equation 7.28 into Equation 7.31 and carries out the integration the result is

$$E(\boldsymbol{r}) = K\left(\frac{2}{\pi}\right)^{1/2} \frac{1}{w(z)}\, e^{-i(kz-\psi(z))} \exp\left[-i\frac{k}{2}\left(\frac{x^2+y^2}{z+i\pi\,w_0^2\,\lambda^{-1}}\right)\right] \qquad (7.32)$$

where

$$\psi(z) = \arctan\frac{\lambda z}{\pi w_0^2} = \arctan\frac{z}{z_0}\,, \tag{7.33}$$

$$w(z) = w_0\sqrt{1+\left(\frac{\lambda z}{\pi w_0^2}\right)^2} = w_0\sqrt{1+\left(\frac{z}{z_0}\right)^2}\,, \tag{7.34}$$

and $z_0 = \pi w_0^2/\lambda$ is called the *Rayleigh range* of the Gaussian beam. Equation 7.34 is an expression for the $1/e$ radius of the spot size of a of a spherical Gaussian wave in a given z plane.

In order to understand Equation 7.32 the properties of a general spherical wave are analyzed and compared to the results of Equation 7.32. The phase of a spherical wave of radius R in the plane z=R is given by

$$\begin{aligned}\phi &= k\,\left[x^2+y^2+R^2\right]^{1/2} \approx k\,R + k\left(\frac{x^2+y^2}{2R}\right). \\ &= k\,z + \frac{\pi}{\lambda R}\,(x^2+y^2)\end{aligned} \tag{7.35}$$

where it has been assumed that the wave is confined near the axis so that z=R is much larger than $(x^2+y^2)^{1/2}$. The phase of a spherical wave on the axis is kz=kR, and the phase in the plane defined by z=R varies from kR at x=0, y=0, z=R to $kR+\pi(x^2+y^2)/\lambda R$. If one considers Equation 7.32 one finds that the final term can be written as

$$\exp\left[-i\frac{\frac{k}{2}(x^2+y^2)}{z+i\pi\,w_0^2\,\lambda^{-1}}\right] = \exp\left[-i\frac{k}{2}\frac{x^2+y^2}{R(z)}\right]\,\exp\left[-\frac{x^2+y^2}{w(z)^2}\right] \tag{7.36}$$

where the radius curvature of the wave is given by

$$R(z) = z\left[1+\left(\frac{\pi\,w_0^2}{\lambda z}\right)^2\right] = z\left[1+\left(\frac{z_0}{z}\right)^2\right]. \tag{7.37}$$

Combining Equations 7.32 and 7.36 yields

$$E(\boldsymbol{r}) = K\left(\frac{2}{\pi}\right)^{1/2}\frac{1}{w(z)}\,e^{-i(kz-\psi(z))}\,\exp\left[-i\frac{k}{2}\left(\frac{x^2+y^2}{R(z)}\right)\right]\,\exp\left[-\frac{x^2+y^2}{w(z)^2}\right]. \tag{7.38}$$

This is the equation for a spherical Gaussian wave with a $1/e$ spot size radius of $w(z)$ and a wave front curvature of $R(z)$. Thus it is seen that a Gaussian plane wave will, due to diffraction, develop into a Gaussian spherical wave as it travels in the z direction. In fact it is at all values of z is a spherical Gaussian wave. It should be noted that the integral of the square of the absolute value of $E(\boldsymbol{r})$ over the plane z is equal to K^2 so that the energy in the Gaussian wave is the same in all z planes. The spherical Gaussian wave is often written as

$$E(\boldsymbol{r}) = K\left(\frac{2}{\pi}\right)^{1/2} \frac{1}{w(z)} e^{-i(kz-\psi(z))} \exp\left[-i\frac{k}{2}\left(\frac{x^2+y^2}{q(z)}\right)\right] \tag{7.39}$$

where q is called the complex radius of curvature of the wave front, which is given by

$$q(z) = z + q_0 = z + i\left(\frac{\pi\, w_0^2}{\lambda}\right) = z + i\, z_0 \tag{7.40}$$

where $q_0 = i\pi\, w_0^2/\lambda = i\, z_0$ is the value of the complex radius of curvature of the wave front at z=0. The inverse of q is given by

$$\frac{1}{q} = \left(z + i\,\frac{\pi\, w_0^2}{\lambda}\right)^{-1} = \frac{1}{R(z)} - i\,\frac{\lambda}{\pi\, w(z)^2}\,. \tag{7.41}$$

The properties of this spherical Gaussian wave require investigation. A spherical Gaussian wave has a waist where $w(z)$ is a minimum. This occurs at z=0 and the beam has a $1/e$ *radius* of w_0 at the waist. The $1/e$ waist *diameter* is $2w_0$ which is also sometimes called the *spot size* at the waist. At other positions $2w$ is called the spot size. At the waist the wave has $R(z)$=∞ and is in fact a plane Gaussian wave. As the wave evolves by Fresnel diffraction the wave becomes a spherical Gaussian wave with a radius of curvature of the wave front of $R(z)$ and a $1/e$ beam radius of $w(z)$. The $1/e$ radius of a Gaussian wave increases from w_0 to $\sqrt{2}w_0$ as it propagates from z=0 to z=z_0. As was previously mentioned the length z_0=$\pi w_0^2/\lambda$ is called the Rayleigh range of the beam. In the far field region where z is much larger than z_0 the beam radius becomes $w \approx \lambda\, z/\pi\, w_0 = w_0\,(z/z_o)$, the radius of curvature of the wave front becomes $R \approx z$, and $\psi \approx \pi/2$. In the far field a Gaussian beam diverges with an half angle given by θ=$w(z)/z$, which approaches θ=$\lambda/\pi\, w_0 = w_0/z_0$ as z approaches infinity. As in any diffraction problem if the waist is small then the diffraction angle, θ, is large and if the waist is large then the diffraction angle is small. The wave front curvature is infinite at the waist, i.e. at z=0, and is also infinite at z=∞. The wave front curvature is a minimum at z=z_0 where it is R=$2z_0$.

The properties of a Gaussian beam at all positions z are now understood so that if the location of the waist and the value of w_0 are known then one can calculate the beam spot size and wave front curvature at any position z. If the location of the waist is at z=0 the Gaussian beam's properties are described entirely by the complex quantity $q(z)$. If one knows q_0 then one can obtain the value of q at the location z_1 from the relation q_1=$q_0 + z_1$, and one can obtain the value of q at z_2 from q_2=$q_0 + z_2$. This leads to the result that $q_2 = q_1 + z_2 - z_1 = q_1 + L$ where L=$z_2 - z_1$.

If a Gasussian beam is incident on a thin lens of focal length f then as the beam passes through the thin lens the radius of the spot size does not change but the curvature of the wave front does change, the change being given by $1/R_o = 1/R_i - 1/f$ where o and i denote the beam just before the thin lens and just after the thin lens. Using Equation 7.41 and $1/R_o = 1/R_i - 1/f$ one finds that the complex curvature of the Gaussian wave after the thin lens is related to the complex curvature before the thin lens by

$$\frac{1}{q_0} = \frac{1}{q_i} - \frac{1}{f} \,. \tag{7.42}$$

This equation is simply the thin lens equation written in terms of the wave front curvature just before and just after the thin lens.

A sample calculation using Gaussian beams may be useful in understanding how one works with Gaussian beams. Consider the following problem. A Gaussian beam has a wavelength λ=500 nm $= 5\times10^{-5}$ cm and a waist radius w_0=0.1 mm =0.01 cm. The waist is located at $z = 0$. What is the beam radius and the wave front curvature at z=10 cm? The Rayleigh range for the beam is $z_0 = \pi\, w_0^2/\lambda$=6.28 cm. The beam radius at z=10 cm is given by (Eq. 7.34)

$$w = w_0\sqrt{1 + \left(\frac{z}{z_0}\right)^2} = 1.88\times10^{-2}\ \text{cm} \tag{7.43}$$

and the wave front curvature is given by (Eq. 7.37)

$$R = z + \frac{z_0^2}{z} = 13.95\ \text{cm}. \tag{7.44}$$

Now suppose the beam passes through a a thin positive lens with a focal length of 5 cm located at z=10 cm. What is the radius and location of the new waist? Since the lens is thin the beam radius immediately after the lens, w^*, is the same as the beam radius just before the lens so that w^*=w=1.88×10^{-2} cm. An asterisk is used to denote quantities after the lens i.e., in the image space. The wave front curvature after the lens, R^*, is determined by $1/R^* = 1/R - 1/f = 1/13.95 - 1/5 = -0.128\ \text{cm}^{-1}$ so that $R^* = -7.79$ cm. The negative sign for R^* indicates a converging beam. It is now necessary to determine the location of the waist and the waist radius after the lens in terms of w^* and R^*. This can be accomplished as follows

$$\frac{z^*}{z_0^*} + \frac{z_0^*}{z^*} = \frac{R^*}{z_0^*} = \left(\frac{w^*}{w_0^*}\right)^2 \left(\frac{z_0^*}{z^*}\right) . \tag{7.45}$$

Equation 7.45 yields

$$\frac{R^*}{(\pi w^{*2}/\lambda)} = \frac{z_0^*}{z^*} = -0.351\ . \tag{7.46}$$

The negative sign indicates that the waist lies downstream from the lens. Using this value for z_0^*/z^* one can calculate w_0^*, which is given by (Eq. 7.34)

$$w_0^* = \frac{w^*}{\sqrt{1+\left(\frac{z^*}{z_0^*}\right)^2}} = 6.23\times 10^{-3}\ \text{cm}. \tag{7.47}$$

Using the value for w_0^* one finds that $z_0^* = \pi w_0^{*2}/\lambda$=2.43 cm, which leads to the result that z=-6.95 cm. It should be noted that the Gaussian beam optics are not the same as geometrical optics. For geometrical optics an object 10 cm from a converging thin lens with a focal length of 5 cm is imaged with magnification 1 at a position 10 cm beyond the lens. For the Gaussian beam the waist is imaged only 6.95 cm beyond the thin lens and with a magnification of the waist radius that is less than 1.

The transmission of a Gaussian beam through an aperture is often an important consideration in various laser problems. The fractional transmission through a circular aperture of radius a is given by

$$\begin{aligned}\frac{I}{I_0} &= \int\int \epsilon\,|E|^2\,c\,dA = \frac{2}{\pi\,w^2}\int_0^a 2\pi\,r\,e^{-2r^2/w^2}\,dr\\ &= 1 - e^{-2a^2/w^2}\ .\end{aligned} \tag{7.48}$$

7.1.3 *Gaussian Optics Stability Criterion*

The stability of an optical cavity based on geometrical optics was discussed in section 7.1, here the stability of an optical cavity is discussed based on Gaussian waves. The criterion for the stability of a Gaussian beam is that an optical cavity is unstable if the $1/e$ radius of the beam becomes infinite at one of the mirrors. Consider a Gaussian wave with a waist located at z=0. The Gaussian wave is such that it has a wave front curvature R_2 at z=z_2 and R_1 at z=z_1. If an optical cavity is such that the mirrors have radii of curvature equal to the curvature of the Gaussian wave then the mirrors will reflect the wave exactly back on itself so that the wave will repeatedly reconstruct itself as it is reflected back and forth between the mirrors. Two such optical cavities are shown in Figure 7.8. The cavities have mirror 2 with a radius of curvature R_2 located at z=z_2 and mirror 1 with radius of curvature R_1 located at z=z_1. As shown mirror 1 can be located so that

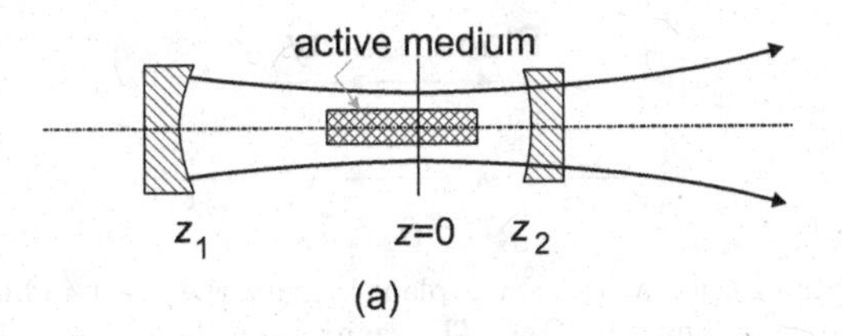

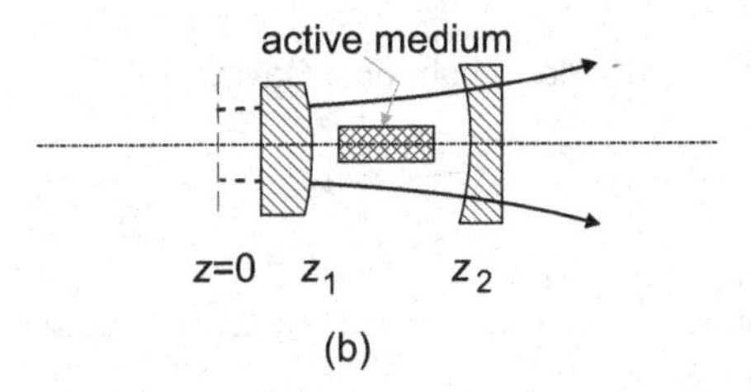

Fig. 7.8 Two stable optical cavities, one with two converging mirrors and the other with one converging and one diverging mirror. In (b) the waist is a virtual waist outside the cavity.

z_1 is either positive or negative. Since the wave front curvature of the a Gaussian beam is $R(z) = z + z_0^2/z$ the radii of curvature for the two mirrors are given by $-R_1 = z_1 + z_0^2/z_1$ and $R_2 = z_2 + z_0^2/z_2$ where $z_0 = \pi w_0^2/\lambda$, where $z_2 - z_1 = L$ and where we have explicitly taken R_1 as negative in the calculations. Taking R_1 as negative implies that z_1 is also negative. Solving for z_1, z_2, and z_0^2 yields

$$\frac{z_1}{L} = -\frac{R_2 - L}{R_2 + R_1 - 2L} = -\frac{g_2\,(1 - g_1)}{g_1 + g_2 - 2g_1\,g_2}\,, \tag{7.49}$$

$$\frac{z_2}{L} = \frac{R_1 - L}{R_2 + R_1 - 2L} = \frac{g_1\,(1 - g_2)}{g_1 + g_2 - 2g_1\,g_2}\,, \tag{7.50}$$

and

$$\begin{aligned}\frac{z_0^2}{L^2} &= \frac{(R_1 - L)\,(R_2 - L)\,[R_1\,R_2 - (R_1 - L)\,(R_2 - L)]}{R_1\,R_2\,[R_2 + R_1 - 2L]^2} \\ &= \frac{g_1\,g_2\,(1 - g_1\,g_2)}{(g_1 + g_2 - 2g_1\,g_2)^2}\end{aligned} \tag{7.51}$$

where $g_1 = 1 - L/R_1$ and $g_2 = 1 - L/R_2$. The $1/e$ radii of the laser beam at the two mirrors are given by

$$w_1^2 = w_0^2\left[1 + \left(\frac{z_1}{z_0}\right)^2\right] = \frac{L\,\lambda}{\pi}\sqrt{\frac{g_2}{g_1\,(1 - g_1\,g_2)}} \tag{7.52}$$

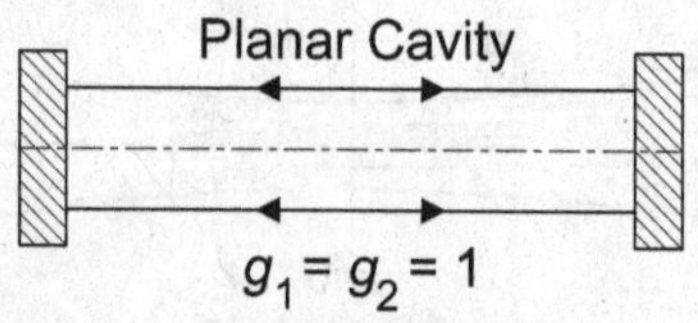

Fig. 7.9 A planar optical cavity. For a planar cavity the waist radius and the beam radii at the mirrors are infinitely large. The cavity is on the border of being unstable.

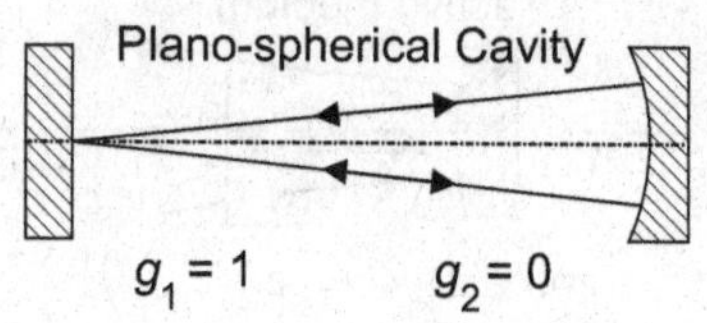

Fig. 7.10 A plano-spherical cavity ($R_1 = \infty$ and $R_2 = L$). The waist is at the plane mirror and has zero radius.

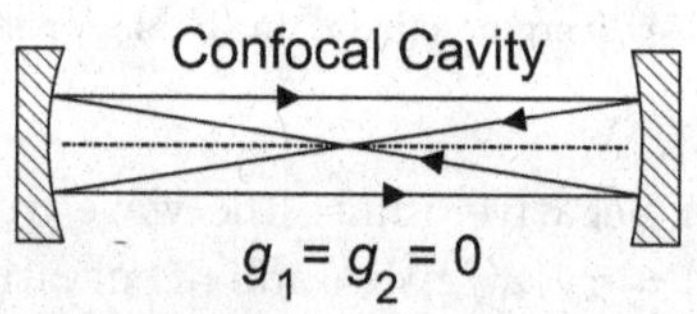

Fig. 7.11 A confocal cavity ($R_1 = R_2 = L$). The beam radius at the waist is $(L\lambda/2\pi)^{1/2}$ and at each of the mirrors is $(L\lambda/\pi)^{1/2}$.

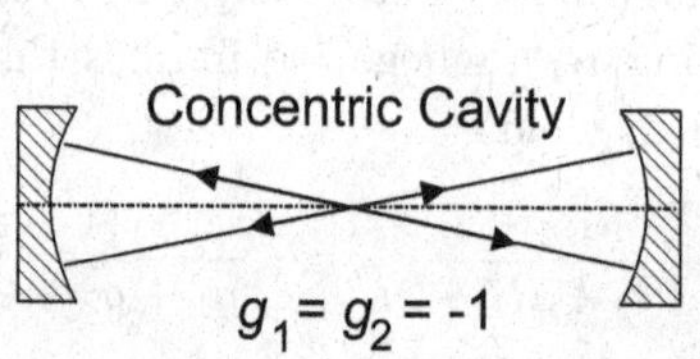

Fig. 7.12 A concentric cavity ($R_1 = R_2 = L/2$). For the concentric cavity the waist radius is zero and the beam radii at the mirrors is infinite.

and

$$w_2^2 = w_0^2 \left[1 + \left(\frac{z_2}{z_0}\right)^2\right] = \frac{L\,\lambda}{\pi} \sqrt{\frac{g_1}{g_2\,(1 - g_1\,g_2)}} \,. \qquad (7.53)$$

The $1/e$ radius of the beam at the mirrors will blow up to infinity at one or both mirrors if either $g_1\,g_2 = 0$ or $g_1\,g_2 = 1$. The optical cavity is therefore stable if

$$0 \leq g_1\,g_2 = \left(1 - \frac{L}{R_1}\right)\left(1 - \frac{L}{R_2}\right) \leq 1 \,. \qquad (7.54)$$

This is exactly the same criterion for stability that was obtained from geometrical optics. Figures 7.9, 7.10, 7.11, and 7.12 show various optical cavities (all the cavities shown are on the boundary between the stable and unstable regions). Some geometrical rays are shown and the figure captions give the Gaussian beam properties. In general it is best to use laser resonators that are inside the region of stability rather than at the boundary in order to avoid small changes in conditions from driving the cavity into the unstable region.

If a Gaussian beam emerges from one optical cavity and one desires to insert the beam into another optical cavity it is important to use coupling optics such that the beam as it enters the next cavity has the correct wave front curvature to match the new cavity.

7.2 Cavity Modes

In the following subsections the allowed frequencies for an optical cavity are analyzed for both the fundamental and higher modes, the modes that can lase are determined, and the effects of the active laser medium on the cavity frequencies is discussed.

7.2.1 *Resonant Frequency of an Optical Cavity*

An optical cavity has an infinite series of resonant frequencies. In Chapter 3 we showed that a cw laser beam must be unchanged in both amplitude and phase after a round trip in an optical cavity (p. 50ff). That the amplitude must be unchanged after a round trip in the optical cavity led to the threshold population condition for laser oscillation. In this section we show that the condition that the phase be unchanged after a round trip in the optical cavity leads to the allowed frequencies for laser oscillation.

The fundamental Gaussian mode is such that the phase of the wave changes as the wave propagates from mirror 1 to mirror 2 by

$$[k\,z_2 - \psi(z_2)] - [k\,z_1 - \psi(z_1)] = k\,L - \arctan\left(\frac{z_2}{z_0}\right) + \arctan\left(\frac{z_1}{z_0}\right) = p\,\pi \tag{7.55}$$

where $z_2 - z_1 = L$ and where p is an integer. The phase change must be equal to $p\,\pi$ where p is an integer so that after a round trip in the cavity the phase is changed by a multiple of 2π leaving the wave unchanged. This leads to the result that the allowed wavelengths are

$$\frac{2\pi}{\lambda_p} = \frac{p\,\pi}{L} + \frac{1}{L}\left[\arctan\frac{z_2}{z_0} - \arctan\frac{z_1}{z_0}\right]. \tag{7.56}$$

where λ_p is the resonant wavelength. The resonant frequencies of the cavity are therefore given by

$$\nu_p = \frac{c}{n\,\lambda_p} = \frac{p\,c}{2n\,L} + \frac{c}{2\pi\,n\,L}\left[\arctan\frac{z_2}{z_0} - \arctan\frac{z_1}{z_0}\right] \tag{7.57}$$

where n is the index of refraction at the frequency ν_p. The difference in frequency between adjacent frequencies of two modes is given by

$$\nu_{p+1} - \nu_p = \frac{c}{2n\,L}\,. \tag{7.58}$$

The difference in frequency between Gaussian modes is identical to the familiar result for a Fabry Perot etalon.

7.2.2 *Higher Order Gaussian Modes*

The discussion in this chapter has so far focused only on the fundamental Gaussian mode of a cavity. There are also higher order Gaussian modes for an optical cavity. The forms of the higher order Gaussian waves are not derived, but merely presented. The wave amplitude for the higher order Gaussian waves in rectangular coordinates is

$$\begin{aligned} E = E_0 \sqrt{\frac{2}{2^{m+n}\,m!\,n!\,\pi}\,\frac{1}{w(z)}}\; H_n\left(\frac{\sqrt{2}\,x}{w(z)}\right)\; H_m\left(\frac{\sqrt{2}\,y}{w(z)}\right) \\ \times \exp\left[-\frac{x^2+y^2}{w(z)^2} - i\,\frac{\pi}{\lambda}\,\frac{x^2+y^2}{R(z)} - \frac{2\pi\,z}{\lambda} - i(m+n+1)\arctan\left(\frac{z}{z_0}\right)\right] \end{aligned} \tag{7.59}$$

where H_n and H_m are Hermite polynomials. The first few Hermite polynomials are

$$\begin{aligned} H_0(x) &= 1 \\ H_1(x) &= 2x \\ H_2(x) &= 4x^2 - 2 \\ H_3(x) &= 8x^3 - 12x \\ H_4(x) &= 16x^4 - 48x^2 + 12\,. \end{aligned} \tag{7.60}$$

As written the Gaussian modes are normalized so that the integral of the absolute square of E is equal to E_0^2. The cavity modes are denoted by the integers n and m. The mode with n=m=0 is called the TEM$_{00}$ mode. The symbol TEM stands for transverse electromagnetic. The 00 mode is identical to the fundamental Gaussian mode that was discussed in Section 7.1.2 of this chapter. The modes with higher n and/or m have an intensity distribution in the xy plane perpendicular to z, direction of propagation of the

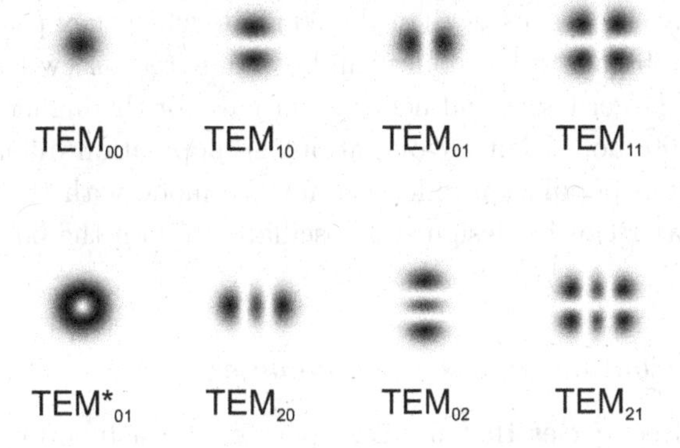

Fig. 7.13 TEM$_{nm}$ intensity distributions. (An inverse image with darkness corresponding to higher intensity.)

beam, that is different from the 00 mode. It depends on the angle in the plane perpendicular to the direction of propagation. In a particular z plane the higher order modes have bright regions separated by dark regions. Some sample TEM$_{nm}$ modes are plotted in Fig. 7.13. For example the 10 mode (n=1, m=0) has a line of zero intensity along x=0 with two bright regions on either side of the line, and the TEM$_{11}$ mode has lines of zero intensity along x=0 and y=0 with bright regions in each quadrant of the particular z plane. One sometimes sees a HeNe laser with an intensity distribution in the xy plane that is a bright ring or donut. This TEM*$_{01}$ intensity distribution is the result of a linear combination of the TEM$_{10}$ and TEM$_{01}$ modes. The transverse modes are described using Cartesian coordinates and Hermite polynomials. For cylidrical coordinates one can also define TEM$_{pl}$ modes using Laguerre polynomials where p corresponds to the number of radial nodes and l to number of angular nodes. The fundamental TEM$_{00}$ modes are identical in the two representations.

The frequency of a $n\,m$ mode is given by

$$\nu_p = \frac{p\,c}{2\,n_i\,L} + \frac{c(n+m+1)\left[\arctan\left(\frac{z_2}{z_0}\right) - \arctan\left(\frac{z_1}{z_0}\right)\right]}{2\pi\,n_i\,L} \tag{7.61}$$

where p is an integer, and the index of refraction at the frequency ν_i is n_i. Usually different TEM$_{nm}$ modes occur as closely spaced groups separated by a larger $c/2L$ frequency spacing.

In general the higher order modes have a larger size in a given z plane than the TEM$_{00}$ mode. It is usually desirable to operate a laser in a

00 mode. Because of their larger size in a given z plane the higher order modes have higher losses than the 00 mode when they pass through an aperture. The higher order modes can be eliminated for cw lasers by using a set up with larger losses and hence lower gain for the higher order modes than for the 00 mode. For cw operation the population difference clamps at the threshold population difference for the mode with the highest gain so that the laser can be designed to oscillate only on the 00 mode in this situation.

7.2.3 *Distribution of Laser Frequencies*

The precise frequencies that a laser operates are a function of both the cavity and the active medium. The cavity supports lasing at discrete modes with frequencies given by Eq. 7.57 (for the TEM_{00} mode) or more generally by Eq. 7.61. Lasing can occur on one or more of these modes if the gain of the active medium is high enough to overcome the losses as illustrated in Fig. 7.14. A *single mode* laser has only one mode lasing at a time whereas a *multimode* laser supports lasing on two or more modes at the same time. These multiple modes can be either different transverse modes or different longitudinal modes separated by $\Delta\nu_p = c/2\,L$. Via the effect of saturation on the $\alpha(\nu)$, lasing on one mode can suppress the gain on other nearby modes thus preventing them from lasing. Whether or not this occurs depends on whether the lineshape is homogeneously or inhomogeneously broadened.

For a homogeneously broadened line centered at ν_0, saturation by a laser operating at frequency ν lowers the gain $\alpha(\nu')$ across the entire line shape as described in Sec. 5.1 and illustrated in Fig. 5.2. In Chapter 6 it was shown that the threshold population difference clamps at a maximum value

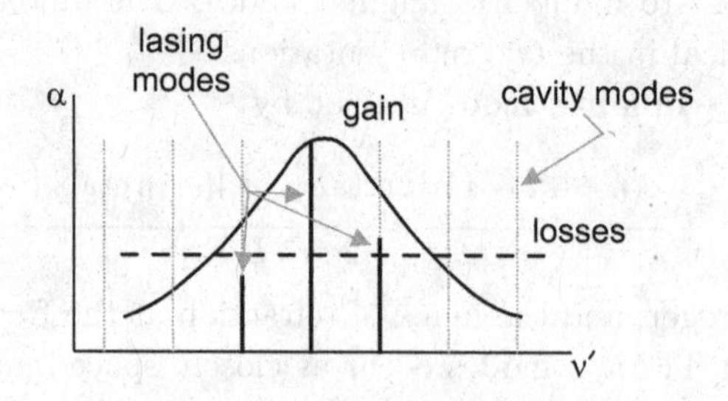

Fig. 7.14 In this illustration, lasing is only possible on the three cavity modes where the unsaturated gain is greater than the losses.

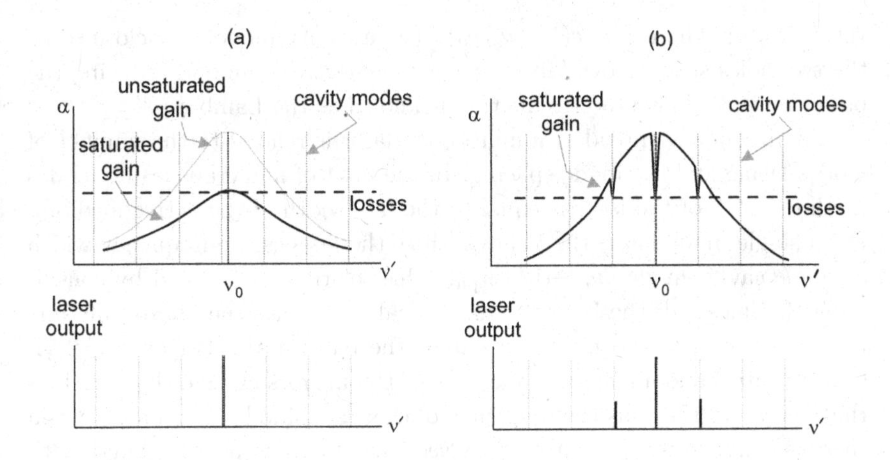

Fig. 7.15 Frequency distribution of lasing modes including saturation of gain medium on (a) homogeneous and (b) inhomogeneously broadened lines. Output is single-mode for homogeneous line, multimode for inhomogeneous line.

of $\Delta n_{\rm th}$ once lasing occurs. Thus, lasing at ν will prevent the population inversion required to lase at any other frequency, and the laser will operate in the single mode configuration at the cavity frequency closest to ν_0 (see Fig. 7.15a). Nonetheless, lasing on higher order TEM modes may still be possible under some circumstances. In particular, the gain clamps only along the path of the lasing (TEM_{00}) mode. Higher order transverse modes have different spatial profiles within the cavity and thus can draw gain from spatial regions of the active medium unaffected by the 00 mode.

For an inhomogeneously broadened line, each cavity mode will interact with different independent homogeneous segments of the velocity distribution within the inhomogeneous envelope. As a result, lasing on one mode does not suppress the gain at other frequencies beyond the width of the spectral hole (Eq. 5.26) and multiple longitudinal modes will be lasing as illustrated in Fig. 7.15b. The line width of gas discharge lasers is dominated by Doppler broadening with typical inhomogeneous widths on the order of a few GHz. The cavity length of these lasers might range from ~10 cm for a simple He-Ne laser to ~1 m for a large frame argon ion laser. The frequency spacing of the cavity modes is equal to $c/2L$=1.5 GHz for L=10 cm to 150 MHz for L=1 m. Thus, with the exception of the shortest cavity length, multiple longitudinal modes fit within the Doppler line width. As the laser beam bounces back and forth in the cavity, each ν_i cavity mode burns *two* holes into the gain profile corresponding to atoms moving with velocities $\pm v$ where $v = c(1 - \nu_i/\nu_0)$. For a cavity mode very close to ν_0 the two holes start to overlap reducing the effective gain and reducing the power output. This power reduction is known as the Lamb dip.

The frequency spread of a multimode laser is related to the number of modes lasing and the mode spacing. In the limit of numerous lasing modes the line width of the laser is equal to the line width of the active medium. For a single mode laser, the bandwidth of the laser is the frequency width of single cavity mode $\delta\nu_i = 1/2\pi\tau_{\text{ph}}$. This relation is obtained by considering the finesse of the Fabry Perot optical cavity. As the photon lifetime increases, ultimately other factors limit the bandwidth. For example, vibrations and random thermal motions of the mirrors change the length of the cavity and the thus the frequency of the lasing mode. Ultimately when these instrumental effects are suppressed, the minimum possible bandwidth is limited by spontaneous emission into the lasing mode as described in Sec. 6.3. Note that there is a subtle difference between a single mode laser and a single frequency laser. In a single mode laser there is one cavity mode lasing within the gain envelope of the active medium, but generally this can occur at any frequency within about a line width of ν_0. Hence the uncertainty in the laser's frequency is on the order of the line width (GHz). Changes in the cavity length due to changes in the temperature or the index of refraction of air can sweep the lasing wavelength back and forth across the line width. In contrast, for a single frequency laser the cavity length is stabilized and the uncertainty in the laser's frequency is on the order of the laser bandwidth (~MHz or better).

7.2.4 *Frequency Pulling and the Electrical Susceptibility of a Laser Medium*

The resonant frequency of an optical cavity containing a laser medium differs from the resonant frequency without the laser medium because of the index of refraction of the medium, which appears in the denominator of the expression for the frequency. The laser medium has a dispersive index near a transition frequency. This affects the resonant frequency of the cavity. In order to understand this effect it is necessary to understand the electric susceptibility and hence the index of refraction. The electric susceptibility is calculated using the Lorentz model for the atom similarly to the development used in Chapter 2 in the discussion of oscillator strengths (p. 31ff). We take N/V to be the number density of atoms in the lower laser level in the laser medium. In the Lorentz model the electron is bound to the atom by a spring with a spring constant $k = \omega_0^2\, m$ and is damped by a damping constant γm where ω_0 is taken as 2π times the atomic transition frequency and m is the electron mass. The equation of motion of the atomic electron in the presence of the electric field of the laser beam, $\mathrm{Re}(E_0\, e^{i\omega t})$, is

$$\ddot{x} + \gamma\,\dot{x} + \omega_0^2\, x = -\frac{e}{m}\,\mathrm{Re}\left(E_0\, e^{i\omega t}\right)\,. \tag{7.62}$$

If one uses $x = \mathrm{Re}(x_0\, e^{i\omega t})$ where x_0 is complex as a trial solution to the equation one finds

$$x_0 = \frac{(e/m)\, E_0}{(\omega - \omega_0)^2 + i\,\omega\,\gamma}\,. \tag{7.63}$$

Since x_0 is large only if ω is nearly equal to ω_0 we make the approximation that $\omega_0^2 - \omega^2 \approx 2\omega(\omega_0 - \omega)$. With this approximation x_0 can be rewritten as

$$x_0 = i\,\frac{e}{m\,\gamma\,\omega}\left(\frac{1}{1 + i\,2(\omega - \omega_0)/\gamma}\right) E_0\,. \tag{7.64}$$

The polarization that results from the laser electric field is given by

$$\begin{aligned} P(\omega) &= \frac{N}{V}\, e\, x = \mathrm{Re}\left[\epsilon_0\,\chi\, E_0\, e^{i\omega t}\right] \\ &= \mathrm{Re}\left[i\,\frac{(N/V)\, e^2}{m\,\gamma\,\omega}\left(\frac{1}{1 + i\,2(\omega - \omega_0)/\gamma}\right) E_0\, e^{i\omega t}\right] \end{aligned} \tag{7.65}$$

where we have made use of the relationship $\boldsymbol{P} = \mathrm{Re}[\,\epsilon_0\,\chi\,\boldsymbol{E}]$ where χ is the electric susceptibility. The electric susceptibility, χ, has both a real part, $\chi'(\omega)$,and an imaginary part, $\chi''(\omega)$ so that $\chi(\omega) = \chi'(\omega) + i\,\chi''(\omega)$. Using the expression for P one finds that

$$\begin{aligned} \chi(\omega) &= \chi'(\omega) - i\,\chi''(\omega) \\ &= i\,\frac{(N/V)\, e^2}{m\,\gamma\,\omega\,\epsilon_0}\left(\frac{1}{1 + i\,2(\omega - \omega_0)/\gamma}\right) \end{aligned} \tag{7.66}$$

$$\chi'(\omega) = \frac{(N/V)\, e^2}{m\,\gamma\,\omega\,\epsilon_0}\left(\frac{2(\omega - \omega_0)/\gamma}{1 + 4(\omega - \omega_0)^2/\gamma^2}\right) \tag{7.67}$$

and

$$\chi''(\omega) = \frac{(N/V)\, e^2}{m\,\gamma\,\omega\,\epsilon_0}\left(\frac{1}{1 + 4(\omega - \omega_0)^2/\gamma^2}\right)\,. \tag{7.68}$$

It is obvious that $\chi' = [2(\omega - \omega_0)/\gamma]\,\chi''$. The real polarization, $\mathrm{Re}\, P$, is given by

$$P(\omega) = \epsilon_0\,\left[\chi'(\omega)\cos\omega t + \chi''(\omega)\sin\omega t\right] E_0\,. \tag{7.69}$$

Thus the polarization is not in phase with the electric field $E_0 \cos\omega t$. The power absorbed per unit volume by the sample is given by the power absorbed by a single atom times the number of atoms per unit volume. The power absorbed by a single atom is given by the force on the atomic electron times the velocity of the atomic electron averaged over the period, T, of the electron's motion. This is equal to

$$\frac{\text{Power}}{V} = \frac{1}{T}\int_0^T \frac{N}{V}\,(e\,E_0\,\cos\omega t)\left(\frac{dx}{dt}\right) dt\,. \tag{7.70}$$

Utilizing the result that $x = P(\omega)\,V/N\,e$ one obtains

$$\frac{\text{Power absorbed}}{V} = \frac{\omega\,\chi''\,\epsilon_0\,E_0^2}{2} = \frac{\omega\,\chi''\,I_\nu}{c} = \frac{N}{V}\,\sigma_{lu}\,I_\nu\,. \tag{7.71}$$

Combining the expression for the power absorbed per unit volume with the expression for $\chi''(\omega)$ one obtains the result that

$$\begin{aligned}\sigma_{lu} = \frac{\omega\,\chi''}{c}\,\frac{N}{V} &= \pi\,r_e\,c\,\frac{\gamma}{(\omega-\omega_0)^2 + (\gamma/2)^2} \\ &= \pi\,r_e\,c\,g(\nu-\nu_0)\,. \end{aligned} \tag{7.72}$$

This result is identical to that obtained in Chapter 2. The expression for σ_{lu} in terms of χ'' can be rewritten to express χ'' in terms of σ_{lu} which yields

$$\chi''(\omega) = \frac{N}{V}\,\frac{c}{\omega}\,\sigma_{lu}\,. \tag{7.73}$$

This expression is correct for a quantum mechanical system provided one includes the possibility of gain as well as absorption. In this case one must write χ'' as

$$\chi''(\omega) = \frac{c}{\omega\,V}\,\left[N_l\,\sigma_{lu}(\omega-\omega_0) - N_u\,\sigma_{ul}(\omega-\omega_0)\right] \tag{7.74}$$

where N_l/V and N_u/V are the number densities of atoms in the lower and upper laser levels respectively and where the absorption and stimulated emission cross sections are the true quantum mechanical cross sections that involve the oscillator strengths of the transition.

The in phase part of the electric susceptibility or dispersion is equal to $2(\omega-\omega_0)/\gamma$ times the out of phase/absorptive part of the electric susceptibility so that

$$\chi'(\omega) = \frac{2(\omega-\omega_0)}{\gamma}\,\chi''(\omega) \tag{7.75}$$

where $\chi'(\omega)$ and $\chi''(\omega)$ are the true quantum mechanical dispersive and absorptive parts of the electric susceptibility.

The expression for $\chi'(\omega)$ contains only that part of the in phase component of the electric susceptibility that arises from the laser transition. There is a part that arises from other atomic transitions. In general one can write $\chi'_T = \chi'_0 + \chi'(\omega)$ where χ'_T is the total in phase part of the electric susceptibility, χ'_0 is the part due to all atomic transitions other than the laser transition, and $\chi'(\omega)$ is the part due to the laser transition. The term χ'_0 is only weakly dependent on the frequency, whereas the term $\chi'(\omega)$ depends strongly on the frequency. The index of refraction is given by $n = [\epsilon/\epsilon_0]^{1/2} = [1 + \chi'_0 + \chi'(\omega)]^{1/2}$. The term $\chi'(\omega)$ causes the index of refraction to have a dispersive shape near the laser transition.

If N_u/V is larger than N_l/V then the frequency of the laser cavity, $\nu_p = p\,c/n\,L + (c/2\pi n\,L)[\arctan(z_2/z_0) - \arctan(z_1/z_0)]$, is shifted from the frequency it would have had in the absence of the laser transition because the frequency depends on the index of refraction. For a laser transition with a population inversion one finds that $\chi''(\omega)$ is negative for all frequencies and one finds that $\chi'(\omega)$ is positive at frequencies above the atomic resonance frequency and negative at frequencies below the atomic resonance frequency. This leads to the result that the resonant frequency of the cavity is shifted toward the atomic resonant frequency and away from the resonant frequency of the cavity in the absence of the population inversion. This effect is called *frequency pulling* and is illustrated in Fig. 7.16.

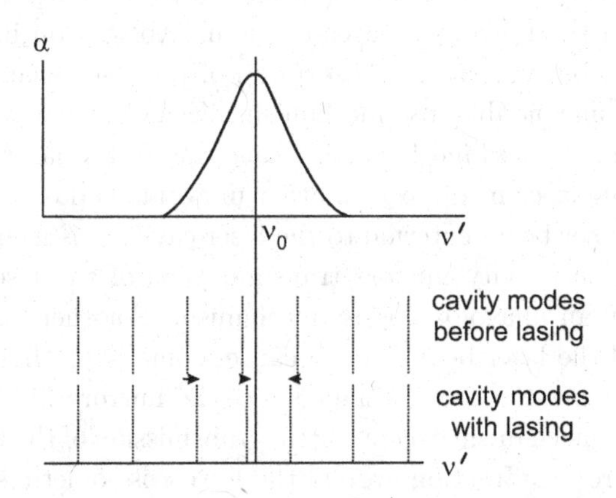

Fig. 7.16 Frequency pulling of cavity modes towards the atomic resonance frequency.

7.3 Cavity Losses

The loss of photons from an optical cavity is represented in earlier chapters by the lifetime of a photon in the cavity

$$\frac{dW_\nu}{dt} = -\frac{W_\nu}{\tau_{ph}} \tag{7.76}$$

where $W_\nu = N_\nu h\nu$ is the total energy in oscillating mode of the cavity. In the absence of any energy input the energy in an optical cavity decays exponentially with a $1/e$ lifetime of τ_{ph}. The analysis of the frequency distribution of a cavity is similar to the analysis of the frequency distribution of the photons in the cavity. Thus the cavity modes are not perfectly sharp, but instead have a cavity response that is Lorentzian with a bandwidth (FWHM) given by

$$\delta\nu_c = \frac{1}{2\pi\,\tau_{ph}} \,. \tag{7.77}$$

There is a quantity called the quality factor of a cavity, Q. The Q factor is defined as 2π times the energy stored in the cavity divided by the energy lost per cycle. Using this definition one finds that

$$Q = 2\pi \frac{W_\nu}{\frac{dW_\nu}{dt}\,T} = 2\pi\,\nu_0\,\tau_{ph} \tag{7.78}$$

where $T = 1/\nu_0$ is the period of the light wave. This leads to

$$\delta\nu_c = \frac{\nu_0}{Q} \,. \tag{7.79}$$

There are several physical mechanisms that result in the loss of energy from an optical cavity. The first mechanism that results in energy loss is absorption in the mirrors or laser medium. Absorption in good dielectric mirrors can be very small. Absorption of the laser beam by the laser medium is an unavoidable loss mechanism. Usually, however, absorption is a relatively minor loss mechanism. Scattering of the laser beam by the mirrors or in the laser medium is another unavoidable loss mechanism. In order to use a laser beam external to the laser cavity it is usually necessary to have one of the cavity mirrors transmit some of the laser beam, and this is certainly an unavoidable loss mechanism. Another loss mechanism is diffraction of the laser beam, which causes some of the light to miss the laser mirrors by passing around the edge of the mirrors. This is often the most severe loss mechanism except for the transmission of the beam through the output mirror. Diffraction around the mirror is sometimes used as the output coupling from a cavity as is discussed in the next section.

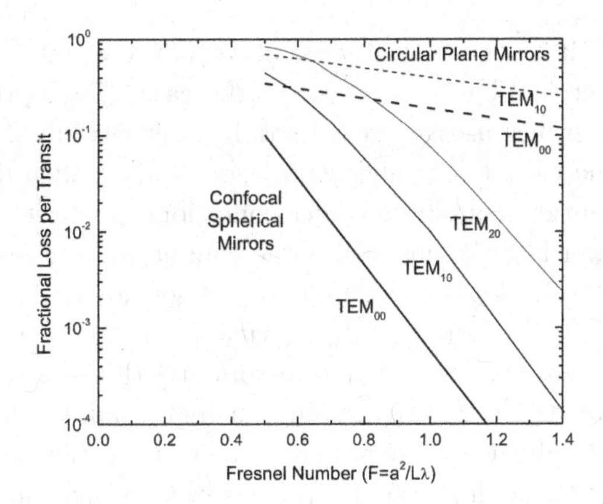

Fig. 7.17 Fractional power lost per cavity transit as a function of the Fresnel number of the cavity mirrors. Line thickness corresponds to different TEM modes, with solid lines for spherical mirror cavity and dashed lines for plane mirrors. Adapted from Fig. 15 of Fox and Li, Bell Sys. Tech. J. **40**, 453 (1961).

The losses from a cavity due to diffraction around the edges of the mirrors were first studied by Fox and Li and by Boyd and Gordon. They used Fresnel diffraction to determine theoretically the losses for different modes of an optical cavity. Their analysis is not reproduced here, but the principal results are shown in Figure 7.17. They calculated the loss per transit through the cavity as a function of the Fresnel number of the mirror, $F = a^2/L\,\lambda$, where a is the diameter of the cavity mirrors. As might be expected they found that the loss per transit in the cavity decreases as the mirror diameter increases and hence as the Fresnel number increases. Furthermore the loss per transit decreases as the mirror diameter increases much faster as a function of the Fresnel number for a cavity using spherical mirrors in a confocal cavity than it does for plane mirrors. In addition they found that the TEM_{00} mode has a smaller loss per reflection than the higher order modes. Their calculations show that for a Fresnel number of 1.0 the diffraction loss from the TEM_{00} mode of a confocal cavity per transit is $5{\times}10^{-4}$, whereas the losses per transit for a circular plane mirror is about 0.17. From these numbers and the other values plotted in Figure 7.17 it is clear that the use of spherical mirrors rather than plane mirrors sharply reduces the loss due to diffraction around the edges of the mirrors.

7.4 Unstable Cavities

In the discussion it has been tacitly assumed that low loss per reflection is a desirable objective. This is not always the case. The optimum output coupling for a laser increases as the gain of the laser medium increases. This means that if one uses a very high gain laser medium then it is necessary to have a high output coupling. Under conditions where it is desirable to extract high power laser beams from large volumes it is necessary that the losses from the cavity be large. The best method for doing this is to use an unstable cavity. For an unstable cavity the primary loss mechanism is diffraction around the edge of the mirrors and this can yield very high loss per reflection while still maintaining a high quality nearly Gaussian beam. If one simply uses a very high transmission through the output mirror to extract the high power the beam quality is much lower than if an unstable cavity is used. The beam extracted from an unstable resonator will necessarily have a large diameter and hence a large Rayleigh length. A pulsed Nd:YAG laser is an example of a laser that requires an unstable cavity. The output power from a Nd:YAG laser can be extremely high while the beam quality is good enough that the laser beam can be focused to a very small spot size.

7.5 Q-Switched Lasers and Mode Locked Lasers

There are many application for two very important techniques that are used to produce laser pulses. The technique of Q-switching is used to produce laser pulses with very high energy, and the technique of mode locking is used to produce a train of very short duration laser pulses. These techniques are both discussed in this section.

As stated Q-switching is a method for producing a very intense laser pulse. This is accomplished by initially pumping the laser medium while keeping the loss rate in the optical cavity high enough that the threshold population difference is not reached and then suddenly reducing the loss rate of the optical cavity. In other words after the pumping process has been running long enough that a large population difference has been obtained in the active medium the "Q" of the optical cavity is changed from the low Q of a lossy cavity to a high Q of a low loss cavity. When the Q is suddenly switched from a low value to a high value then the laser is far above threshold for the new larger Q of the optical cavity. The result is the initiation of laser action by the spontaneous photons that are already present in the modes that are now able to oscillate. The population difference is rapidly reduced by the emission of a high energy laser pulse. Another way of describing this Q switching is in terms of the photon lifetime of the cavity. Initially a large population difference is built up while keeping the photon lifetime short by making the optical cavity lossy so that the threshold population difference is not achieved. When the photon lifetime is suddenly lengthened then the system is far above the new threshold population difference with the longer photon lifetime and a large energy laser pulse is produced. Typically a Q switched laser pulse lasts 1-20 ns. A Q switched ruby laser can easily produce pulses with energies with more than 1 J per pulse. Q switching is commonly used with ruby lasers, Nd:YAG lasers, and other high power lasers.

The two most commonly used methods for producing the Q switching are by the use of a saturable dye absorber or by the use of an electro-optic switch. A saturable dye absorber has a high absorption coefficient for low light intensities but has a low absorption coefficient when the light intensity is high enough that the dye is saturated. With a saturable dye absorber as the laser pumping process begins the absorber prevents laser action. When the spontaneous emission reaches a level where the dye absorber saturates then the photon lifetime is suddenly reduced and a high energy laser pulse is produced.

The use of a saturable absorber is easily accomplished, but does not permit one to trigger the Q switched pulse at a predetermined time. Precise triggering of the Q switched pulse can be obtained with electro-optic devices. One method for setting up a Q switched laser using an electro-optic device can work follows. In addition to the active medium the laser cavity contains a linear polarizer and the electro-optic device. The device consists of an electro-optic crystal and the voltages necessary to drive the electro-optic crystal. The electro-optic crystal is operated with voltages such that it acts as a quarter wave plate when an appropriate voltage is applied to the crystal. The linearly polarized light coming from the linear polarizer is changed into right circularly polarized light by the quarter wave plate. After reflection from the end mirror of the cavity the light is changed into left circularly polarized light. The left circularly polarized light is converted into linearly polarized light as it passes back through the electro-optic crystal. The returning light beam, however, has a linear polarization that is orthogonal to the polarization passed by the linear polarizer. Thus the system is very lossy. When the voltage to the electro-optic crystal is reduced to zero the electro-optic crystal does not alter the polarization of the light beam and the Q of the optical cavity is greatly increased. Therefore a large energy laser pulse is generated when the voltage to the electro-optic crystal is switched to zero. The use of an electro-optic crystal is not as simple as the use of a saturable dye absorber, but it does permit one to trigger the laser pulse at a given time whereas the saturable absorber simply generates a pulse when the light intensity in the cavity reaches a high enough value. There are other methods for producing Q switched pulses from a laser.

Mode locking is a method for the production of a train of very short laser pulses. If a laser is operated with a very short duration pulse bouncing back and forth in the optical cavity of a laser the output of the laser is a train of very short pulses. The time necessary for a pulse to make a round trip in a laser cavity is equal to $2n\,L/c$ where L is the length of the cavity and n is the average value of the index of refraction in the cavity. Thus the time between the very short pulses in the output pulse train is $2n\,L/c$. The difference in frequency between the longitudinal modes of a laser is equal to $c/2nL$, which is the inverse of the time between the pulses in the train of pulses. The train of pulses can be thought of as being due to a Fourier sum of optical waves at the frequencies of the modes of the laser cavity. The individual modes in the Fourier sum must each have a particular phase in order to produce a pulse that circulates in the optical cavity of the laser. The mode phases are said to be "locked" in order to produce the optical pulse that is bouncing back and forth in the optical cavity. This is the origin of the term mode locking. Mode locking can be obtained by modulating the properties of the optical cavity so that a pulse that bounces back and forth in the cavity is naturally formed.

There are a number of techniques for accomplishing mode-locking. One method of producing mode locking of a laser is to insert a saturable dye absorber in the cavity. The saturable absorber causes the laser to operate in a pulsed mode. When the light intensity is high then the absorber does not greatly attenuate the beam whereas when the light intensity is low the absorber does greatly attenuates the beam. The saturable absorber in effect acts as a shutter that is open only when a high intensity pulse is passing through it. Mode locking can be used with either a homogeneously or inhomogeneously broadened laser medium. The result that there must be many modes lasing at the same time in order to form the Fourier series of a pulse might be thought to be impossible. This is not, however, the case. The modulation of the laser beam by the saturable absorber introduces side bands at enough mode frequencies and with the correct phases for the different modes in order to form the optical pulse.

One common application of mode locking is to mode lock a pump laser. The mode locked pump laser is then used to pump a tunable laser. The output from the tunable laser is a series of very short pulses. It is possible to produce tunable laser pulses as short as tens of fempto-seconds (10^{-15} s) or less with a mode locked pump laser. The very short tunable laser pulses can be use to study time resolved spectroscopy on a very fast time scale. It is possible to amplify the tunable laser pulses so that one can obtain quite high intensity tunable very short duration mode locked pulses. Many lasers can be mode locked. It is possible to produce mode locked laser action with the use of electro-optic techniques as well as with saturable absorbers.

7.6 Role of the Optical Cavity in Laser Operation

After completing a long and mathematical chapter one might be a bit confused and ask is why is an optical cavity useful in the operation of a laser? There are several reasons why optical cavities are useful for laser operation. In Chapter 3 the threshold population density difference in a laser was determined to be (Eq. 3.14)

$$\Delta n_{\text{th}} = \frac{8\pi\, \nu_0^2\, \Delta\nu}{(c/n)^3} \frac{\tau_u}{\tau_{\text{ph}}} = -\frac{8\pi\, \Delta\lambda}{\lambda_0^4} \frac{\tau_u}{\tau_{\text{ph}}}. \tag{7.80}$$

The longer the photon lifetime, τ_{ph}, the lower the threshold population difference. For a laser with a cavity of length L and with mirrors with reflectivities R_1=1.00 and R_2=0.95 so that mirror 2 has a transmission of T_2=0.05 the photon lifetime is $\tau_{\text{ph}} \approx 2L/cT_2 = 40(L/c)$ whereas if mirror 2 has a reflectivity of 0.99 and therefore a transmission of 0.01 then the photon lifetime is $\tau_{\text{ph}} = 200(L/c)$ i.e. the photon lifetime is increased by a factor of 5 and the threshold population difference is decreased by a factor of 5. In addition to decreasing the threshold population difference a longer photon lifetime results in a narrowing of the intrinsic frequency width of the modes, which is given by $1/(2\tau_{\text{ph}})$.

In Chapter 6 it was shown that the output power of a cw laser, P_L, is given by Eq. 6.32 to be $P_L = (r-1)(h\nu\, p/\tau_{\text{ph}})$ where $r = R_{gu}/R_{\text{th}}$ and p is the number of modes within the line width. In this expression R_{gu} is the rate, at which atoms or molecules are pumped into the upper laser level per unit volume of the laser and $R_{\text{th}} = \Delta n_{\text{th}}\, A_{ul}$. One might think that the laser output power decreases as the photon lifetime increases since the above equation for P_L has τ_{ph} in the denominator. This is, however not the case. The threshold population density can be rewritten as $\Delta n_{\text{th}} = V/(p\, A_{ul}\, \tau_{\text{ph}})$

so that the expression $P_L = (r-1)(h\nu\, p/\tau_{\rm ph})$ becomes

$$P_L = \left[\frac{R_{gu}}{\Delta n_{\rm th}\, A_{ul}} - 1\right] h\nu \frac{p}{\tau_{\rm ph}} = R_{gu}\, V\, h\nu - h\nu \frac{p}{\tau_{\rm ph}}\,. \tag{7.81}$$

If the photon lifetime is large enough that $h\nu\, p/\tau_{\rm ph}$ is small compared to $R_{gu}\, V\, h\nu$ then $P_L \approx R_{gu}\, V\, h\nu$. Thus for a very long photon lifetime the output laser power is equal to the rate of excitation into the excited laser level per unit volume times the laser volume times the energy per emitted photon. For a good cavity (long $\tau_{\rm ph}$), almost every atom excited into the upper level u produces a photon into the output laser beam.

The design of the laser cavity affects many other laser parameters beyond the photon lifetime. A very important function of a laser cavity is to enable the output laser beam to have a well defined Gaussian beam profile. Without the cavity the beam profile would have a rapid divergence and would not be capable of being focused to a very small spot size. Gaussian beams are important in many ways. The transmission of laser beams over great distances either in free space or in optical fibers requires high quality Gaussian laser beams. For example, in lunar laser ranging experiments a laser beam is retro-reflected off the surface of the moon. Including dispersion in the earth's atmosphere, the beam size at the moon is approximately 2 km. Not exactly small on a human scale, but tiny on the scale of the Earth-moon distance (3.8×10^5 km). At the other extreme, single mode optical fibers and optical data storage applications (CD/DVD/Blu-ray) all require spot sizes $\ll 2$ μm.

Other laser applications require extremely high power per unit area. Again these applications would be impossible without high quality laser beams. For very high gain lasers an unstable cavity (i.e. one that has a short photon lifetime) enables one to produce a near Gaussian beam with a very high output power. Some lasers such as the N_2 laser have a high enough gain that for applications that do not require transmission over a long distance or focusing to a very small region it is possible to use only a simple flat mirror at one end of the laser and a flat piece of quartz for the second output 'mirror' in order to produce a useful laser. Lasers of this type can be used to pump dye lasers.

In Chapter 5 it was shown (Eq. 5.30) that the optimum transmission for the output mirror of a laser is $T_{\rm opt} = \left[(\alpha_0\, \alpha')^{1/2} - \alpha'\right] 2L$ where α_0 is the gain per round trip inside the cavity and α' is the loss per round trip. A well designed optical cavity enables one to minimize α' and hence increase both the optimum output transmission and optimum output power of the laser.

Finally, it should be noted that the circulating power inside the cavity is much greater than the output power. The output power of a laser is equal to the laser power going in one direction inside the laser cavity times the transmission of the output mirror i.e. the power inside the laser cavity is given by P_L/T_2 where T_2 is the transmission of the output mirror of the cavity. If T_2=0.01 the power inside the laser cavity is about 100 times the output power. This is sometimes important since experiments that require a high power optical beam incident on a target can be done by inserting the target into the laser cavity provided the target does not introduce large additional losses. Examples include sensitive (i.e. very weak) laser absorption measurements and the intracavity second harmonic generation of 532 nm light from a Nd:YAG laser using a non-linear crystal (Chapter 13).

Summary of Key Ideas

- Using both geometrical and Gaussian wave optics, a laser cavity is stable against small changes in the mirror separation if

$$0 \leq g_1\, g_2 = \left(1 - \frac{L}{R_1}\right)\left(1 - \frac{L}{R_2}\right) \leq 1 \qquad (7.54)$$

  where $g_1 = 1 - L/R_1$ and $g_2 = 1 - L/R_2$ and where L is the separation of the mirrors and the R's are the mirror radii. Some common cavity configurations are shown in Figs. 7.9-7.12 on page 150. Diffraction losses are much lower in cavities using spherical mirrors than plane mirrors (Sec. 7.3).
- It is useful to describe the propagation of laser beams using Gaussian optics. For the TEM_{00} fundamental mode, and assuming a minimum waist radius w_0 located at z=0, the beam radius at the position z is

$$w(z) = w_0\sqrt{1 + \left(\frac{z}{z_0}\right)^2} \qquad (7.34)$$

  where $z_0 = \pi\, w_0^2/\lambda$ is the *Rayleigh range* (the distance for the beam to double in size from the minimum).
- The spacing between longitudinal modes in a laser cavity is given by

$$\nu_{p+1} - \nu_p = \frac{c}{2n\,L}\,. \qquad (7.58)$$

  With an active medium, the resonant frequencies of a cavity are pulled toward the resonant frequency of the laser medium (Sec. 7.2.4).
- Unstable cavities are often used to extract near diffraction limited beams from very high gain lasers (Sec. 7.4).

- A Q-switched laser is a laser that has a low Q (i.e. a short photon lifetime) and builds up a large population difference without lasing until the Q is suddenly increased so that the population is far above threshold and a giant laser pulse is emitted (Sec. 7.5). A mode locked laser produces a series of very short output pulses.

Suggested Additional Reading

B. E. A. Saleh and M. C. Teich, *Fundamentals of Photonics*, John Wiley and Sons (1991).

A. E. Siegman, *Lasers*, University Science Books (1986).

A. Yariv, *Quantum Electronics* 3rd *Ed.*, John Wiley and Sons (1989).

William T. Silfvast, *Laser Fundamentals*, Cambridge University Press (1996).

A. L. Schawlow and C. H. Townes, "Infrared and Optical Masers" *Phys. Rev.* **112**, 1940 (1958).

A. G. Fox and T. Li, "Resonant Modes in an Maser Interferometer" *Bell Sys. Technol. J.* **40**, 453 (1961).

J. P. Boyd and G. D. Gordon, "Confocal Multimode Resonator for Millimeter Through Optical Wavelength Masers" *Bell Sys. Technol. J.* **40**, 489 (1961).

A. E. Siegman, "Unstable Optical Resonators" *Appl. Optics* **13**, 353 (1974).

D. C. Sinclair and W. E. Bell, *Gas Laser Technology*, Holt, Rinehart, and Winston (1969).

R. Paschotta, *Encyclopedia of Laser Physics and Technology*, Wiley-VCH (2008).

G. P. Karman, G. S. McDonald, G. H. C. New and J. P. Woerdman, "Laser optics: Fractal modes in unstable resonators" *Nature* **402**, 138 (1999).

William Koechner, *Solid-state Laser Engineering, 2nd ed.*, Springer Verlag (1988).

N. Hodgson and H. Weber, *Laser Resonators and Beam Propagation, 2nd ed.*, Springer (2005).

Problems

1. An optical cavity for use at λ=500 nm is constructed using spherical mirrors with radii of curvature R_1=15 cm. and R_2=30 cm. The separation of the mirrors is L=20 cm. Calculated g_1 and g_2. Is the cavity stable or not?

2. An optical cavity for use at λ=500 nm is constructed using two mirrors with radii of curvature R_1=R_2=20 cm and with a cavity length of 20 cm. Is the cavity stable?

3. For the cavity of Problem 2 calculate the $1/e$ radius of the Gaussian beam in the 00 mode at the waist and at each mirror.

4. For the cavity of Problem 2 calculate the far field divergence with the assumption that the divergence is not altered as the beam is transmitted out of the cavity. How large is the beam ($1/e$ radius) at a distance of 10 m? at the distance equal to the distance from the earth to the moon?

5. If a cavity is designed as a confocal cavity but the mirrors have slightly different focal lengths it is possible for the cavity to become unstable if the cavity length changes slightly. Explain how this is possible. If the cavity is designed to be near the confocal condition but slightly inside the region of stability the cavity can not become unstable if the cavity length changes slightly. Explain this statement.

6. A TEM_{00} mode Gaussian beam with λ=500 nm has a waist at z=0. The $1/e$ radius of the waist is 0.05 mm. What is the Rayleigh range of the beam? What is the $1/e$ radius of the beam at z=20 cm? What is the curvature of the wave front at z=20 cm? What is the far field divergence of the beam?

7. At z=20 cm the Gaussian beam of Problem 6 passes through a thin lens with a focal length of 5 cm. What is the curvature of the wave front of the beam just after the lens? What is the Rayleigh range of the beam after the lens? Where is the new waist and what is the $1/e$ radius of the beam at the new waist?

8. For the optical cavity of Problem 2 how large must the mirrors be so that the Fresnel number for the mirrors is 1.0?

9. An optical cavity is to be constructed by using two mirrors each with a focal length of 20 cm. What is the maximum length of a stable cavity?

10. Sketch an optical cavity that is constructed by using one converging and one diverging mirror. Discuss the location of the waist for such a cavity. Can a stable cavity be constructed by using two diverging mirrors?

11. An optical cavity for use at λ=500 nm is constructed using two converging mirrors with R_1=15 cm and R_2=25 cm. The mirrors are located so that L=10 cm. Is the cavity stable? Where is the waist located? What is the value of the Rayleigh range, the $1/e$ radius at the waist, and the $1/e$ radius at each mirror?

12. Sketch a Q switched laser cavity using a saturable dye absorber. Sketch a Q switched laser cavity using a electro-optic device. Explain the operation of both in your own words.

Chapter 8

Diode Lasers

Solid state semiconductor diode lasers are the most widely used and important lasers. Diode lasers are important for technological applications in the areas of optical communications and optical data storage. For these applications the diode laser is used essentially as an on-off modulated laser by turning the current through the diode laser on or off. Diode lasers are used in consumer electronics such as DVD and CD players and recorders, for bar code read out, and for many other applications. The wavelength of a diode lasers can be varied either by changing the current through the diode laser or by changing the temperature of the diode laser. Diode lasers are also used for scientific applications where their tunable character is essential as well as for some technological application where the tunable character is important. Diode lasers have become extremely important for applications where it is important to be able to tune the lasers to a particular atomic or molecular transition. These applications include laser spectroscopy, laser optical pumping, laser cooling of atoms and molecules, trapping of atoms, and producing Bose Einstein condensates. They are also used in research on the development of quantum computers.

All lasers can be tuned within the gain envelope of the active medium, i.e. the line width of the laser transition. Many lasers such as the CO_2 laser and the He-Ne laser are, however, tunable over only a small wavelength range due to the narrow line width and are therefore considered as fixed frequency lasers. Other types of lasers such as titanium-sapphire lasers and dye lasers are tunable over a relatively large wavelength range and are called tunable lasers. The line width of diode lasers is much wider than that of fixed frequency lasers, permitting a diode to be tuned over a range of wavelengths. Diode lasers can be tuned by changing the temperature of the diode. Typically the center wavelength of the diode laser transition changes by about 0.3 nm/°C. This frequency shift as a function of the temperature results from a combination of the thermal expansion of the diode, changes in the index of refraction of the diode with temperature, and changes in the distribution of the mobile carriers in the diode as a function of the temperature. The center frequency of a diode laser also varies when the current through the laser is changed. This effect is largely due to the change in the temperature of the diode since a change in the current through the diode laser changes the power dissipated in the diode laser. This chapter discusses how diode lasers work, and how they can be set up to operate as narrow bandwidth tunable lasers.

8.1 Solid State Theory

For many lasers the emission of light comes from individual atoms or ions. In diode lasers, however, light emission occurs from 'mobile' electrons injected into a semiconductor crystal. The most important diode lasers are made using sandwiches of GaAs between layers of GaAlAs or sandwiches of GaInAsP between layers of InP. In order to understand diode lasers it is necessary to understand how solid state semiconductors function.

8.1.1 *Electrons in a Crystalline Solid*

Understanding semiconductor diodes requires understanding the quantum mechanical behavior of electrons in a crystalline solid. In quantum mechanics, electrons have a wavelength $\lambda = h/p$ where h is Planck's constant and p is the linear momentum of the particle. When an electron moves in a region of constant potential energy the linear momentum and hence the wavelength of the electron is constant, and the associated wave is a simple sine wave. In one dimension the form of the electron wave in a constant potential is

$$\Psi = A \sin\left(\frac{2\pi\, x}{\lambda} + \theta\right) = A \sin\left(\frac{2\pi\, px}{h} + \theta\right) . \tag{8.1}$$

This wave function is a solution to the differential equation

$$\frac{d^2\Psi(x)}{dx^2} = -\frac{4\pi^2 p^2}{h^2}\,\Psi(x) . \tag{8.2}$$

If one considers the possibility that the potential energy may vary with the position so that the linear momentum also varies with the position then one has the relationship

$$p^2(x) = 2m\,[E - V(x)] \tag{8.3}$$

where E is the total energy of the particle, $V(x)$ is the potential energy of the electron, and $p^2(x)/2m$ is the kinetic energy of the electron. Substituting Equation 8.3 into Equation 8.2 one obtains

$$\frac{d^2\Psi(x)}{dx^2} + \frac{8\pi^2 m}{h^2}\left[E - V(x)\right]\Psi = 0\,. \tag{8.4}$$

Equation 8.4 is called Schrödinger's equation, and the solutions for, Ψ, are called the wave functions of the electron. The quantity $|\Psi(x)|^2\,dx$ is the probability of finding the electron between x and $x + dx$. Since the probability of finding the electron somewhere must be unity the wave function is normalized such that the integral of $|\Psi(x)|^2$ over all space is equal to one. Although this discussion of the Schrödinger equation is brief and incomplete it illustrates how the wavelength of the electron and the Schrödinger equation are connected.

To a first approximation the effect of a crystalline solid is to present a region of nearly constant potential energy for an electron. Thus the solid acts as a potential well in which the electron moves. This approximation is especially useful in understanding a metallic conductor. Consider the approximation where an electron in a metallic conductor moves in a one dimensional potential well with potential energy $V(x) = 0$ inside the well and $V(x) = V$ outside the well. The depth of the potential well, V, is called the inner potential of the metal. Typically the inner potential of a metallic solid is of the order of 10 eV. The inner potential arises from the attractive forces of the positively charged ion cores of the atoms that make up the metallic solid. For an electron outside the metal this force is largely compensated by the electrons of the solid, and the primary force on an electron outside the solid is due to the image charge. For an electron in the solid the electrons do not provide complete shielding and an electron experiences the inner potential.

If an electron has a total energy E, that is less than V, as measured from the bottom of the potential well then the kinetic energy, $K = E - V$, of the electron is positive only inside the well, and is negative outside the well. The wave function for an electron decreases exponentially in regions where the kinetic energy is negative. This requires boundary conditions on the Schrödinger equation so that wave functions are possible only for particular discrete values of the energy, which are called the energy eigenvalues. This is similar to the vibrations of a string where only certain discrete values of the resonant frequencies are possible. For solutions for $E \leq V$ the approximation that the wave function is zero outside the potential well is made. The solutions to the wave equation in this approximation are sine

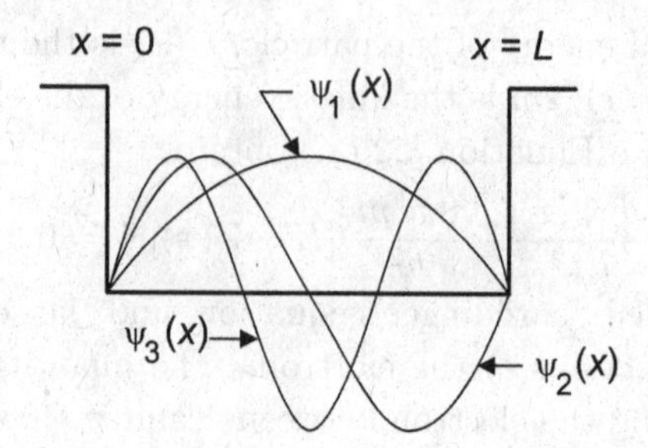

Fig. 8.1 A potential well and the three lowest energy electron wave functions corresponding to the potential well.

waves within the region of the potential well with the boundary condition that the wave function is equal to zero at the boundaries, which are taken as $x = 0$ and $x = L$. The boundary conditions lead to the solutions

$$\Psi(x) = A \sin \frac{n \pi x}{L} . \tag{8.5}$$

These wave functions correspond to wavelengths given by

$$\lambda_n = \frac{h}{p_n} = \frac{2L}{n} \tag{8.6}$$

where n is a positive integer. The allowed wavelengths correspond to the allowed energies

$$E_n = \frac{p_n^2}{2\,m} = \frac{n^2\,h^2}{8m\,L^2} . \tag{8.7}$$

The integer n is the quantum number labeling the allowed states for an electron in a potential well. Figure 8.1 shows schematically a potential well associated with a solid and the wave functions for the three lowest energy levels. These wave functions are intended to represent the electrons outside the last closed shell of the atoms making up the metallic conducting solid. In a metal the electrons outside the last closed shell are called the conduction electrons.

For a real crystalline solid of macroscopic dimensions the potential well contains a very large number of electrons, one to four electrons outside the last closed shell per atom in the solid. At absolute zero of temperature these electrons fill up the levels of lowest energy with two electrons one with spin up and one with spin down in each quantum level. If the well contains N electrons the quantum number of the highest filled level is $N/2$. The energy of this level is called the Fermi energy, E_F, which is given by $E_F = N^2 h^2/32mL^2$. The quantity $\phi = V - E_F$, which is called the work function

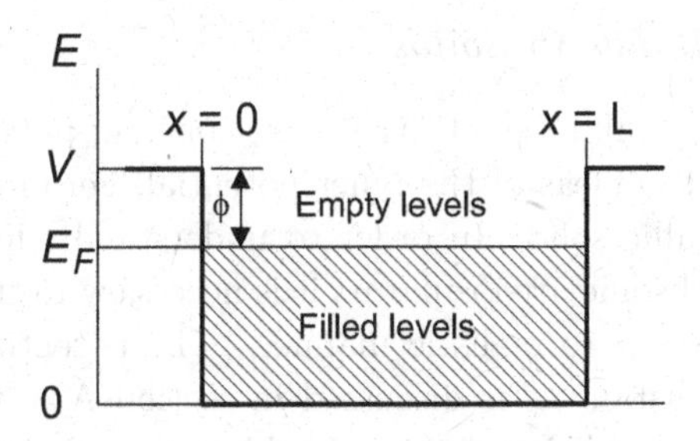

Fig. 8.2 The energy levels of the conduction electrons of a metal as described by the electron in a box model. V is the inner potential, E_F is the Fermi energy, and ϕ is the work function.

of the metal, is the lowest energy required to remove an electron from the solid. These ideas are illustrated in Fig. 8.2. For most metals the Fermi energy is in the range two to ten eV and the work function is usually smaller than the Fermi energy. Although the one dimensional model is obviously an oversimplification of the energy levels of a metallic solid the Fermi energy calculated using it is of the correct order of magnitude. Copper has one free electron per atom and a nearest atom separation of 0.25 nm. The value of E_F using the one dimensional model is $E_F = (h^2/32m)(N/L)^2 = 1.5$ eV where m has been taken as the free electron mass. The energy levels for the electrons in the potential well are quantized, but for a one dimensional potential well the energy separation of the levels is very small being of the order of magnitude 10^{-7} eV and for a three dimensional potential well the energy separation of the levels is of the order of magnitude 10^{-21} eV. For most purposes the energy levels of a crystalline metallic solid can be considered as a continuum.

The discussion of the wave functions for an electron in a potential well was very simplified. The electrons were treated as standing waves, and the time dependence of the wave functions was ignored. For some purposes it is useful to treat the electrons as running waves with a wave function given by

$$\Psi_k(\boldsymbol{r}) = A\, e^{i(\boldsymbol{k}\cdot\boldsymbol{r} - \omega t)} \tag{8.8}$$

where $\omega = 2\pi E/h$ and where $\boldsymbol{k}$ is the wave vector of the electron wave. The wave vector is a vector that has a magnitude equal to $2\pi/\lambda$ and points in the direction that the wave is running.

8.1.2 *Energy Bands in Solids*

The picture of a crystalline solid as a simple constant potential well enables us to understand the ideas of the inner potential, Fermi energy, and work function of a metallic solid. In order to understand why some materials are conductors and some are insulators it is necessary to treat the potential well of a solid in a more realistic manner. The potential well in a solid is not simply a constant as is discussed in section A of this chapter, but instead it has the periodicity of the crystal lattice. An electron that is near or penetrates into an ion core has a lower potential energy than one that is between the ion cores. In Fig. 8.3 the periodic character of the potential well is illustrated. The ion cores are centered at the potential minima. The ion cores are separated by the length a_0. The potential well is therefore periodic in space with a period a_0. In one dimension the potential energy of an electron therefore satisfies the relationship $V(x) = V(x + n\,a_0)$ where n is an integer. In the earlier analysis the electrons were treated as standing waves with the spacing between the nodes of the electron wave given by L/n. When L/n and a_0 are not related in some particular way the electron is as likely to be found in high energy as in low energy parts of the potential well and the energy of the electron is determined by the average value of

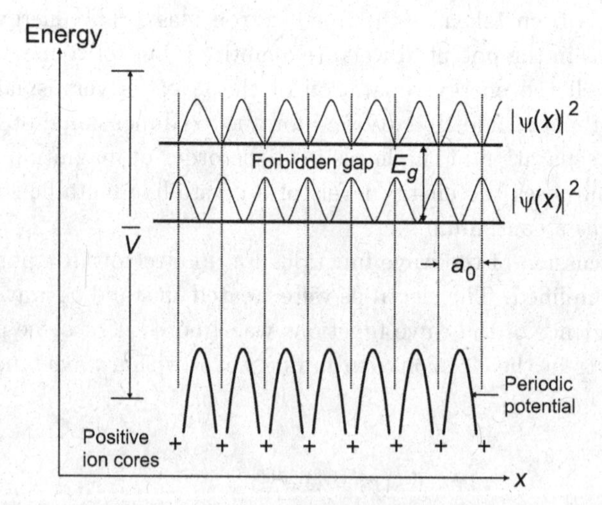

Fig. 8.3 A schematic diagram representing the periodic potential energy of an electron in a crystalline solid. The periodic potential energy is produced by the rows of ion cores in the solid. Also shown are the values of $|\Psi(x)|^2$ for the high and low energy standing wave functions at the band edge. In a plot of $\Psi(x)$ alternate maxima of $|\Psi(x)|$ would be of opposite signs.

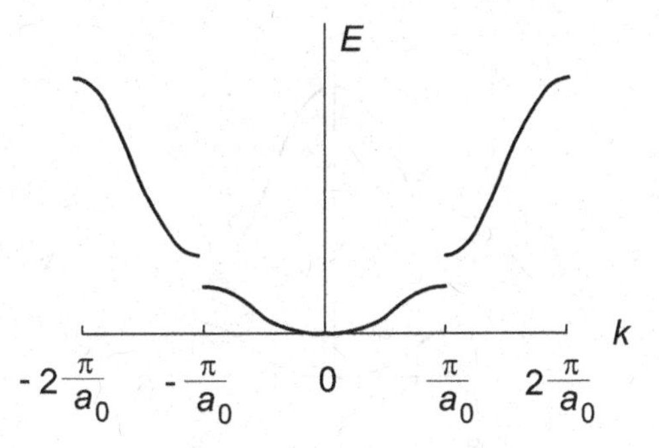

Fig. 8.4 The energy of an electron in a periodic potential as a function of k. Except near the band gaps the energy of the electrons is given by $E = k^2h^2/8\pi^2m$. The states near the top and bottom of the band gap can not be represented by running waves. They are the standing waves shown in Figure 8.3.

the potential well. If, however, $L/n\,a_0 = 1$ then the standing electron wave that has its nodes located at the position of the ion cores will have a very different energy than the electron wave that has its nodes located midway between the ion cores. The energy difference between these waves with the same wavelength is called the energy band gap, E_g. This is illustrated in Fig. 8.3. The simple model of the electron as a free particle in a box would predict that the energies of these electrons should be the same. There are no allowed energy levels within the forbidden band gap. The value of E_g is large for solids that have large amplitudes of the periodic part of the potential well and is small for solids that have small amplitudes of the periodic part of the potential well.

Another way of thinking about the origin of the band gap is to consider a running electron wave in the crystal lattice. The running wave is partially scattered from each ion core. For most running waves the scattered waves from different ion cores do not have a constructive phase relationship and no strong scattered wave builds up. If, however, the electron wavelength, λ, is equal to $2a_0$ then the waves that are back scattered from different ion cores are in phase and a strong scattered wave builds up. It is easy to see that $\lambda = 2a_0$ is the same condition as $L/n = a_0$ that was obtained using the standing wave picture. In Fig. 8.4 the energy of the electrons with wave vector k, $E = E(k)$ is plotted as a function of $k = 2\pi/\lambda$. The first band gap occurs at $k = \pm\pi/a_0$ and higher band gaps occur for $k = \pm n\,\pi/a_0$. The region that contains a given band is called a Brillouin zone. The lowest Brillouin zone extends from $k = -\pi/a_0$ to $k = \pi/a_0$. The plot of E seen in Fig. 8.4 is not the one usually encountered. Rather one usually sees a

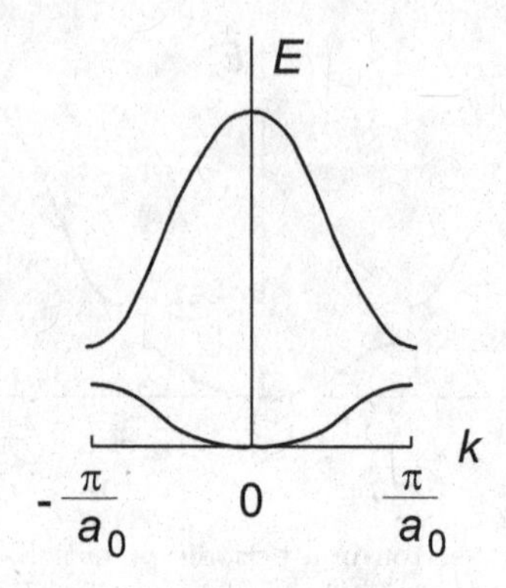

Fig. 8.5 The energy of an electron in a periodic potential as a function of k in the reduced zone representation.

picture known as the "reduced zone scheme". The reduced zone scheme is illustrated in Fig. 8.5. The energy levels from the zones with the absolute value of k greater than π/a_0 are folded back into the first Brillouin zone as is shown. Why these pictures are essentially the same is addressed briefly. In one dimension the translational symmetry of the crystal requires that the charge density in the solid be periodic with a period a_0. This means that the wave function must satisfy the relationship

$$\Psi(x + a_0) = C\Psi(x) \qquad |C|^2 = 1\,. \tag{8.9}$$

If the wave function in the crystal is the same after N displacements of length a_0 (i.e. assuming there are periodic boundary conditions) then

$$\Psi(x + N\,a_0) = C^N\,\Psi(x) = \Psi(x) \tag{8.10}$$

so that $C^N = 1$. There are N solutions for C. They are given by $C = e^{i2\pi n/N}$ where n is an integer. This leads to the result that

$$\Psi_k(x + a_0) = e^{i\,k\,a_0}\,\Psi_k(x) \tag{8.11}$$

where $k = 2\pi n/Na_0$ and where n is the integer that determines the allowed values of k for the electrons in a crystalline solid. It is clear that $k^*(x) = k(x)$ where $k^* = k + 2\pi/a_0$. The wave function for an electron with wave vector k is identical to the wave function for the wave vector $k^* = k+2\pi/a_0$. With this it is clear that the energy bands can all be represented in the reduced zone scheme as illustrated in Fig. 8.5. The reduced zone picture is usually the one used in discussions of diode lasers.

Another way of looking at the existence of energy bands in solids is to think about the energy bands as arising from the atomic energy levels of the

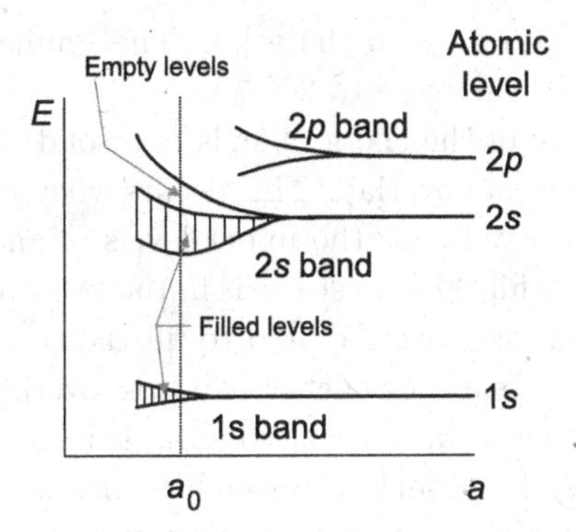

Fig. 8.6 The splitting of the 1*s* and 2*s* bands of lithium as the atoms are brought together toward a_0, the atomic separation in the metal. Note that the filled 2*s* levels are lower in energy than the 2*s* atomic level. This is the source of most of the binding energy of the metal. The shape and positioning of the curves is only qualitatively correct.

atoms in the solid as the atoms are assembled to form the solid. This way of looking at the band structure of a solid is called the tight binding method. Perhaps the simplest solid to consider is Li metal. A Li atom has the two electrons in the 1*s* level and has one electron in the 2*s* level. Consider a large number of Li atoms assembled on a crystalline lattice as they are in Li metal but with a much larger lattice spacing. The widely separated atoms have energy levels that are the same as isolated atoms. Now consider what happens as the lattice spacing shrinks to the lattice spacing in the Li metal. As the electron clouds in the atoms begin to overlap there is an interaction that splits the energy levels that are degenerate at large separations into a large number of closely spaced discrete energies. In the limit where there are a very large number of Li atoms these closely spaced energy levels form a band of allowed energies. The width in energy of the band increases as the overlap of the wave functions increases. Figure 8.6 shows the band structure of Li as a function of the separation of the atoms. The final result is essentially the same as was obtained from the periodic potential. There are allowed bands separated by forbidden gaps. For Li metal the 1s band is completely full, the 2*s* band is half full, and the higher $2p, 3s, 3p, \ldots, nl$ bands are empty. This atomic picture of the band structure of a solid makes clear how many electrons are required to fill a band. The interaction of the atoms that results in a band does not alter the number of energy levels but instead just alters the energy of the levels. Since the number of levels is not changed it is clear that the number of electrons that are needed to fill the band is identical to the number of electrons needed to fill the corresponding atomic levels of the N atoms in the solid. The Pauli principle limits the number of electrons in the band to $2N$.

The atomic picture of the energy bands in a solid also makes it easy to understand how bands can overlap. This occurs when the lower levels from one band have an energy below the upper levels in another band. In this situation the electrons fill the lowest levels in the two overlapping bands. In this situation $4N$ electrons are required to fill the two overlapping bands. Of course it is not necessary for energy bands to overlap. If the separation of the atomic energy levels of an isolated atom is larger than the spreading of the levels then as the atoms are assembled into a crystalline solid the energy bands of the solid do not overlap. The overlapping of bands does not occur in a one dimensional solid, but is important for real three dimensional solids.

8.1.3 *Conductors, Insulators and Semiconductors*

How the energy levels in a solid are split into allowed bands and forbidden gaps has been discussed. The electrons in the solid can only go into the states in the allowed bands. The outer electrons in a solid are in states that are similar to the states of an electron in a potential well which is the size of the solid. This means that the outer electrons of the solid are associated with the entire crystal and are not confined to the region about the atom from which they originated. Thus far only what the electronic energy levels of a crystalline solid are has been discussed. Another important problem is what is the probability that a given level is filled? So far it has been assumed that the electrons in a solid fill the lowest energy levels with two electrons in each state one with spin up and the other with spin down. This is true at the absolute zero of temperature. At higher temperatures there are some filled states up to energies a few times kT above the uppermost filled level at $T = 0$ K and there are a some unfilled states below that level. The expression for the probability that a state is filled is given by the Fermi-Dirac distribution function

$$f = \frac{1}{e^{(E-E_F)/kT} + 1} \tag{8.12}$$

where f is the probability that a state of energy, E, is occupied if the system is at the absolute temperature T and where E_F is the Fermi energy. The maximum value of f is 1 rather than 2 since it is common to regard the two Pauli spin states as distinct. A plot of f as a function of the energy is shown in Fig. 8.7 for a particular value of the temperature. As shown

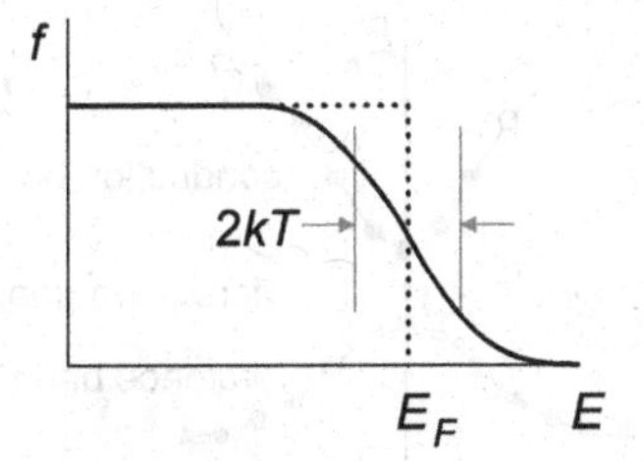

Fig. 8.7 The Fermi Dirac distribution function $f = 1/[\exp(E - E_F) + 1]$. The function has an inversion symmetry through the point where the curve crosses the line $E = E_F$. The dashed line shows f for T=0.

the value of the function, f, for energies well below the Fermi energy is approximately equal to 1, and the value of f for energies far above the Fermi energy is approximately equal to 0. The change from 1 to 0 occurs in an energy range approximately given by $2kT$ and centered about the Fermi energy. When the electron density is high enough or the temperature is low enough that Fermi statistics are important then the electrons are said to form a degenerate Fermi gas.

Whether a material is an electrical conductor or insulator is determined by how the electron energy levels in the material are filled. As will be discussed an electrical conductor is a material that has a partially filled conduction band, and an electrical insulator is a material that has a filled valence band and an empty conduction band. This is illustrated schematically in Figure 8.8.

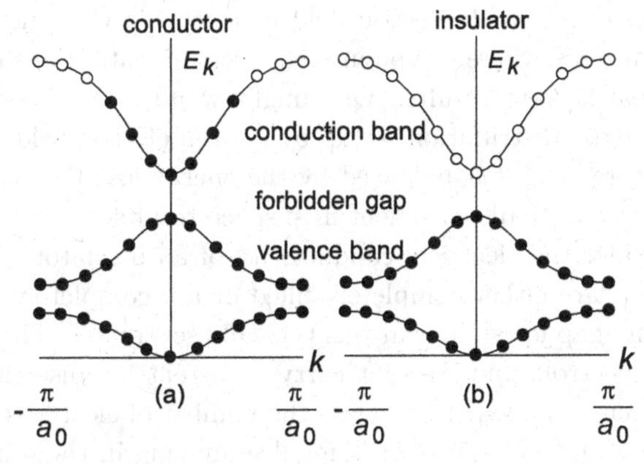

Fig. 8.8 A schematic representation of the filled levels of an insulator and a conductor. The filled circles represent filled levels and the white center circles represent unfilled levels. For the conductor the conduction band is partially filled, and for the insulator the conduction band in completely empty.

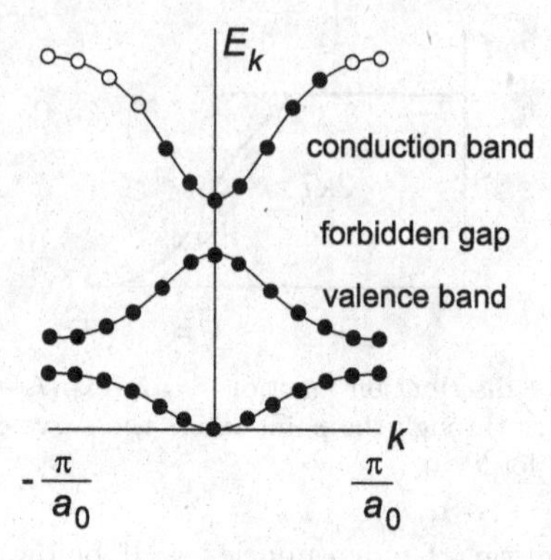

Fig. 8.9 The filled energy levels of a conductor that is carrying an electrical current. Note that there are more electrons with positive k than with negative k.

Now consider the electrical conductivity of a metallic conductor. A conductor has a partially filled conduction band. An electrical current is carried by running waves. For zero electric field there are as many electrons with wave vector $\boldsymbol{k}$ as with wave vector $-\boldsymbol{k}$ and therefore no net electrical current. A current is obtained only when there is a different number of filled states for electrons with wave vector $\boldsymbol{k}$ than with wave vector $-\boldsymbol{k}$. The situation where there are more electrons in states with wave vector $\boldsymbol{k}$ than in states with wave vector $-\boldsymbol{k}$ is discussed. In order to achieve this situation energy must be provided to depopulate the states with wave vector $-\boldsymbol{k}$ and to populate those with wave vector $\boldsymbol{k}$. Because the states are closely spaced a weak electric field can provide this energy. This is shown in Figure 8.9 where the populated electron states are shown so that a net electrical current results. One might wonder what determines the particular electron distribution set up by a given electric field. The answer is that the energy input is balanced by the energy loss through scattering so that a steady state displacement in $\boldsymbol{k}$ space results.

Next consider the electrical conductivity of an insulator. An insulator has bands that are either completely filled or are completely empty. The forbidden band gap for an insulator is typically several eV. The filled bands have mobile electrons but can not carry a current because the states are all full and there is no way to increase the number of electrons moving in a given direction and to decrease the number moving in the opposite direction. In order to increase the flow of electrons in a given direction one must excite an electron into the empty conduction band from the filled valence band. This typically requires an excitation of a few eV that can not be provided by either a weak electric field or by thermal fluctuations. As will be discussed in the next section of this chapter a pure semiconductor is similar to an insulator in that there is at the absolute zero of temperature a filled valence band and an empty conduction band, but for a semiconductor the forbidden band gap is relatively small so that it is possible to excite some electrons into the conduction band by removing some electrons from the valence band so that the semiconductor can carry an electrical current.

How the electron band picture of a crystalline solid explains the electrical conductivity of both metallic conductors and insulators is now understood. It should be noted that the overlapping of bands can play an important role in determining whether a solid is a conductor or an insulator. For example magnesium has two $3s$ electrons in its outermost occupied shells in the isolated atom. Thus one might think that magnesium would be an insulator since the $3s$ band would be full. This is not the case because the $3s$ band overlaps with the $3p$ band providing the necessary high density of empty states for a conductor.

8.1.4 *Semiconductors*

As explained previously a pure semiconductor is a material that has a filled valence band and an empty conduction band at the absolute zero of temperature similar to an insulator. The pure semiconductor, however, has a forbidden band gap small enough that appreciable thermal excitation of electrons out of the valence band and into the conduction band occurs at room temperature. For GaAs the forbidden band gap is about 1.43 eV at room temperature. A pure semiconductor, which is also called an intrinsic semiconductor, has an electrical conductivity that is less than the conductivity of a metallic conductor and that increases rapidly with the temperature. At thermal equilibrium the number of electrons in the conduction band is equal to the number of unoccupied states in the valence band. Current is carried by the electrons in both bands. It is nevertheless common to discuss the current carried by the valence band electrons as though it were carried by the excess positive charge created when an electron is excited to the conduction band. This excess positive charge is called a hole. Experiments such as the Hall effect show that a hole moves in an electric or magnetic field just as if it had a positive charge equal in magnitude to the electron charge. Thus it is usual to discuss the electron current in the conduction band and the hole current in the valence band.

The discussion of the band structure of a semiconductor such as GaAs has been greatly simplified. The bands in real semiconductor materials are not simple parabolic functions of k except near the edges of the forbidden gaps, but are instead complicated functions of the $\boldsymbol{k}$ vector in three dimensions. Also both the electrons in the conduction band and the holes in the valence band act as though they have an effective mass that is different from the mass of a free electron. One must take into account these and other considerations if one wishes to analyze semiconductor lasers in detail.

In addition to pure semiconductors it is common to dope semiconductors with impurities in order to obtain useful characteristics. Doped semiconductors are called extrinsic semiconductors. For a semiconductor such as GaAs the valence of Ga is 3 and the valence of As is 5. The material GaAs is called a 3-5 compound. A GaAs crystal has on the average four valence electrons per atom. Impurity atoms are used that replace either a Ga atom or an As atom substitutionally at a lattice site. The impurity atoms should have a valence different from the atom they are replacing. If the impurity has one more electron per atom than the atom it replaces then the impurity will provide an ion core with a charge one unit greater than the ion core it replaces and an extra electron. This ion core together with the additional electron of the impurity atom forms a weakly bound system similar to the electron and proton of a hydrogen atom. The impurity system is called a donor impurity, and the corresponding energy levels are called donor levels. A donor impurity is easily ionized producing an electron in the conduction band. On the other hand if the impurity atom has one fewer valence electrons than the atom it replaces it acts as a positive ion core that can attract an electron from the valence band to form a weakly bound system called an acceptor impurity, and the corresponding energy levels are called acceptor levels. When an acceptor is formed it leaves behind a hole in the valence band. Semiconductors that are doped with donor impurities are called n-type semiconductors and semiconductors that are doped with acceptor impurities are called p-type semiconductors.

At very low temperatures an extrinsic semiconductor has a very low conductivity. The donor levels are filled and there are no electrons in the conduction band, and the acceptor levels are empty leaving no holes in the valence band. As the temperature increases the donors become ionized and electrons are released into the conduction band, and the acceptor levels become populated leaving holes in the valence band. The electrons in the conduction band and the holes in the valence band cause the extrinsic

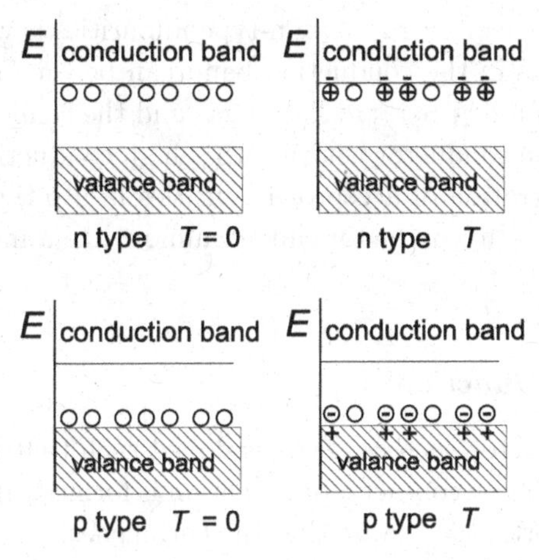

Fig. 8.10 The localized impurity levels in n- and p-type semiconductors at $T=0$ and at room temperature T. At low temperatures the impurities are neutral. At room temperature an electron or hole is released.

semiconductor's conductivity to increase. At relatively low temperatures the donors are largely ionized and the acceptors are largely filled so that the conductivity of an extrinsic semiconductor is larger at room temperature than the conductivity of an intrinsic semiconductor. The conductivity of an extrinsic semiconductor does show a temperature dependence due to excitation of electrons across the band gap just like an intrinsic semiconductor does, but for an extrinsic semiconductor the fractional change in the conductivity with temperature is not as large as for an intrinsic semiconductor since the extrinsic semiconductor has a much larger conductivity than an intrinsic semiconductor. Figure 8.10 shows schematically both n-type and p-type semiconductors at absolute zero and at room temperature. At $T = 0$ in an n-type semiconductor the donor levels are half filled since each level contains one electron, and the Fermi level is therefore at the energy of the donor impurity level. As the temperature increases to room temperature the electrons in the donor impurities are released into the conduction band so that the donor levels are depopulated. This implies that the Fermi level has decreased and is below the donor level at room temperature. In the same way the Fermi level in a p-type semiconductor is at the acceptor level at $T = 0$ but at room temperature the Fermi level is above the acceptor level. At room temperature the resistivity of an extrinsic semiconductor is typically a few ohm-cm while the resistivity of an intrinsic semiconductor is much higher and may depend on the density of unavoidable impurities in the material. If the concentration of n-type impurities is very high then the density of electrons in the conduction band can become high enough that the electrons form a degenerated Fermi gas and the semiconductor acts almost like a metallic conductor with a rather high conductivity. In the same way if the concentration of acceptor impurities in a p-type semiconductor is high enough the semiconductor can act almost like a metallic conductor.

8.1.5 *The p-n Junction*

When a semiconductor such as GaAs is doped such that it has p-type and n-type material in close contact a p-n junction is formed. In discussing a p-n junction it is important to realize that the Fermi energy, which is also called the Fermi level, is an important thermodynamic property of the semiconductor. In equilibrium the Fermi level of a system is the same throughout the system. For a p-n junction this might seem to be impossible since the Fermi levels of an isolated p-type semiconductor are very different from the Fermi level of an isolated n-type semiconductor. In order to equalize the Fermi level throughout a p-n junction some mobile electrons from the n part of the junction spill over into the p part of the junction or some holes from the p part of the junction spill over into the n part of the junction. This produces a double layer with a net positive charge in the n part of the p-n junction and a net negative charge in the p part of the p-n junction. There is therefore an electric field directed from the n part to the p part of the p-n junction. This leaves a region depleted of mobile carriers, which is called the depletion layer. This is illustrated in Figure 8.11. There are almost no mobile carriers in the depletion layer. It should be emphasized that the electric field due to the double layer arises even though there is no applied field. The electric field due to the double layer is due entirely to the redistribution of charges within the material and is similar to the electric field that produces a contact potential difference.

Due to the electric field in the depletion layer the energy bands vary as functions of the position through the p-n junction even though the Fermi level is a constant throughout the p-n junction. The variation in energy of the bands produces an energy barrier, E_B, to the free movement of electrons or holes across the depletion layer. This barrier to the free movement of electrons or holes across the depletion layer is also spoken of as being due to the electric field in the depletion layer.

Because the depletion layer is depleted of mobile carriers it acts as an

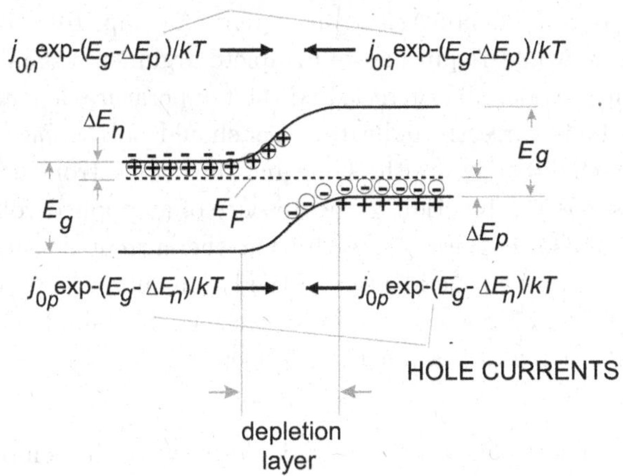

Fig. 8.11 A schematic diagram of a p-n junction. There is no voltage applied across the junction. The electron currents going each direction across the junction are equal. The hole currents going each way across the junction are also equal.

insulating layer in a semiconductor. Any voltage applied across the semiconductor appears primarily across the depletion layer. One might think that as an insulating layer the p-n junction would prevent the flow of current. This is of course not the case, and although the flow of current is altered it is not completely prevented. The depletion layer is not completely devoid of mobile carriers, but instead the number of mobile carriers is merely reduced since the Fermi Dirac distribution function is reduced as the energy is further above the Fermi energy. Next the conduction of electrons across a p-n junction will be discussed. A discussion of the conduction of holes across a p-n junction will be analyzed subsequently. When no voltage is applied across a p-n junction the probability of an electron crossing from the n-doped material into the p-doped material is proportional to the probability that the electron is thermally excited to an energy greater than the energy of the barrier. The energy barrier to the flow of electrons is given by $E_B = E_g - \Delta E_p$ where E_g is the band gap and ΔE_p is difference in energy between the Fermi level and the top of the valence band in the p-doped material. If $E \geq E_B$ then usually $E - E_F \gg kT$ so that the Fermi Dirac distribution function is approximately given by $f = \exp[-(E - E_F)/kT]$. In this situation the electron current density in either direction is given by

$$j_n = j_{0n}\, e^{-(E_g - \Delta E_p)/kT} \tag{8.13}$$

where j_{0n} is a constant that varies only slowly with the temperature and where the subscript n denotes electron current. A complete derivation of the current density would require more complete arguments than were given, but the result is essentially correct, and the temperature dependence of the current density is correctly indicated. As should be the case the current densities in each direction are the same and the net electron current density flowing across the p-n junction in the absence of an applied voltage is zero. Equation 8.13 also indicates correctly that the current density flowing in each direction would not be identical if the Fermi energy were not the same in both the p- and n-doped parts of the junction. In a similar way the hole current density across a p-n junction is given by

$$j_p = j_{0p}\, e^{-(E_g - \Delta E_n)/kT} \tag{8.14}$$

where j_{0p} is a constant that varies only slowly with the temperature and ΔE_n is the difference between the Fermi level and the bottom of the conduction band in the n-doped material. The hole current densities in each direction across the p-n junction also are equal in magnitude so that there is no net hole current density across the junction. Note that the electron and hole currents across the p-n junction balance separately and do not balance each other.

The doping of an extrinsic semiconductor results in there being a much higher density of mobile carriers of one type than of the other type in the doped material. In an extrinsic semiconductor the current is primarily carried by the carriers with the highest density which are called the majority carriers. In n-type material the electrons are the majority carriers, and in p-type material the holes are the majority carriers. In n-type material the holes are called the minority carriers, and in p-type material the electrons are called the minority carriers.

If an emf is used to apply a voltage across the p-n junction the situation is altered and a net current density flows across the junction. Figure 8.12 illustrates schematically the situation where a voltage is applied across the p-n junction. The applied voltage shifts the Fermi level in the p-doped material with respect to the Fermi level in the n-doped material. This changes the height of the barrier for the forward current i.e. for electrons from the n-doped material to enter the p-doped material or holes from the p-doped material to enter the n-doped material. The barrier for the reverse current is unchanged. This leads us to the result that there is a net electron current density across a p-n junction that is given by

$$j_n = j_{0n} e^{-(E_g - \Delta E_p)/kT} \left(e^{eV/kT} - 1 \right) \tag{8.15}$$

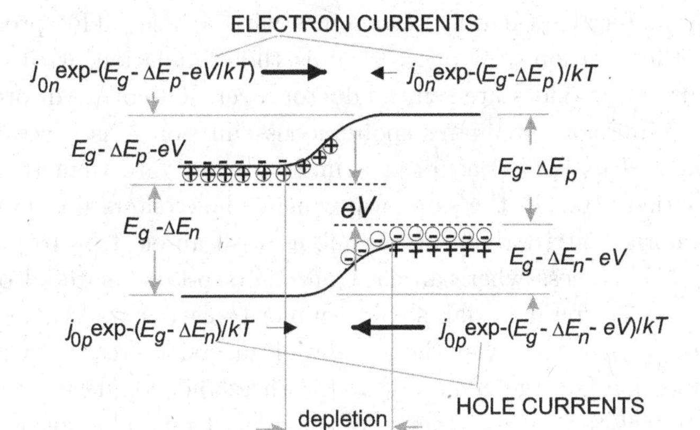

Fig. 8.12 The energy levels of a p-n junction with a voltage applied across the p-n junction. The junction is forward biased. The flow of electrons and holes are such that current is carried from right to left.

where V is the applied voltage. In the same way there is net hole current density across the p-n junction that is given by

$$j_p = j_{0p} e^{-(E_g - \Delta E_n)/kT} \left(e^{eV/kT} - 1 \right) . \tag{8.16}$$

The net current through a junction at a constant temperature will therefore vary with the applied voltage as $I = I_0[\exp(eV/kT) - 1]$. In the forward biased situation (V positive) the current through a junction increases exponentially whereas in the back biased situation (V negative) the current saturates at $-I_0$. For a p-n junction I_0 is typically a very small current. Note that I_0 depends exponentially on the inverse of the temperature. The forward current can be large. It can be shown that the forward biased current through a p-n junction is primarily carried by the majority carrier of the more heavily doped material. Thus if the n side of a p-n junction is more heavily doped than the p side then most of the current through the p-n junction is carried by electrons when the junction is forward biased. On the other hand if the p side of the p-n junction is more heavily doped than the n side then most of the current through the junction is carried by holes when the junction is forward biased. A p-n junction is forward biased when the voltage between the p and n doped material is positive, and a junction is back biased when the voltage between the p and n doped material is negative.

When an electron from n-type material passes through the junction and enters the p-type material it becomes a minority carrier. An electron in p-type material ultimately recombines with a hole. This process can result in the creation of a photon. It is the stimulation of this process that occurs in a solid state semiconductor laser. Of course in order that stimulated emission dominates spontaneous emission it is necessary that the stimulated emission occurs at a much greater rate than the natural recombination rate. In the same way when a hole enters n-type material it is a minority carrier that can recombine with an electron to produce a photon. The process where an electron enters p-type material or where a hole enters n-type material is called minority carrier injection. As with any laser it is necessary that the population inversion created by minority carrier injection is large enough that the threshold population is reached before laser action begins to occur. When an electron and a hole recombine with the emission of a photon one has the restriction that the initial linear momentum of the electron must be equal to the momentum of the final electron plus the momentum of the photon plus any momentum given to the crystal lattice. The linear momentum of the electron and the photon are related to the their wave vectors. The $\boldsymbol{k}$ vector of the photon is typically about 10^5 while the $\boldsymbol{k}$ vector of the electrons is typically ten time larger. An analysis of the process of recombination with the emission of a photon leads one to conclude that in the reduced zone picture that the electron does not significantly change its $\boldsymbol{k}$ vector so that the transitions are called vertical transitions. This discussion has assumed that the maximum in the valence band energy lies directly above the minimum in the conduction band energy in $\boldsymbol{k}$ space. A semiconductor where this is the situation is called a direct band gap material. GaAs is a direct band gap semiconductor.

One final comment is appropriate. An essential difference between a laser like the ruby laser and a semiconductor laser is that in a ruby laser the electrons involved in the laser action are localized near the ruby ion whereas the electrons involved in laser action in a semiconductor are mobile electrons that can move about the semiconductor crystal freely.

8.2 Semiconductor Diode Lasers

8.2.1 *GaAs Double Heterojunction Lasers*

Semiconductor diode lasers are made using a wide variety of semiconductor materials. The most important semiconductor diode lasers are those based on the III-V compound semiconductors. For the most part, this chapter focuses on diode lasers that utilize compounds made from Ga, As, and Al

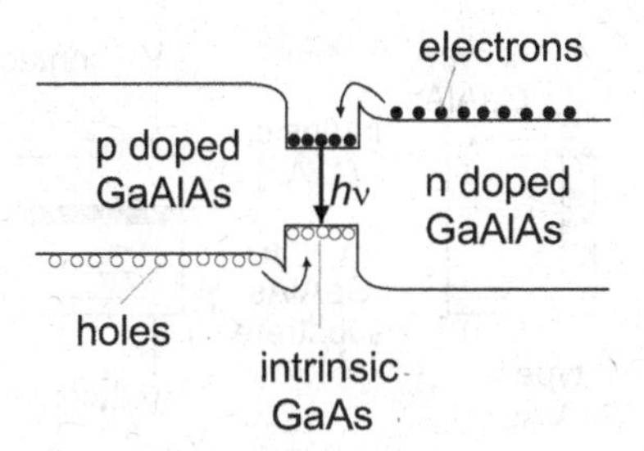

Fig. 8.13 A schematic diagram of a double heterostructure diode laser. Electrons and holes are injected into the intrinsic GaAs potential well where they are stimulated to recombine.

but some other materials are discussed at the end of this section. Semiconductor diode lasers can be constructed using GaAs and $Ga_{1-x}Al_xAs$ where the subscripts indicate that a fraction x of the Ga is replaced by Al. The band gap in $Ga_{1-x}Al_xAs$ is somewhat larger than the band gap in GaAs. Although laser action can be obtained with a single junction diode most semiconductor lasers are formed using a double heterojunction (also known as double hetrostructure lasers). This double heterojunction structure and its energy band structure for forward biased current drive are shown schematically in Fig. 8.13. The double heterojunction is constructed with intrinsic GaAs in the center of a sandwich with extrinsic $Ga_{1-x}Al_xAs$ on either side of the intrinsic GaAs. The $Ga_{1-x}Al_xAs$ on one side of the intrinsic GaAs is heavily p-doped and on the other side it is heavily n-doped.

Diode lasers are pumped by the current through them. The double heterojunction is forward biased when the applied voltage difference between the p- and n-doped $Ga_{1-x}Al_xAs$ is positive. In the forward biased situation holes are injected into the intrinsic GaAs from the p-doped $Ga_{1-x}Al_xAs$ and electrons are injected into the intrinsic GaAs from the n-doped $Ga_{1-x}Al_xAs$. The electrons that are injected into the intrinsic GaAs can not diffuse across the intrinsic region and escape into the p-doped $Ga_{1-x}Al_xAs$ because the larger band gap causes the electrons to be trapped in the potential well formed by the intrinsic GaAs. In the same way the holes are also trapped in the intrinsic GaAs. The trapped electrons and holes diffuse around in the trap until they recombine with each other. The recombination produces a photon a high percentage of the time. For GaAs the wavelengths emitted are typically in the range 650-880 nm. The emitted photons can stimulate further recombination thereby producing laser action. A given GaAs laser can not lase over the entire wavelength range of 650-880 nm, but instead lases only over a relatively narrow range

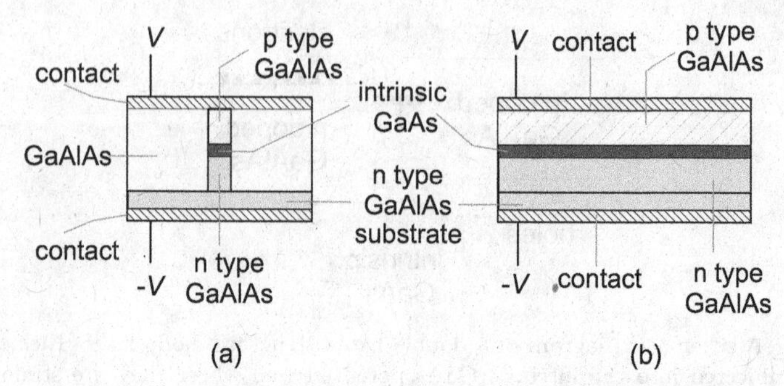

Fig. 8.14 (a) A cross section diagram of a buried heterostructure GaAs diode laser. The active laser medium is shown in black. The laser beam is emerging perpendicular to the image. (b) A cross section diagram of the same laser but with the laser beam emerging along the laser medium axis.

of frequencies at a given temperature. Since the laser wavelength is determined by the GaAs band gap one might wonder how manufacturers can obtain lasers that lase at such different wavelengths. The answer to this question is that the band gap in GaAs can be changed by strain, and the GaAs crystal can be subjected to different strains by the way it is deposited onto the GaAlAs layers.

An important consideration in the laser action in semiconductor double heterojunctions is that the photons produced be confined to the intrinsic GaAs layer. In *index-guided* lasers this confinement occurs because the index of refraction of the heavily doped n-type or p-type $Ga_{1-x}Al_xAs$ is smaller than the index of refraction of the intrinsic GaAs layer. The index of refraction is smoothly graded at the transition between the two materials. The smooth gradation of the index of refraction causes photons that reach the boundary between the intrinsic layer and one of the extrinsic layers to be refracted back into the intrinsic layer. The smooth gradation of the index of refraction between GaAs and $Ga_{1-x}Al_xAs$ occurs because the Al concentration varies smoothly. Typically the Al concentration varies from 0 in the GaAs to 0.4-0.6 in the $Ga_{1-x}Al_xAs$. *Gain-guided* lasers lack the gradations in the refractive indexes, and simply confine the output Gaussian beam by the physical dimensions of the GaAs gain material. While simpler to construct, gain-guided lasers are much less efficient than index-guided lasers.

As seen in Figure 8.14 the active GaAs region is completely surrounded by $Ga_{1-x}Al_xAs$ so that the laser radiation emitted in 'any' direction in the GaAs is refracted back into the GaAs when it reaches a GaAs-$Ga_{1-x}Al_xAs$

boundary (assuming the angle of incidence is greater than the minimum angle of total internal reflection). Also the mobile electrons and holes are confined to a relatively small rectangular parallelepiped of GaAs so that the threshold current is relatively low. A typical GaAs laser has a threshold injection current of 10-30 mA.

Double heterojunction lasers have many advantages that make them uniquely useful. They are manufactured using the well developed semiconductor production techniques so that they can be produced and sold at moderate prices. The population inversion is produced by minority carrier injection at moderate currents (5-50 mA) using commonly available transistor circuitry. In addition the pump current can be modulated thereby modulating the light output of the laser. The modulation can be very rapid (20 GHz) for use in communications systems. The efficiency of the laser is high with up to 50% of the electric power being converted into laser light power. The lasers are small in size. These properties make the double heterojunction lasers useful for optical communications transmission and optical data handling. They are also useful for applications such as grocery store checkout bar code scanners. Finally the wavelength of the light emitted by the double heterojunction lasers is easily varied so that these lasers have many uses for atomic and molecular spectroscopy. The operation of a tunable GaAs laser is discussed in section 8.2.4 of this chapter.

8.2.2 *Laser Cavity and Mode Structure*

The discussion of the basic double heterostructure laser in the previous section was generally limited to describing the layers of semiconductor materials along the axis of the p-n junction. A device composed of a simple plane of GaAs lacks confinement of either the current density or the laser radiation in the directions perpendicular to the width of the GaAs layer. Instead, GaAs lasers are typically fabricated as a buried structure as shown in Figure 8.14 where the active layer is surrounded on all sides by low index of refraction materials.

The optical cavity of a GaAs laser is formed by cleaving the semiconductor crystal. The reflectance of light transmitted from a material with index of refraction n to air (at normal incidence) is equal to

$$R = \left(\frac{n-1}{n+1}\right)^2 . \tag{8.17}$$

For a glass-air interface ($n_{\text{glass}} \approx 1.5$), this leads to the familiar value of $R = 0.04$, but the index of refraction of GaAs is much higher ($n_{\text{GaAs}} \approx 3.6$)

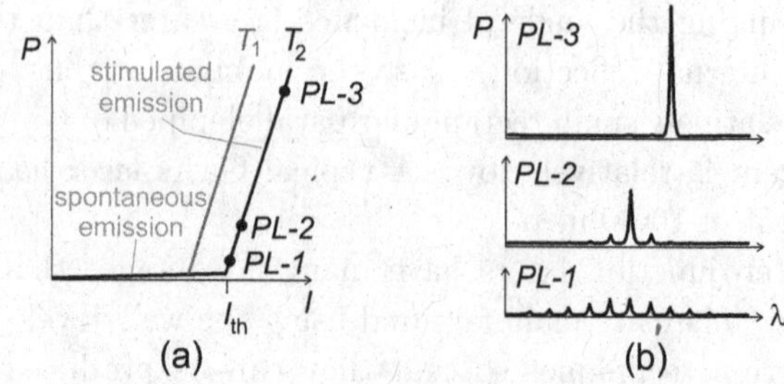

Fig. 8.15 (a) Power output of a diode laser as a function of injection current for two different diode temperatures ($T_2 > T_1$). (b) Dependence of output spectrum on power for three different power levels. Note shift in wavelength and transition towards single-mode operation. This is for a nominal 'single-mode' diode laser. For a 'multi-mode' diode, a few longitudinal modes may still be lasing at high powers.

producing a much larger reflectance of $R \approx 0.32$. Diode lasers are high gain devices so that it is possible to use the bare cleaved surfaces as the cavity mirrors. Nonetheless, a reflective coating is generally applied to the back surface of the diode to increase the photon lifetime in the active GaAs laser region and increase the total power. In fact in order to obtain a high output power an antireflection coating is sometimes put onto the other cleaved surface. It is helpful if the antireflection coating is not perfect so that there is a small amount of feedback to help the laser action. Recall from the discussion on p. 192 that the 'sides' of the optical cavity for an index-guided GaAs laser also relies the low index of refraction of the doped $Ga_{1-x}Al_xAs$ that refracts the light back into the GaAs.

Typical dimensions of the GaAs active layer in a double heterojunction laser range from $(0.1-0.2) \times (3-10) \times (100-1000)$ μm. For a cavity length of $L = 250$ μm, the longitudinal mode spacing (Eq. 7.58) is equal to $\Delta\nu = c/2nL = 1.2\times10^{11}$ Hz, or $\Delta\lambda \approx 0.3$ nm. This spacing is much less than the width ~ 5 nm gain curve so lasing on multiple longitudinal modes is generally possible. This is most evident at low operating currents right above threshold. The line broadening of the recombination process is homogeneous, so at higher currents, lasing is restricted to the longitudinal mode(s) closest to the peak of the gain curve as illustrated in Fig. 8.15. Note that the peak of the gain curve, and thus the emission wavelength, also shifts with the injection current. Depending upon the dimensions of the active layer in the directions perpendicular output beam, operation on multiple transverse modes is likely. Single mode TEM_{00} output can be achieved for small dimensions. Since the active layer of double heterojunction lasers is already very thin (0.1 μm), this restriction mainly applies to the width of the device.

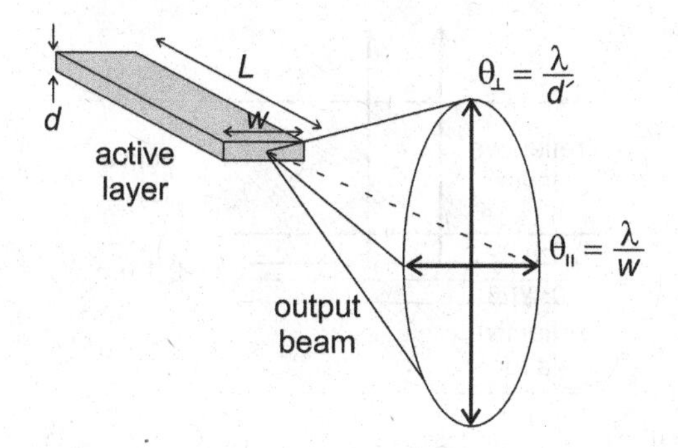

Fig. 8.16 Angular divergence of diode laser output.

The small size of the active layer/wave guide also leads to substantial diffractive spreading of the Gaussian beam in the far field (Sec. 7.1.2). For example, if the width of the active layer in a 780 nm diode is $w = 5\ \mu$m, the far field divergence along that axis is equal to $\theta_w = \lambda/w = 9°$. Further, owing to the different dimensions of the active layer/wave guide along the two directions perpendicular to the beam direction, the angular divergence along the two directions are not equal so the output beam of diode lasers, even those operating in the TEM_{00} mode, is generally not circular but oval shaped as illustrated in Fig. 8.16. Typical divergence angles for 560-800 nm diode lasers for the directions parallel and perpendicular to layered structure are $\theta_{||} \sim 10°$ and $\theta_\perp \sim 30°$. Note that the 30° divergence angle along the vertical axis is not due to the thickness of the active layer d, but the size of the beam as it emerges from the edge of the cavity, d' which is typically about ~1.5 μm. A lens can be used to collimate the beam. A circular beam can be created by using an anamorphic prism pair that magnifies the size of the beam passing through them in one dimension while leaving the size unchanged in the other beam dimension (similar to the prismatic beam expander shown on page 292). A circular TEM_{00} output beam can also be obtained by passing the output of the diode laser through a single-mode optical fiber.

8.2.3 *Other Types of Diode Lasers*

8.2.3.1 *Surface Emitting Diode Lasers*

The light emission in the double heterostructure lasers described in Sections 8.2.1 and 8.2.2 emerges along an axis parallel to the semiconductor

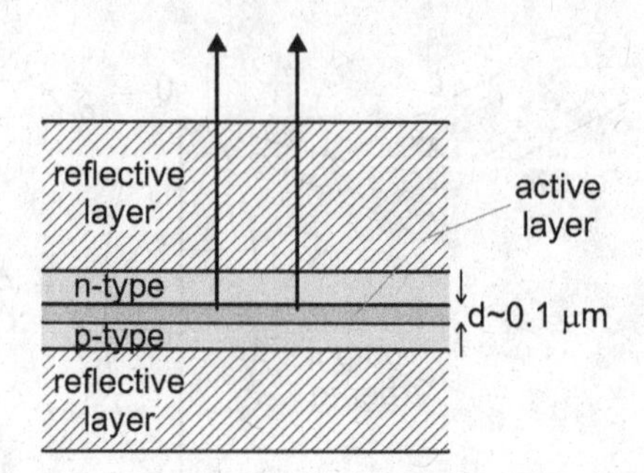

Fig. 8.17 Schematic layout of a vertical cavity surface emitting laser VCSEL. Not to scale, contacts and substrate layers are omitted for clarity.

layers and are thus known as *edge emitting lasers.* Diode lasers can also be constructed so the emission occurs along an axis perpendicular to the semiconductor layers. Such lasers are called *vertical cavity surface emitting lasers* (VCSEL). A schematic representation of the layered structure of a VCSEL is shown in Fig. 8.17. Since the length of gain medium in a VCSEL is equal to the short thickness of the active layer ($L \approx 0.1$ μm), the single pass gain of the laser is much less than in an edge emitter ($L \approx 100$ μm). As a result, highly reflective layers (> 99%) are deposited on either side of the gain medium to increase the photon lifetime. The reflective layers are made up of alternating layers of high and low index of materials with thicknesses equal to a quarter the wavelength of light and are known as a distributed Bragg reflectors (DBR).

8.2.3.2 *Quantum Well Lasers*

For most GaAs lasers the active region is on the order of 100 nm in thickness. For some applications there are advantages in making the active GaAs region as small as 5-10 nm in thickness. In this situation the mobile electrons or holes are confined in a potential well that is small enough that the allowed quantum energy levels for these mobile carriers are no longer a near continuum as they are for a macroscopic size solid, but they are instead discrete. In this situation where the mobile electrons are in discrete energy levels the gain coefficient can be very large, and the threshold injection current can be extremely low. These lasers are called *quantum well* lasers. A typical threshold current or operating current for a quantum well laser is about a factor of ten smaller than for a ordinary double heterojunction laser because the linewidth and hence the density of states is smaller for the quantum well laser than for an ordinary diode laser. Figure 8.18 shows

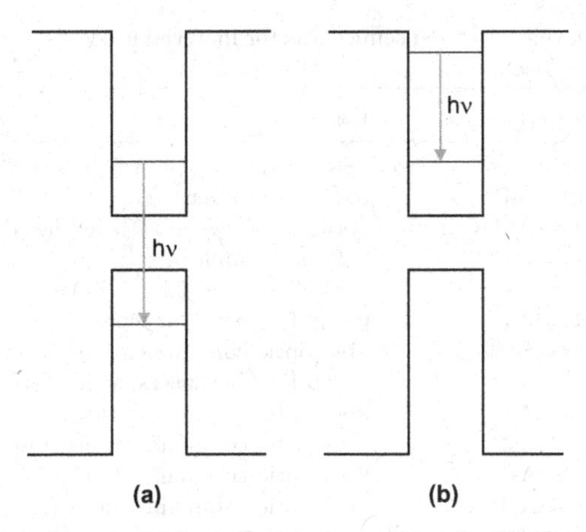

Fig. 8.18 (a) A quantum well laser with only a single level in both the conduction band and the valence band; laser emission is between the two levels. (b) A quantum well laser where there are two levels in the conduction level and the laser emission is between these two levels.

two different quantum well lasers. In one there are two quantum levels, one in the conduction band and one in the valence band, with the laser emission between levels in the two bands, and the other with the two quantum levels in the conduction and with the laser emission between the levels in the same band. There are also situations where there are a large number of distinct levels with emission between the various levels. Quantum well lasers can be constructed both as edge-emitters and VCSELs. Since the active region is quite narrow, the output power of a single quantum well is rather low. To overcome this, *multiple quantum well* (MQW) lasers are fabriacated with multiple identical wells.

8.2.3.3 *Other Laser Materials*

Although the discussion has centered on GaAs this is not the only material used to create semiconductor lasers. Table 8.1 lists the semiconductors used in a number of commonly available diode lasers. For example, the semiconductor material $In_xGa_{1-x}As_{1-y}P_y$ can be operated as a laser in the wavelength range from 1.1-1.6 microns. The entire field of semiconductor diode lasers is currently one where research is very actively altering and improving the status of the available lasers and in the future semiconductor lasers are sure to be available at wavelengths and at power levels where they are not commercially available today. Diode lasers, and lasers that can be pumped by diode lasers such as Nd:YVO_4, glass fiber, and Ti:sapphire

Table 8.1 Wavelengths and semiconductor material used in some of the most common diode lasers.

λ (nm)	Material	Use
405	InGaN	optical storage (Blu-ray/DVD-HD)
635	AlGaInP	red laser pointer
650	GaInP/AlGaInP	optical storage (DVD), red laser pointer
670	AlGaInP	bar code scanner
785	GaAlAs	optical storage (CD)
808	GaAlAs	pump for solid state lasers
850	InGaAs/GaAs	fiber optic communications
980	InGaAs	pump for fiber lasers, solid state lasers
1064	AlGaAs	fiber optic comm., 'substitute' for Nd:YAG
1310	InGaAsP	fiber optic communications (O-band)
1550	InGaAsP	fiber optic communications (C-band)
1625	InGaAsP	fiber optic communications (L-band)

lasers have now displaced most of the earlier types of gas lasers which are covered in Chapter 11. The use of diode lasers for pumping other lasers has been made possible by the development of small, highly reliable diode lasers that produce very high powers (1 W or greater). High power diode lasers or diode arrays are further addressed in Section 8.2.5.

8.2.4 *Tunable Diode Lasers*

Diode lasers are of great importance for use in physics, chemistry, and engineering research as *tunable* lasers for use in spectroscopic applications. In the previous section it was stated that GaAs diode lasers operated in the wavelength range 650-880 nm. This should not be interpreted as meaning that a given laser can operated at all wavelengths in this range. A given laser can only operate over a much smaller wavelength range. A typical upper level lifetime for a diode laser is 10^{-9} s, and a typical bandwidth is 5×10^{12} Hz or about 10 nm. In order to cover the entire range 650-880 nm it is necessary to utilize several different lasers with different center frequencies for each laser. The center frequency of a diode laser depends on the composition of the semiconductor material. In the heterostructure GaAs diode laser discussed in the previous section the active medium was taken to be pure intrinsic GaAs. If an atom similar in chemistry and size such as P is substituted for a small fraction of the As then the center frequency of the laser is shifted. The shift depends on the concentration of the P. Thus by varying the composition of the lasing material the center frequency of the laser can be varied. It is possible to purchase diode lasers with different

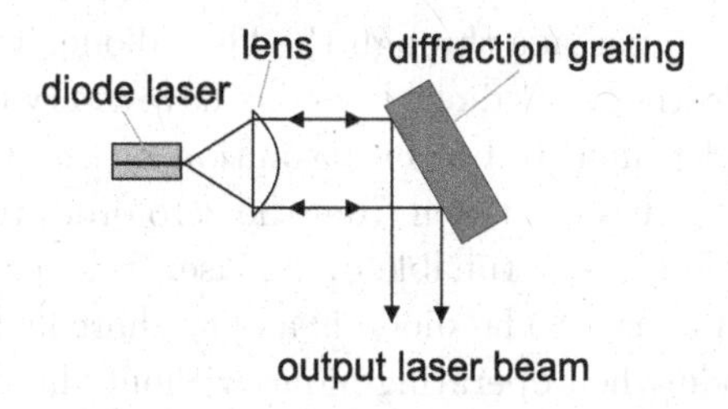

Fig. 8.19 A schematic diagram of a tunable diode laser setup (Littrow configuration).

center frequencies. There are wide variety of diode lasers with different center frequencies. Normally one can only purchase a diode with a frequency near the desired frequency. After one obtains a diode laser with a center frequency near the frequency desired it is usually necessary to adjust the center frequency of the laser to obtain the exact center frequency desired. This can be done as follows. The center frequency of a diode laser also depends on the current through and the temperature of the diode (see Fig. 8.15). In order to vary the frequency of a diode laser by a significant amount one must change either the temperature of the diode or change the current through the diode that changes the diode's temperature. Changing the diode temperature changes both the length of the diode and its index of refraction so that the diode frequency changes. By changing the temperature or current through the diode (or both) one can obtain the frequency desired.

There are two principal methods of setting up a tunable diode laser. Figure 8.19 shows a schematic diagram for the first method for setting up a tunable diode laser. The diode laser has a reflective coating at one surface and an antireflection coating at the other surface. A lens is used to collimate the laser beam. The output beam from a laser is divergent due to diffraction since it is emerging from a source of small size (Sec. 8.2.2). The lens typically has a speed of about $f/1$ and is positioned about one focal length from the antireflection coated surface from which the laser beam emerges. The diffraction grating is used in a Littrow mounting so that the wavelength, λ, of the light that is diffracted back into the active material satisfies the condition

$$n\,\lambda = d\,(\sin\theta + \sin\theta_d) = 2d\,\sin\theta \qquad (8.18)$$

where n is the order number, d is the grating spacing, θ is the angle of incidence which is also equal to the angle of diffraction, θ_d. The light of wavelength λ that is diffracted back into the active medium is amplified during its round trip passage through the laser diode. In the steady state the light emitted by the GaAs diode laser can have a very narrow bandwidth that is determined primarily by the resolution of the diffraction grating. The output from the laser is taken from the zero order reflection.

In order to understand a tunable diode laser it is necessary to consider the modes of such a laser. The diode is a very short high gain amplifying medium. The diode when operating alone without the diffraction grating has longitudinal modes that are widely separated since the modes are separated by $c/2nl$ where l is the very short length of the diode laser. These modes are separated by a few tenths of a nm. For example for a length of $l = 0.5$ mm and an index of refraction of 3.5 the mode separation is $\Delta\nu = c/2nl = 8.6 \times 10^{10}$ Hz which leads to $\Delta\lambda = 0.2$ nm. The diode laser alone typically operates on one or two modes of the diode that are closest to the center frequency of the gain curve. The complete laser has an optical cavity consisting of the path from the reflective surface on the backside of the diode to the diffraction grating. The optical cavity of the complete laser has a length L that is much larger than l so that the longitudinal modes of the optical cavity are much closer together than are the modes of the diode alone. For example if one takes the length of the optical cavity using the diffraction grating as $L = 10$ cm and the average index of refraction as 1 (since most of the cavity in now in air) then the mode separation is $\Delta\nu = c/2nL = 1.5 \times 10^9$ Hz, which is much smaller than the mode separation of the diode alone. The complete laser operates on one or several of the longitudinal modes of the complete optical cavity that are within the frequency width of the mode of the diode alone that is nearest the center of the gain curve. The output from the diode laser is commonly obtained by taking the zeroth order reflection from the diffraction grating. It is also possible to extract an external beam by the use of a beam splitter located between the diode and the diffraction grating. Many other cavity designs for tunable diode lasers using a diffraction grating are also used. Some of these are discussed in Sec. 12.2.1 in reference to tunable dye lasers.

Diode lasers of this type can be tuned over a range of about one nm by varying the angle of diffraction of the grating and at the same time changing the current through the diode laser. In order to tune the laser over a larger range it is necessary to vary the temperature of the diode. Typically a diode laser's wavelength varies by about 0.3 nm/°C due to thermal changes in the length and the index of refraction of the GaAs laser material. Thus a change in the temperature of 60°C will change the wavelength by 18 nm. It is possible to change the temperature of a diode

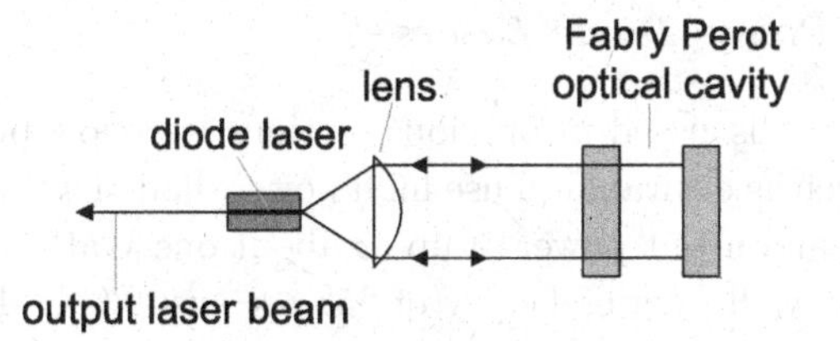

Fig. 8.20 A schematic diagram of a tunable diode laser setup that uses a Fabry Perot etalon for the tuning element.

laser by the use of Peltier coolers. One can also heat the diode to change its temperature. It is, however, the author's experience that if the diode is heated above room temperature by more than about 25°C the lifetime of the diode is significantly shortened. Tunable diode lasers of this type are used for a wide variety of atomic and molecular physics experiments where it is necessary to set a laser to the wavelength of an atomic or molecular transition.

The second method of setting up a tunable diode laser is shown schematically in Fig. 8.20. In this setup the diode laser has antireflection coatings on both surfaces. The light from the diode is collimated by a lens and is incident on a Fabry Perot cavity. The mirrors for the Fabry Perot cavity are coated so that they have a relatively low reflectivity. There are wavelengths where the light incident on the Fabry Perot cavity is completely transmitted. At these wavelengths there is no reflected wave. There are other wavelengths where the Fabry Perot cavity has only a low transmission. At these wavelengths there is relatively high reflected wave. The reflected wave is amplified as it passes through the diode laser. The laser oscillates at a single wavelength corresponding to the wavelength with a high reflected intensity nearest the maximum of the diode gain curve. It is possible to vary the wavelength of a diode laser operating with this type of setup by changing the spacing of the Fabry Perot cavity and the current through the diode at the same time in such a way that the change in wavelength of the of the cavity matches the change in wavelength of the diode gain curve.

Tunable diode lasers are used in physics, chemistry, and engineering research for applications as varied as absorption spectroscopy or atom trapping. Diode lasers are among the most important research lasers. With further development of diode lasers it is clear that the applications of diode lasers in basic research will continue to expand.

8.2.5 *High Power Diode Lasers*

The diode lasers discussed in previous sections were low power lasers. For many purposes it is desirable to use high power diode lasers. A single diode laser can have an output power of up to about one Watt. Diode lasers can have an output wall plug efficiency of 25% or even 30%. This means that they can take 4 W of electrical power from the wall plug and convert it into 1 W of light. This is an extremely high efficiency compared most to other possible light sources. Also diode lasers have a very long lifetime provided they are not exposed to sudden power surges, too high temperatures, or other environmental problems. Diode lasers can be used directly for high power applications, and they have significant advantages over flash lamps for pumping solid state lasers (known as DPSS - diode pumped solid state lasers) or glass fiber lasers. For pumping solid state lasers diode lasers have higher efficiency, longer lifetimes and higher reliability than flash lamps. Probably the only real disadvantage of using diodes for pumping solid state or fiber lasers rather than flash lamps is the higher initial cost for the high power laser setup.

Single diode lasers can be made to operate with output powers up to about 1 W. If 1 W is sufficient power to pump another laser then the system can be set up as shown schematically in Fig. 8.21. In Fig. 8.21(a) the solid state laser is pumped with the pump light entering the solid state laser active medium from the end. In Figure 8.21(b) the pump laser enters the solid state laser active medium from the side. For pumping from the end of the solid state laser rod the absorption coefficient of the laser rod should be such that the pump beam pumps the active ions through out the entire rod. For pumping from the side it is important that the absorption coefficient due to the active ions be such that the pump light is absorbed

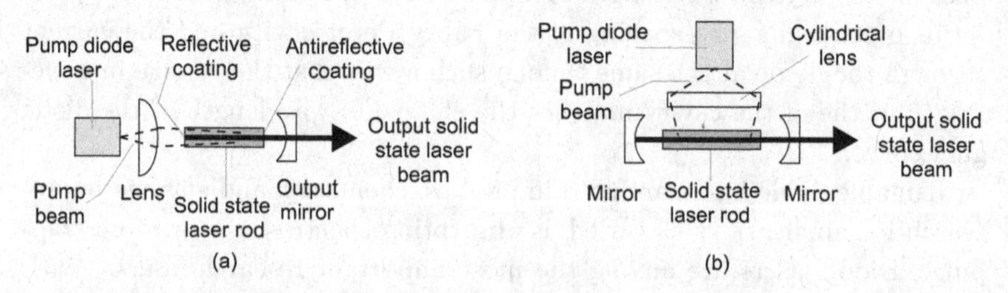

Fig. 8.21 (a) The use of a high power diode to pump a solid state laser crystal from the end. (b) The use of a high power laser to pump a solid state laser crystal from the side.

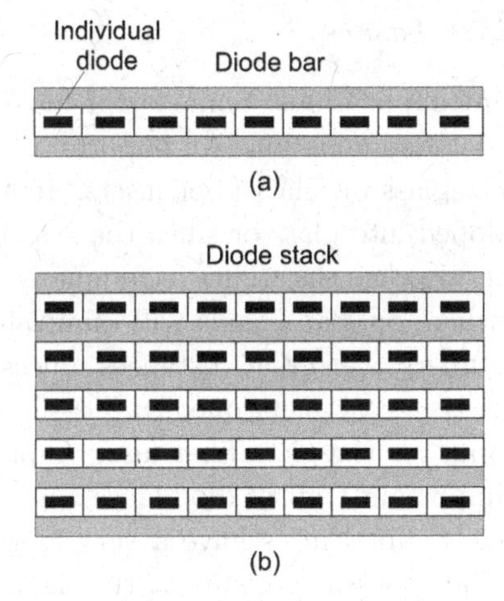

Fig. 8.22 (a) A schematic diagram of a diode bar as seen looking into the lasers. The black rectangles are the laser outputs. The grey area is for cooling the diodes. (b) A schematic diagram of a diode stack.

going through the diameter of the rod. Although the figure shows the side pumping from only one side it is possible to pump the laser from opposite sides or even from more than two directions.

For applications that require a pump power greater than about 1 W it is necessary to use more than one diode laser. Diode bars are produced with many laser diodes on a single semiconductor chip. Figure 8.22(a) illustrates schematically a laser diode bar. The dark rectangles represent the diode lasers with the laser beam emerging out of the page. Figure 8.22(b) illustrates a laser diode stack. Diode laser bars and stacks can emit hundreds or even kilowatts of laser power. For these high output powers it is necessary to provide active cooling for the bars or stacks. Water cooling is often used to maintain the bars or stacks at a stable temperature that is low enough that the semiconductor diodes are not damaged. Obviously a diode bar or a diode stack is suited for side pumping of a long solid state laser crystal or a glass fiber laser. The diode bar should be selected to be roughly the same length as the laser crystal or laser glass fiber. The use of a diode bar to pump a laser crystal or laser glass fiber results in a high power output from the solid state laser or glass fiber laser. Laser bars and stacks can be purchased with fiber coupling from the output of the individual lasers.

8.2.6 *Uses of Diode Lasers*

Diode lasers have many applications. Some of the most important applications are in the field of communications An important applications of diode lasers is to pump Er^{3+} glass or silica fiber lasers. It has been found that when Er^{3+} ions are doped into glass or silica the system can amplify light signals at appropriate wavelengths. This technique is now widely used in telecommunications where optical signals will diminish as they are transmitted through silica fibers over great distances unless they are amplified along the way. Diode lasers, as mentioned, are easily modulated and are used for the signal input for the glass fiber laser amplifiers. In this application another very important trait of diode lasers is critical. Diode lasers operated at reasonable temperatures have a very long operating lifetime. Without the long lifetime constant repairs to the signal lasers would make long distance communications much more difficult and expensive. Indeed the modern internet might be impossible. Applications such as streaming video would be very difficult.

Many consumer electronics also use diode lasers. CD players use a λ = 785 nm diode, DVD players a λ = 650 nm diode, and Blu-ray/HD DVD players and recorders use a λ = 405 nm GaN laser diode. Again the long lifetimes of laser diodes is essential, and it is very impressive how long devices such as CD players operate in regular use.

As mentioned diode lasers are regularly used for scientific experiments where the diodes are often used to excite particular atoms or molecules into a desired excited level. Of special note, widely available 780-785 nm laser diodes used for CD players/recorders can be modified to work at the 780.24 nm wavelength of the rubidium resonance line ($5\,^2S_{1/2} - 5\,^2P_{3/2}$). This has promoted Rb to be a choice atom for optical pumping as well as atom trapping and cooling experiments since the widespread availability of low cost diode lasers starting in the late 1980s. Prior to this, the choice alkali atom for many similar studies had been sodium since the D_1 and D_2 lines (589.59 and 589.00 nm) had wavelengths near the peak efficiency of the Rhodamine 6G dye widely used in tunable dye lasers as described in Chapter 12.

8.3 Combining Laser Beams

There are a number of applications of high power lasers. For example industrial uses such as the cutting and machining of metal or military applications. For these applications CO_2 lasers, fiber lasers, or solid state lasers are commonly used. Both fiber and solid state lasers are usually pumped by diode lasers. For all applications diode lasers have the advantages of high efficiency, low cost, high reliability, compact size and a relatively wide wavelength availability. Nonetheless, the output of a single diode laser is limited. To produce a high power, the output from large number of individual units needs to be combined together.

This section discusses methods related to combining laser beams. An excellent reference on the problems of combining laser beams has been written by T. Fan (see Suggested Additional Readings). The problems in combining beams are somewhat different for incoherent laser beams and for coherent beams. The term incoherent as used in this context requires a bit of explanation. Each individual laser beam has a significant longitudinal and transverse coherence. The beams, however, may have different wavelengths. Any one particular beam has no coherence with the other beams. These beams are called incoherent beams. If the beams have the same wavelength and the same relative phase at all times they are called coherent beams. The discussion begins with the problems in combining incoherent beams. This is referred to as wavelength combining of beams.

The output beam quality from a single diode laser is usually not particularly good. The use of a lens to combine the individual output beams from a diode bar or stack into a single beam results in a large output power, but an even poorer quality output beam. A high beam quality is required for applications such as laser machining, laser welding, or any applications where high power focused into a very small spot is needed. Hence, the raw output of a diode bar array is generally not useful in these applications. One way around this problem is to use the combined output of the diode bar to pump another laser that has good beam quality. For example, a diode-pumped Nd:YAG laser. While this is certainly a feasiable option, this pumping is never completely efficient. This method naturally results in a lower efficiency than producing a beam of good quality directly using diode lasers.

The problem of combining the output of two or more diode lasers into a high quality beam is a longstanding, important, and difficult problem. The objective is to combine two or more laser beams so that they are going in the same direction along the same axis and have nearly the same diameter. This means taking two laser beams and deflecting them so that they overlap and are going in the same direction. It should be clear that each of the diode laser beams must have a bandwidth that is small compared

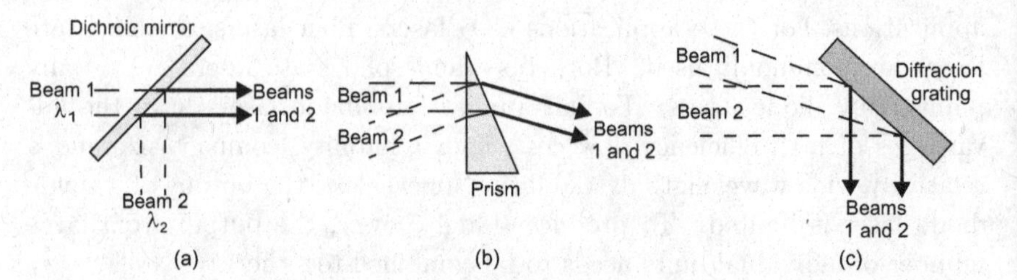

Fig. 8.23 Three possible ways to combine laser beams of *different* wavelengths. (a) the use of a dichroic mirror to combine the two laser beams. The dichroic mirror must pass without reflection beam 1 and reflect beam 2 in order for the system to work perfectly. (b) The use of a prism to combine two laser beams. (c) The use of a diffraction grating to combine two laser beams.

to the diode gain bandwidth in order that the beam quality of each beam has high quality. Otherwise the individual diode lasers will have a poor beam quality and it is not possible to take a large number of low quality beams and combine them into a single high quality beam. The beams can be deflected in various ways including using dichroic mirrors, prisms or diffraction gratings. Figure 8.23 illustrates schematically three possible methods for combining laser beams. In order for any of these methods to work well the wavelengths from each of the laser diodes must be stable. For example if one uses a diffraction grating to combine the beams the wavelengths of the two beam and the angles of incidence of the two beams must be such that the two beams are deflected in the same direction. If one of the diode laser beams changes its frequency then the two beams will no longer be deflected in the same direction as the other beam. In the same way if a prism is used to deflect the beams into the same direction then if the wavelength of one of the beams changes then the beams are no longer deflected into the same direction. In the case of the use of the dichroic mirrors if one of the lasers changes its wavelength the output intensity is altered because the reflection coefficient of the dichroic mirror is wavelength dependent. If one desires to wavelength combine a large number of diode lasers then the problem of maintaining the correct wavelength for each of the lasers is a major problem.

A clever solution to this problem has been found by Daneu *et al.* and Chann *et al.* and is illustrated in Fig. 8.24. In this figure two diode lasers are shown, and the output of the two lasers is also shown. The beams from two lasers are focused by a lens onto a diffraction grating (not shown are additional micro- lenses that collimate the output of the individual diode lasers). The lens transforms the small lateral displacement Δx of the two

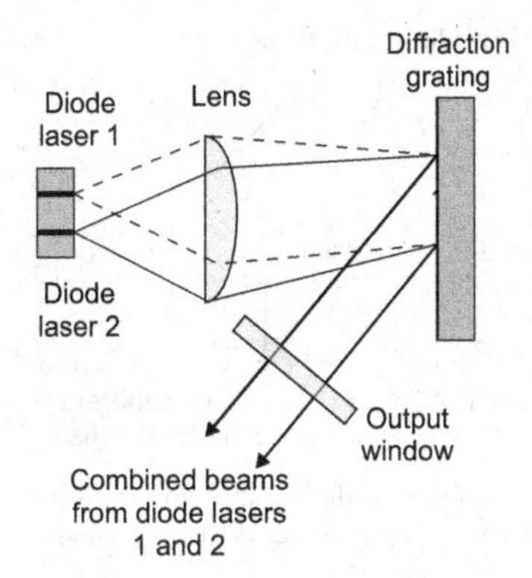

Fig. 8.24 A method to combine many laser beams. Two lasers are shown operating with different wavelengths and with different angles of incidence on the diffraction grating so that they are diffracted in the same direction. The output window also serves to reflect a small part of each beam back into the laser diode it originally came from where it is amplified as it double passes the laser. This forces each laser to operate at the correct wavelength for beam combining.

lasers into different angles of incident onto the grating, θ_i. If the two lasers operate at slightly different wavelengths λ_i, the two beams can be made to be diffracted at the same angle λ_{out}, and will overlap and be going in the same direction. Assuming the diffraction grating is operating in first order, this requires

$$\lambda_i = d\left(\sin\theta_i + \sin\theta_{\text{out}}\right) \tag{8.19}$$

where d is the grating spacing. The question is how to get the diodes to lase at the correct wavelength and to remain at this wavelength. This problem is solved by using an output window that reflects a small fraction of the each laser beam back on itself. The reflected beams are diffracted back to the individual laser from which the beam started. This feedback is reflected from the end mirror of the diode laser and amplified as it double passes the laser. This causes the laser to lase at exactly the right wavelength, λ_i, to be diffracted toward the output window. This feedback is much like feedback for the tunable dye laser discussed in Section 8.2.4. Although Fig. 8.24 shows only two diodes it is possible to combine a large number of diode laser beams in this manner. A typical laser diode bar has a string of 10-100 individual diodes with a total width of about a 1 cm. Assuming the lens is placed one focal length f away from both the diode array and the diffraction grating, $d\theta = dx/f$ and differentiation of Eq. 8.19 yields

$$\Delta\lambda \approx d\,\cos\theta\left(\Delta x/f\right)\ . \tag{8.20}$$

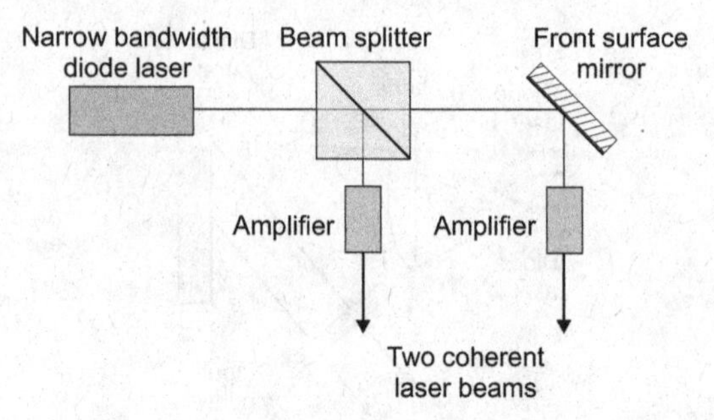

Fig. 8.25 The use of a beam splitter followed by two amplifiers to produce a two coherent beams, however, the two laser beams may be shifted in phase with respect to each other.

Thus, for a lens with $f = 15$ cm, a $1/d = 1800$ groove/mm diffraction grating, and $\approx 50°$ angle of incidence θ the wavelengths of two lasers at opposite ends of the diode bar are required to differ in wavelength by 24 nm. Alternatively, if instead of using a diode bar, the output from hundreds or thousands of individual fiber coupled diodes are incident on the diffraction grating so that the feedback from the output window is reflected back onto the particular fibers and hence into the particular diode laser, from which they were emitted, the beam will be amplified as it double passes the diode laser. The output of the diodes is combined to produce a high output power. Using this technique the high output power from a large number of diode lasers can be combined and can have high beam quality and brightness. The combined beams can be focussed into a very small spot. The wavelength combined beams do not have good longitudinal coherence, but they do have excellent transverse coherence.

This is an appropriate point to discuss briefly the concept of beam brightness, B. The brightness or radiance of a laser is defined as the power divided by the minimum spot size times the square of far field divergence angle. For a Gaussian beam the minimum spot size is $\pi\omega_0^2$ and the square of the far field divergence is $\lambda^2/\pi\omega_0^2$ so that the brightness of a Gaussian beam is given by $B = P/[(\pi\omega_0^2)(\lambda^2/\pi\omega_0^2)] = P/\lambda^2$. The brightness or radiance of a beam is a measure of the power density at the focus of the beam. For a non-Gaussian beam the brightness is reduced from that of a Gaussian beam by an amount that depends on the quality of the beam.

The next problem to be addressed is how to combine two or more coherent beams. The first thing required is the question of how to produce two coherent beams. One can produce a coherent beam using a narrow bandwidth diode laser or another type of narrow bandwidth laser. The output

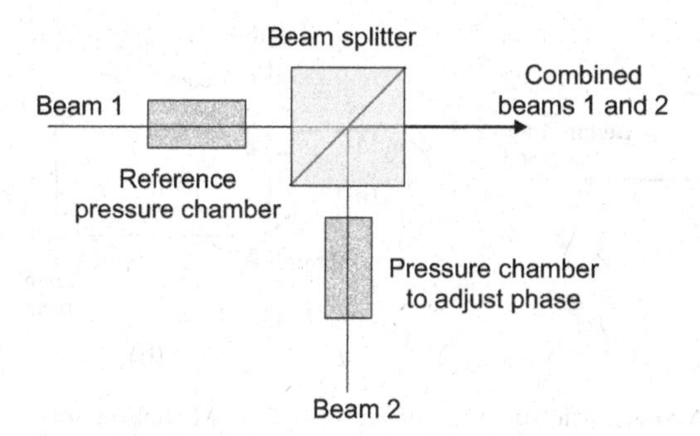

Fig. 8.26 The use of a beam splitter to combine two laser beams. The reference pressure chamber is to avoid the need to have the pressure in the adjustable pressure chamber be varied by too great an amount in order to combine the two laser beams constructively.

of the narrow bandwidth laser can then be split using a beam splitter. The two beams emerging from the beam splitter are coherent. Often the output from the beam splitter is weaker than desired for some application. If so, the two beams can be passed through amplifiers resulting in two intense coherent beams as illustrated in Fig. 8.25.

In order to combine these two coherent beams constructively one can use a beam splitter as shown in Fig. 8.26. In order for the combined beams to interfere constructively it is necessary that the two waves emerging from the beam splitter have the same phase. The waves emerging from the two amplifiers may not have exactly the same phase and since phase shifts in the optical wave will occur as the beams pass through the glass in the beam splitter, on reflection, and in the two different amplifiers it is necessary to have some adjustment in the phase of one of the laser beams. This can be done in a number of ways. One possible method of adjusting the phase of one of the beams is to pass the beam through a different optical path length than the other beam. This can be done in a variety of ways one of which is by passing the beam through a cell where the gas pressure can be changed thereby changing the index of refraction as illustrated in Fig. 8.26. More typically, an interferometer is used for the combining of two coherent beams. For example, the use of a Mach Zender interferometer to combine two coherent laser beams is illustrated schematically in Fig. 8.27(b). Note that some variation in the phase of one beam is needed to assure constructive interference in the output beam. For example, if the two beam splitters are dichroic then it is possible to have one laser beam go along one path and the other beam go along the other path. In this case a small variation in the phase (i.e., by varying the index of refraction) along one path can be used to force constructive coherent in the output beam.

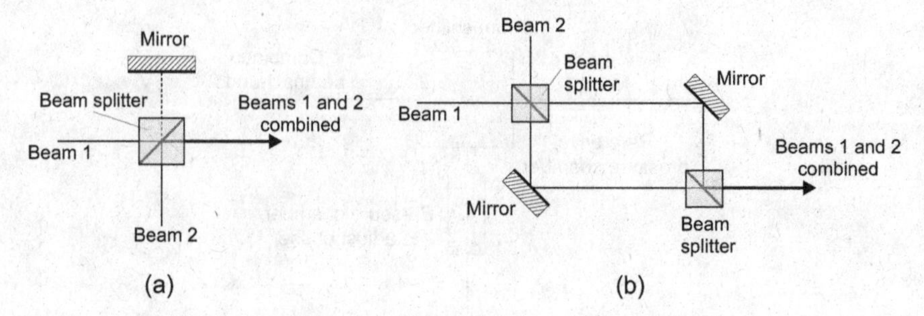

Fig. 8.27 (a) A schematic diagram of the use of a Michelson interferometer for the combining of two coherent laser beams. The mirror is adjusted so the two beams combine constructively in the output channel. (b) The use of a Mach Zender interferometer to combine two coherent laser beams.

Summary of Key Ideas

- Energy levels in a solid are split into allowed bands separated by forbidden gaps. Insulators have a filled valence band and empty conduction band. Conductors have a partially filled valence band. Semiconductors have an empty conduction band at absolute zero, but the forbidden band gap is small enough that some electrons are excited into the conduction band at room temperature (Sec. 8.1.2-8.1.3).
- Pure GaAs is an intrinsic semiconductor with a direct band gap of 1.43 eV. GaAs can be doped to produce extrinsic semiconductors. An n-type semiconductor is one with donor impurities that has energy levels that are easily ionized forming positive ions and populating the conduction band. A p-type semiconductor has acceptor impurities that easily form negative ions leaving holes in the valence band (Sec. 8.1.4).
- A p-n junction is formed when adjacent parts of the same semiconductor are doped with p- and n-type impurities. When a p-n junction is biased so that current flows the free electrons or holes can flow into the junction where they recombine resulting in the emission of light. The process where an electron enters a p-type material (or a hole enters a n-type) is called minority carrier injection (Sec. 8.1.5). A semiconductor diode laser uses minority carrier injection to produce a population inversion at moderate currents (Sec. 8.2.1).
- By changing the composition of dopants to the semiconductor material, GaAs diode lasers can be constructed with center wavelengths in

the range 650-880 nm. Other semiconductor materials are used to obtain emissions at other wavelengths (Sec. 8.2.3). By varying the diode current and temperature, the center wavelength of an individual laser can be shifted by ~5 nm (Sec. 8.2.4). An external cavity diode laser can produce a very narrow band width laser with a tunable wavelength (Sec. 8.2.4).

- Diode lasers have many uses including optical communication and optical data storage (CD, DVD) (Sec. 8.2.6). Diode laser bars and stacks are obtained by combining individual diodes in one and two dimensional arrays. High power lasers are used as a pump for solid state lasers (DPSS) (Sec. 8.2.5). Multiple laser beams can be combined both for coherent and incoherent beams (Sec. 8.3).

Suggested Additional Reading

Charles Kittell, *Introduction to Solid State Physics, 8th Ed.*, John Wiley and Sons (2004).

W. Koechner and M. Bass, *Solid-State Lasers: A Graduate Text, 6th Ed.*, Springer (2003).

William T. Silfvast, *Laser Fundamentals*, Cambridge University Press (1996).

A. E. Siegman, *Lasers*, University Science Books (1986).

D. Sands, *Diode Lasers*, CRC Press (2004).

L. A. Coldren, S. W. Corzine, and M. L. Mashanovitch, *Diode Lasers and Photonic Integrated Circuits*, John Wiley and Sons (2012).

J. Buus, M.-C. Amann, and D. J. Blumenthal, *Tunable Laser Diodes and Related Optical Sources, 2nd Ed.*, Wiley-IEEE Press (2005).

B. E. A. Saleh and M. C. Teich, *Fundamentals of Photonics*, John Wiley and Sons (1991).

T. Y. Fan, "Laser beam combining for high-power, high-radiance sources" *IEEE Journal of Selected Topics in Quantum Electronics* **11**, 567 (2005).

V. Daneu, A. Sanchez, T. Y. Fan, H. K. Choi, G. W. Turner, and C. C. Cook, "Spectral beam combining of a broad-stripe diode laser array in an external cavity" *Opt. Lett.* **25**, 405 (2000).

B. Chann, R. K. Huang, L. J. Missaggia, C. T. Harris, Z. L. Liau, A. K. Goyal, J. P. Donnelly, T. Y. Fan, A. Sanchez-Rubio, and G. W. Turner, "Near-diffraction-limited diode laser arrays by wavelength beam combining" *Opt. Lett.* **30**, 2104 (2005).

Problems

1. What wavelength does the GaAs forbidden band gap of 1.43 eV correspond to? A GaAs diode laser operates at $\lambda = 800$ nm at room temperature. Estimate the temperature at which the laser will operate at 788 nm.

2. If a GaAs laser operates at $\lambda = 800$ nm and if the GaAs sandwich is 10 μm across estimate the divergence of the laser beam in the far field.

3. Calculate the wavelength of an electron at the Fermi surface in a one dimensional solid with an internuclear spacing of 0.25 nm and with one valence electron per atom.

4. It was stated that the average separation of energy levels in a three dimensional solid was of the order of magnitude of 10^{-21} eV. Over what distance does an electric field of 3 V/m have to act to impart an energy of 10^{-21} eV to an electron?

5. For pure GaAs the forbidden band gap is 1.43 eV. If the Fermi energy is midway between the bottom of the conduction band and the top of the valence band what is the probability that a state at the bottom of the conduction band is occupied at room temperature?

6. For the one dimensional solid of Problem 3 calculate the lowest value of the $\boldsymbol{k}$ vector for which there is a band gap.

7. At thermal equilibrium the population of the quantum states is determined by the Fermi Dirac distribution function. Using this, discuss why the mobile carrier density in the depletion layer is so much less than the mobile carrier density in the bulk material.

8. Sketch the design of a tunable diode laser for use at $\lambda = 800$ nm that uses a diffraction grating in the optical cavity. Select a lens with a suitable focal length and a diffraction grating with a suitable number of grooves per cm for your design. Discuss what considerations must go into the tuning of the laser. In particular discuss how one can change the temperature in order to change the wavelength of the laser.

9. If a GaAs layer is 10 nm in width calculate the quantum energy levels of the quantum well. What is the energy separation of the first excited level

and the ground level? You may assume that the effective electron mass in the solid is equal to the free electron mass.

10. If a GaAs diode laser is 100 μm in length and if the index of refraction of GaAs is 3.64 what is the separation of the modes of the cavity formed between the two ends of the semiconductor material?

Chapter 9

Solid State Impurity Ion Lasers

There are a number of fixed frequency solid state lasers that make use of impurity ions doped into insulating crystals or glass as the active medium. This chapter discusses four such lasers: the ruby laser, the Nd:YAG laser, the Nd:glass laser, and glass or silica fiber lasers. In addition this chapter discusses one tunable impurity ion laser, the Titanium:sapphire laser (the Ti:sapph laser). Solid state lasers have several important features that make their use desireable. The density of the doped impurity ions can be quite high and is adjustable during crystal growth. The output powers from impurity ion lasers can be very high. Because of the high concentration of impurity ions the laser crystals can be of moderate size, typically a few centimeters in length. Most solid state impurity ion lasers can be pumped by diode lasers. The excitation density can be controlled by selecting a crystal with the appropriate concentration of impurity ions and by adjusting the optical excitation power. Solid state impurity ion lasers do, however, have a few disadvantages. If the solid state material is not uniform in its properties including uniform doping then the wave front of the laser beam will be distorted. In addition for very high power operation if the solid state material absorbs even a small fraction of the power in the laser beam then it may heat up and change its properties. If the heating is severe or very non-uniform then the solid state material may fracture and a new crystal must be obtained.

In general the fields due to the crystal or glass affect the energy levels of the doped ions so that the energy levels are different from those of a free ion. While it is reasonable to discuss in some detail the energy levels of atoms like He or Ne in a text on lasers a treatment of the splitting and shifting of the energy levels of impurity ions is a subject that would require too great length. Thus the energy levels of the impurity ions important for lasers are merely discussed without describing in detail how these energy levels have arisen from the energy levels of the free ions.

9.1 Ruby Laser

The first laser that was constructed was a ruby laser. Ruby, which is also called corundum, is a sapphire crystal containing Cr^{3+} impurities. The Cr^{3+} impurity ions form the active material for the laser. Sapphire is an Al_2O_3 crystal, which has a hexagonal crystal structure. The Cr^{3+} impurity ions replace the Al^{3+} ions substitutionally. The Cr^{3+} ions absorb both in the blue and green regions but not the red region of the spectrum so that a ruby crystal has its characteristic red or pink appearance. The Cr^{3+} ion number density for ruby a laser is typically $1 - 2 \times 10^{19}$ cm^{-3}. Ruby crystals are grown commercially to large sizes and with a wide range of Cr^{3+} impurity ion densities.

The relevant energy levels of the Cr^{3+} ion in a ruby crystal are shown in Figure 9.1. The ground level of the ions is denoted by $^4A_2$ and although it is shown as a single level it is in fact made up of two energy levels that are separated 0.38 cm^{-1}, an energy too small to be seen on the scale of the energy level diagram of Figure 9.1. There are two broad energy levels called $^4F_2$ and $^4F_1$ centered about $16{\times}10^3$ and $25{\times}10^3$ cm^{-1} above the ground level. Absorption from the ground level into these two broad levels is responsible for the absorption by ruby in the green and blue regions of the spectrum. Both of the two broad levels decay rapidly via non-radiative transitions. The decay rate for these non-radiative decays is about 2×10^7 s^{-1}. The

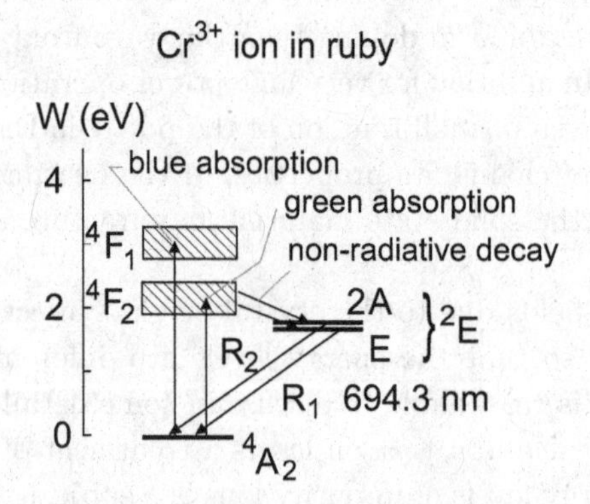

Fig. 9.1 The energy levels of the Cr^{3+}ion in a ruby crystal. The energy is given in eV above the ground level of the Cr^{3+} ion.

non-radiative decay results in the population of Cr^{3+} ions in a closely spaced pair of levels denoted by $2\bar{A}$ and $\bar{E}$ and separated by 29 cm^{-1}. The pair of levels $2\bar{A}$ and $\bar{E}$ is denoted by ^2E. The $2\bar{A}$ and $\bar{E}$ levels decay radiatively to the ground level with a spontaneous lifetime of about 3 ms. The transitions from the $2\bar{A}$ and $\bar{E}$ levels to the ground level are called the R_2 and R_1 transitions respectively. Laser action can occur on both the R_1 and R_2 transitions, but usually laser action occurs primarily on the R_1 transition at 694 nm. The radiative decay rate of the upper laser level, the $\bar{E}$ level, is 330 s^{-1}.

The ruby laser is essentially a three level laser. It is necessary to populate the upper laser level more highly than the lower or ground level for a three level laser. Despite this difficulty the first laser to be constructed was in fact a ruby laser. The population inversion in a ruby laser can be produced by absorption of light from a pump flash lamp exciting the Cr^{3+} ions into the two broad $^4F_2$ and $^4F_1$ levels that then decay rapidly to the upper laser level. The broadband absorption cross sections are large enough that it is possible to produce either cw or pulsed laser action in ruby. The ruby laser is most commonly operated in a pulsed mode.

There are several ways to construct a ruby laser. In the simplest design the ends of the ruby crystal rod are polished and coated with a highly reflecting material. Thus there are no external mirrors and the ruby rod with the reflecting ends constitute the optical cavity. The ruby rod is inserted along the axis of a spiral flash lamp. The flash lamp is activated by a rapid discharge that results from dumping the output from an energy storage capacitor through the flash lamp. The output light from the flash lamp occurs for a duration of about 5×10^{-4} s. The duration of the light from the flash lamp is much shorter than the lifetime of the upper laser level so that almost all the excited Cr^{3+} ions end up in the ^2E levels. The ruby rod is heated by the absorbed light. For high power operation this can be a severe problem. It is common to flow a cooling fluid past the ruby rod in order to carry off the excess heat.

More sophisticated ruby lasers use external mirrors. In a ruby laser with external mirrors the ends of the ruby rod are orthogonal to the laser beam axis, are polished flat and, are coated with a non-reflective dielectric coating in order to eliminate reflections from the ends of the rod. With these lasers the flash lamp is usually in the form of a straight rod which lies parallel to the ruby rod. The flash lamp rod is located along one of the two foci of an elliptical cylindrical reflector and the ruby rod is located along the other so that essentially all the light emitted by the flash lamp passes

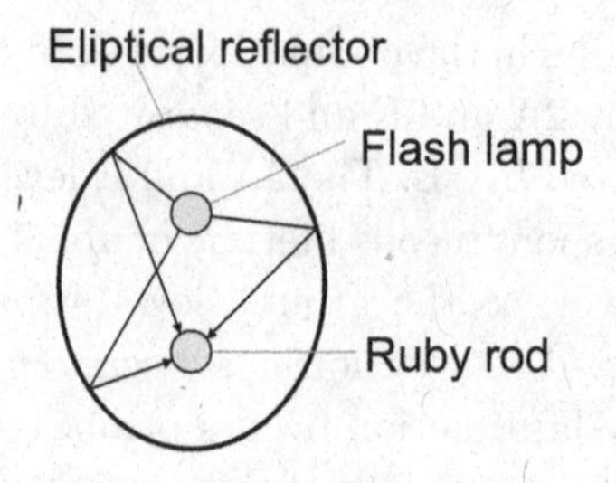

Fig. 9.2 A schematic diagram of a cross sectional view of the reflector used in the excitation of a ruby laser.

through the ruby rod where it is absorbed. Figure 9.2 shows a schematic diagram of a the cylindrical elliptical reflector for use with a ruby laser.

Ruby lasers can be operated at extremely high output powers in a Q-switched pulsed mode. As described in Section 7.5 of Chapter 7, Q-switching is a method used to produce very short bursts of high power oscillation from a laser by rapidly increasing the "Q" of the optical cavity. Initially the photon lifetime of a laser cavity is kept short so that even though a large population inversion is produced the system does not reach threshold. Then the photon lifetime is suddenly switched so that it is much longer, the threshold population difference is very quickly reduced, and the system is suddenly far above threshold. Induced emission rapidly depletes the population inversion producing an intense burst of photons into the optical cavity in a time comparable to the photon lifetime. The simple ruby laser with the ends coated to act as mirrors can not be Q-switched, but the advanced cavity design using external mirrors enables one to use intracavity devices with a ruby laser to obtain Q-switching. With Q-switching it is possible to produce laser pulses with an energy of 1-20 Joules per pulse and with pulse durations of about 10^{-8} s.

Let us briefly consider the operational requirements for producing a population inversion large enough to reach threshold in a typical flash lamp pumped ruby laser. Consider a 5 cm long ruby crystal, 5 mm in diameter with 2×10^{19} Cr^{3+} ions/cm^3. One end of the ruby rod has a high reflector with R_1=1, and the output coupler end has R_2=0.96. From Eq. 3.24 the photon lifetime of this cavity is $\tau_{\text{ph}} \approx 2nL/cT_2$. The index of refraction of ruby is n=1.76. So $\tau_{\text{ph}} = 15$ ns. The threshold population difference required for lasing can be calculated using Eq. 3.14,

$$\Delta n_{\text{th}} = \frac{8\pi\,\Delta\lambda}{\lambda_0^4}\,\frac{\tau_u}{\tau_{\text{ph}}}\,. \tag{9.1}$$

The linewidth for transition increases almost quadratically with the crystal temperature. At room temperature, it is approximately 11 cm^{-1} which converts to $\Delta\lambda$=0.53 nm. Note that the wavelength λ_0 in the denominator is the wavelength in the medium, $\lambda_0 = 694\ \text{nm}/1.76 = 394$ nm. From the 330 s^{-1} radiative decay rate of the $\bar{\text{E}}$ level, we find $\tau_u = 3\times10^{-3}$ s. So

$$\Delta n_{\text{th}} = \frac{8\pi\,\Delta\lambda}{\lambda_0^4}\,\frac{\tau_u}{\tau_{\text{ph}}} = \frac{8\pi\,(0.53\ \text{nm})}{(394\ \text{nm})^4}\,\frac{3\times10^{-3}\ \text{s}}{15\ \text{ns}} = 1\times10^{17}\ \text{cm}^{-3}\,. \qquad (9.2)$$

The required number of excited $\bar{\text{E}}$ atoms is much greater than Δn_{th} since this is the population difference between an the upper laser level and the lower level which in this case is the ground level,

$$\Delta n_{\text{th}} = \left(n_u - \frac{g_u}{g_l}\,n_l\right)\,. \qquad (9.3)$$

Accounting for the populations in all the relevant Cr^{3+} ion levels,

$$n_{\text{tot}} = n(^4\text{F}_2) + n(^4\text{F}_1) + n(2\bar{\text{A}}) + n(\bar{\text{E}}) + n(^4\text{A}_2)\,. \qquad (9.4)$$

Since the $^4\text{F}_1$ and $^4\text{F}_2$ levels rapidly decay to the ^2E levels, the population in these levels are essentially zero. The relative populations of the $2\bar{\text{A}}$ and $\bar{\text{E}}$ levels rapidly reach thermal equilibrium. The laser action is typically out of the $\bar{\text{E}}$ level, which is is slightly lower in energy, than the $2\bar{\text{A}}$ level. Since the small 29 cm^{-1} energy difference between the two levels is much less than $kT \approx 200\ \text{cm}^{-1}$ at room temperature $n(2\bar{\text{A}}) \approx n(\bar{\text{E}})$. With these approximations $n_{\text{tot}} \approx 2\,n(\bar{\text{E}}) + n(^4\text{A}_2)$. Substituting this into Eq. 9.3 along with the statistical weights of upper level $g_u = 2$ and lower level $g_l = 4$, one obtains

$$n(\bar{\text{E}}) = \frac{1}{4}\,n_{\text{tot}} + \frac{1}{2}\,\Delta n_{\text{th}}\,. \qquad (9.5)$$

The populations in the $\bar{\text{E}}$ and $2\bar{\text{A}}$ levels are maintained in rapid thermal equilibrium, so that as the $\bar{\text{E}}$ level begins to decay it is fed by the $2\bar{\text{A}}$ level and both levels are depleted as the laser action occurs.

Let us now examine the requirements on the flash lamp required to pump this sample ruby laser. Because ruby is a 'three level laser', it is necessary for the flash lamp to pump more than half the Cr^{3+} ions into the excited level in a time short compared to the lifetime of the excited ^2E levels. The pulse duration of the flash lamp should be much shorter than the the lifetime of the $\bar{\text{E}}$ level. A pulse duration of about 5×10^{-4} s is satiafactory for this purpose. The primary concern is that the absorbed energy is large enough to excite more than half the ions into the upper levels. The pulse energy absorbed by the Cr^{3+} ions must exceed the product of the number density of Cr^{3+} ions/cm^3 times volume of the ruby crystal times the energy to excite the ions into the $^4F_1$ and $^4F_2$ levels. For the 5 cm long 5 mm diameter ruby crystal with an ion density of $2 \times 10^{19}\text{cm}^{-3}$ the result is that the flash lamp must have an output power greater than about 16 J. Flash lamps are readily available with output powers great enough to deliver significantly greater energies than this in a time short compared to the lifetime of the $\bar{E}$ level so that the ruby laser can be driven far above threshold with the result that powerful pulsed laser action can be obtained.

Ruby lasers can be used for applications where a high energy per pulse is important. For example ruby lasers have been used for optical probing of the atmosphere. Experiments might include looking at the reflected light from water droplets or other particulate material in the atmosphere. From the delay times between the initial pulse and the reflected light one can learn the distance to the water droplets or other particulate matter. These types of experiments are called LIDAR (for light detection and ranging) experiments. Ruby lasers are also used as the light source for Thomson scattering experiments in plasma physics, and are used for the light source for making holograms.

9.2 Neodymium (Nd:YAG) Lasers

A number of widely used high power solid state lasers use neodymium ions (Nd^{3+}) doped in a host medium. The host material can either be a crystal such as yttrium aluminum garnet (YAG), $Y_3Al_5O_{12}$, yttrium orthovanadate, YVO_4, yttrium lithium fluoride, $LiYF_4$, or an amorphous material such as glass. The energy level structure of the Nd^{3+} ions in the various hosts are similar, the relevant energy levels of the Nd^{3+} ion in a YAG

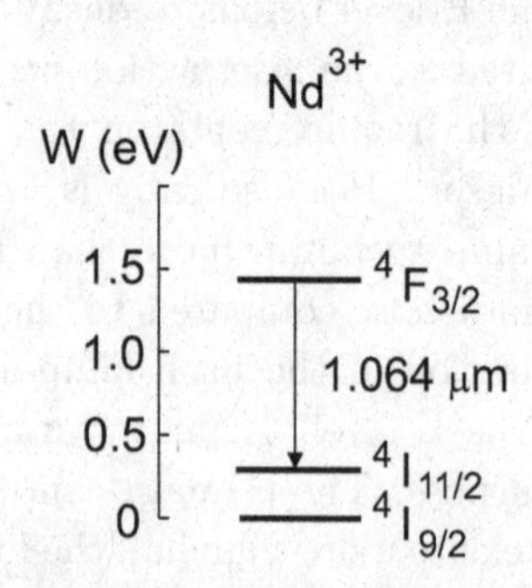

Fig. 9.3 The energy levels of the Nd^{3+} ion in a Nd:YAG crystal. The energies are given in eV above the ground level of the Nd^{3+} ion.

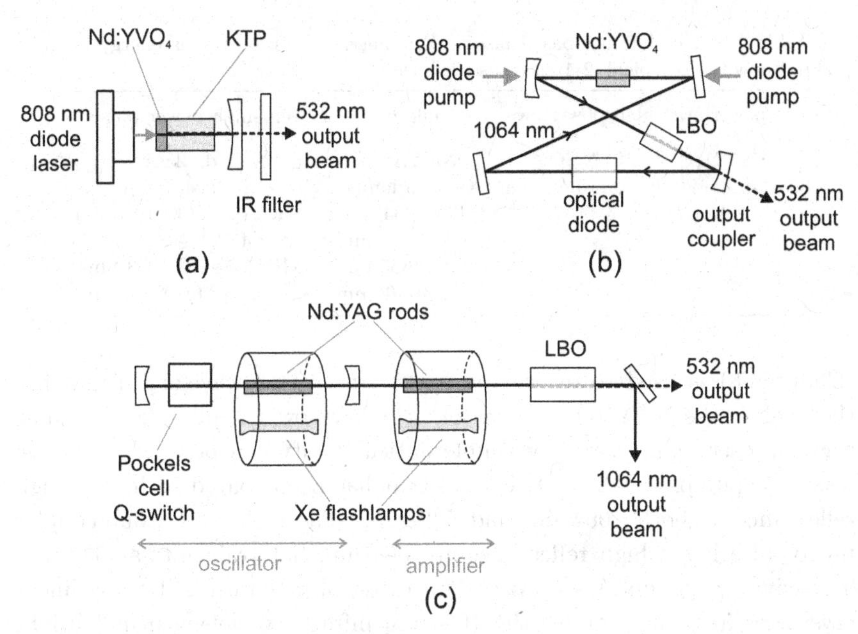

Fig. 9.4 Three Nd-based lasers: (a) a low-power diode pumped green laser pointer, (b) a high-power cw diode pumped ring laser, (c) flash lamp pumped pulsed Nd:YAG laser. In all three lasers a non-linear crystal (KTP or LBO) is used to frequency double the 1064 nm Nd^{3+} output to a 532 nm green beam.

crystal are shown in Figure 9.3. While the particular choice of host material effects (i) the absorption bands of the light used to pump the laser, (ii) the wavelength of the laser transition, and (iii) the linewidth of the transition (Sec. 4.4), a discussion of the Nd:YAG laser permits an understanding of all the Nd^{3+} ion based lasers.

The Nd:YAG laser is a four level laser with a very high gain. The ground level of the Nd ion is denoted by $^4I_{9/2}$. The Nd:YAG crystal can be optically pumped using either a diode laser or flash lamp. Nd:YAG has two broad absorption bands (not shown in Fig. 9.3) in the 730-760 nm and 790-820 nm wavelength ranges. Absorption of pump light is used to excite various levels that decay rapidly to the level denoted by $^4F_{3/2}$. The laser transition is between the $^4F_{3/2}$ and the $^4I_{11/2}$ levels and has a wavelength of 1.0641 μm. The $^4I_{11/2}$ lower laser level is essentially always unpopulated due to rapid relaxation.

Nd:YAG and Nd:YVO_4 lasers can be operated either cw or pulsed. Figure 9.4(a,b) illustrates two types of cw diode pumped solid state (DPSS) lasers using neodymium ion as the active medium. An 808 nm laser diode

Table 9.1 Neodymium-based lasers. For pulsed lasers, output is energy per pulse with a typical 2-4 ns pulse length.

low-power cw	high-power cw	pulsed	very-high power pulsed
$Nd:YVO_4$	$Nd:YVO_4$	Nd:YAG	Nd:Glass
808 nm diode	808 nm diode bar	Xe flash lamp	Xe flash lamp
≤5 mW	1-20 W	1 J @ 1064 nm	Nova: 100 kJ @ 1053 nm
		0.5 J @ 532 nm	45 kJ @ 351 nm
		0.2 J @ 355 nm	NIF: 3 MJ @ 1053 nm
		0.05 J @ 266 nm	1.5 MJ @ 351 nm

(Chapter 8) is used to excite the Nd^{+3} ions. This wavelength matches the peak of the $Nd:YVO_4$ absorption. For low-power applications, such as a green laser pointer, a very simple optical cavity can be used. A diode laser end pumps a $Nd:YVO_4$ crystal that has been coated to have a high reflectance at both 1064 nm and 532 nm. The dichroic output coupler mirror also has a high reflectance at 1064 nm, but transmits at 532 nm. The cavity also contains a potassium titayl phosphate (KTP) non-linear crystal to frequency double the 1064 nm infrared wavelength into bright green light at 532 nm (see Sec. 13.1). For higher power applications (Watts versus mW), a $Nd:YVO_4$ crystal is pumped by a number of diode lasers. The high power DPSS laser illustrated in Figure 9.4(b) uses a ring-cavity design with a lithium triborate crystal, (indicated as LiB_3O_5 or LBO) for second harmonic generation. For cw operation, high power DPSS lasers are very efficient and reliable compared to Ar-ion lasers. For example, it requires around 30 kW of electrical power for an Ar ion laser to generate a 5 W output beam at 514.5 nm, whereas less than 100 W are needed by a $Nd:YVO_4$ DPSS laser to generate a 5 W output at 532 nm.

Figure 9.4(c) illustrates the general design of a pulsed Nd:YAG laser using a flash lamp for pumping. Diode lasers can also be used for pumping a Nd:YAG laser. Some Nd:YAG lasers have two stages, an oscillator stage and an amplifier stage. The mirrors that form the optical cavity for the oscillator are located externally to the Nd:YAG rod. The ends of the Nd:YAG rod are orthogonal to the laser beam, are polished flat, and are coated with a non-reflecting dielectric. The Nd:YAG rods are optically pumped using a co-linear xenon flash lamp located in an elliptical reflector as illustrated in Fig. 9.2. The cavity in Figure 9.4(c) also contains a Pockels cell (electro-optical switch) used for Q-switching (Sec. 7.5). It is common to either double or triple the frequency of the laser light in order to obtain a beam with a wavelength of 532 nm or 355 nm respectively. Pulsed Nd:YAG lasers usually operate with a repetition rate of 10-30 Hz and with an output energy of 0.1-1 Joule per pulse.

Nd:YAG lasers are commonly used for applications where high output powers are required. Table 9.1 summarizes a few types of neodymium-based lasers. It is also possible to obtain laser action from Nd^{3+} ions in glass. There are limits on the physical size of Nd:YAG crystals that can be easily grown, but very large Nd:Glass slabs can be fabricated. These lasers can operate with extremely high output powers and are used in applications such as laser driven fusion where the highest possible output powers are required. For example, the National Ignition Facility (NIF) laser uses 192 beam lines to produce a 1.85 MJ pulse at a frequency tripled wavelength of 351 nm. During the short duration of the pulse, the output power of this laser is close to 500 TW. For very high power Nd:Glass lasers it is essential to have a very high quality beam with a uniform wave front. Small hot spots can be amplified to a high enough power that the glass slab can be cracked. To prevent this, spatial filters are placed between the numerous amplifier stages.

9.3 Glass and Silica Fiber Lasers

Glass or silica fiber lasers are extremely important lasers for commercial use. A glass or silica fiber is very long and thin piece of glass or silica. Fiber lasers are used for communications, for applications requiring high power, and for applications requiring short pulses. Fiber lasers consist of a long glass or silica fiber with ions doped into the glass. The ions are pumped into a level with a large enough population inversion that laser action occurs. For communications a commonly used ion is Erbium, Er^{3+}, doped into the fiber. The Er^{3+} ions can be pumped at 980 nm or 1480 nm and radiate at 1530 nm. Other ions that can be doped into glass or silica fibers to provide an active gain medium include Neodymium (Nd^{3+}), Ytterbium (Yb^{3+}), Thulium (Tm^{3+}), and Praseodymium (Pr^{3+}). The laser light is confined to the fiber by total internal reflection. The lasing fiber is always surrounded by a pump cladding. The pump light is generated using multi-mode diode lasers. The diode pump light can enter the laser fiber and pump the Er^{3+} into the lasing level, One might think that the pump cladding should be concentric with the laser fiber, but this geometry allows the pump light to pass down the pump cladding with too little interaction with the fiber to satisfactorily pump the Er^{3+} ions. The pump cladding can have a variety of possible shapes. A commonly used pump cladding is rectangular in shape, which reflects enough pump light into the laser fiber. A polymer protective cladding surrounds the fiber laser and pump cladding to prevent damage. Fiber lasers can be used as either amplifiers or oscillators. For communications an electrical signal is converted into a modulated optical signal by a diode laser and this optical signal is injected into a fiber laser where it is amplified. The output of the fiber laser is coupled into an extremely long glass fiber through which the signal is transmitted to a distant receiver. In many applications a large number of fibers are bundled together to support the tramsmission of numerous different signals.

There are two types of glass fibers, single mode and multi-mode fibers. Single mode fibers are typically 8-10 μm in diameter, and multi-mode fibers may be 50 μm or more in diameter. A single mode fiber supports only a single transverse mode. Single mode fibers have great advantages over multimode fibers and hence are used for most applications. For large distance communications optical amplifiers are used to replenish losses in the fiber. The amplifier is a relatively short length of fiber that is doped and pumped so that it acts as an *optical amplifier*. For extremely long distance signal transmission a repeater may be needed. A *repeater* is a device that converts the optical signal into an electrical signal, cleans the signal, converts the electrical signal back into an optical signal, and then launches the optical signal through a glass fiber amplifier into another length of optical cable. With modern glass fibers the absorption is very low. As a consequence the primary limitation on the length of fiber that can used before a repeater is needed is often not determined by the absorption of the glass fiber, but is instead due to the chromatic dispersion of the glass fiber. Chromatic dispersion is the spreading of different Fourier component wave lengths in a pulse as the pulse travels along the fiber. Chromatic dispersion is due the wavelength dependence of the index of refraction of the glass.

Wavelength-division multiplexing is used to increase the bit rate transmission. Wavelength-division is the use of different channels, each channel reserved for a different wavelength. Modern glass fibers can support the rates transmission of an enormous number of bits per second for distances of hundreds of km for each channel. At the end of the optical cable a receiver converts the optical signal into an electrical signal that is suitable for use in computers, television sets and other electronic devices.

Some glass fiber lasers can produce high power, and some glass lasers can be used as Q-switched lasers. It is possible to make glass fiber ring lasers. Figure 9.5 shows schematically a high power laser. A signal is generated from a diode laser and is injected into the fiber laser. The fiber

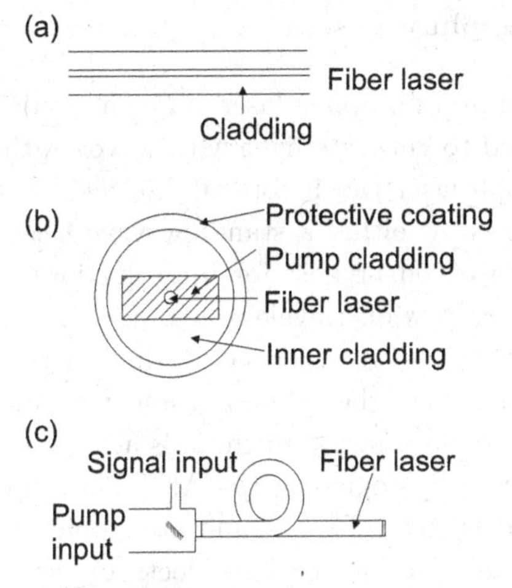

Fig. 9.5 (a) A glass or silica fiber with a doped core; (b) a fiber laser in cross section; (c) a schematic of the complete glass fiber laser.

laser is pumped by multimode diode lasers. Figure 9.5(a) shows the glass fiber laser. Figure 9.5(b) shows the fiber laser in cross section. The pump cladding is indicated as rectangular in shape, but other non-concentric geometries are used. Figure 9.5(c) indicates the entire laser set up. Powers as high as a kW can be obtained using a single fiber. Very high powers may necessitate the use of a large core fiber. This may result in multi-mode operation rather than single mode operation.

There are many reasons to utilize fiber lasers and fiber transmission for various applications. These include: (i) The fibers can be coiled up into a compact laser for use in a laboratory. (ii) The fiber lasers can have very high wall plug to light efficiency and hence fiber lasers can operate with low pump powers. (iii) Diffraction limited use is relatively easy to obtain. (iv) Fiber lasers can have very high output powers. (v) Glass is an electrical insulator and does not act as an antenna thereby making the fibers insensitive to electrical noise. (vi) The cost of glass fibers is much lower than copper wire. (vii) Most importantly glass fibers have a broad bandwidth and can transmit an enormous amount of information in a single fiber. There are many applications of fiber lasers in communications, in industry and in laboratory research.

9.4 Titanium:sapphire Lasers

The Titanium:sapphire (Ti:sapph) laser is a commonly used tunable solid state laser. It is used to generate light with wavelengths in the range 670-1150 nm. A Ti:sapph laser can be operated either cw or pulsed. The cw lasers can be operated as either a standing wave laser or a ring laser. A sapphire crystal is a excellent host for impurity ions that lase since it is transparent over a very wide wavelength range. It is, also, resistant to thermal damage due to its high thermal conductivity. Sapphire (Al_2O_3) is used for the host Cr ions in the ruby laser and for Ti ions in the Ti:sapph laser. The Ti:sapphire laser active medium is a sapphire crystal with some Ti^{3+} ions substitutionally replacing the Al^{3+} ions. The Ti^{3+} ion is confined by the crystal lattice. The ground electronic level of the Ti^{3+} ion is denoted as $^2T_2$, and the first excited electronic level is denoted as ^2E. Both the ground electronic level and the first excited electronic level have many excited vibrational levels with very small separations between the vibrational levels as shown in Fig. 9.6. Absorption of light supplied by a pump laser excites $^2T_2$ ground vibrational level into the ^2E highly excited vibrational levels. The highly excited vibrational levels of both the $^2T_2$ and the ^2E levels rapidly undergo non-radiative decay to vibrational levels within kT of the lowest vibrational level. The radiative decay of the low lying vibrational levels of the ^2E level into excited vibrational levels of the $^2T_2$ level forms the laser transition. The lifetime of the ^2E electronic level decreases as the temperature of the Ti:sapph crystal increases because the non-radiative decay rate increases when the temperature increases. It

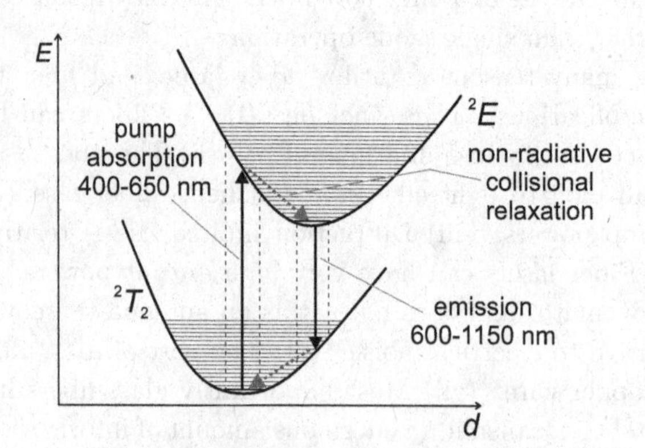

Fig. 9.6 Energy levels of a Titanium:sapphire laser. Horizontal lines denote vibrational energy levels.

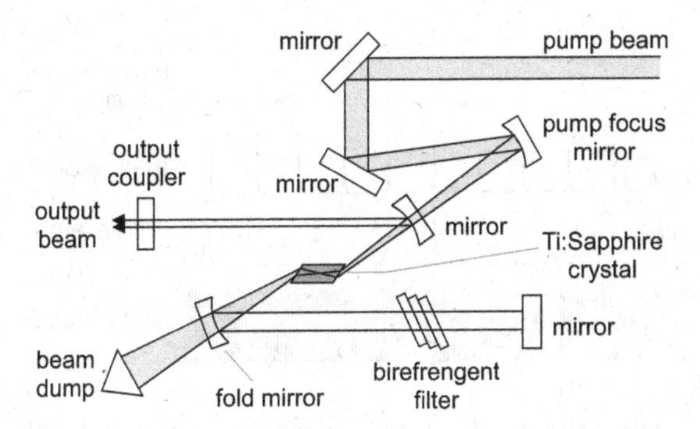

Fig. 9.7 A schematic diagram for a standing wave Ti:sapph laser. Note that the pump laser beam enters the laser cavity parallel to the output beam via a dichroic mirror and the pump beam is dumped out of the cavity through a dichroic mirror.

is therefore desirable to keep the Ti:sapph crystal cool in order to obtain laser action. Due to the small separations between the vibrational levels, the output wavelength is continuously tunable. There are no higher electronic levels of the Titanium ion in the sapphire crystal and hence there is no loss of gain due to further excitation or absorption by electronic levels, but sometimes there are losses due to absorption by Ti^{4+} ions that may be present in a Ti:sapph crystal.

The Ti^{3+} ions absorb wavelengths from about 400 to 650 nm. For pulsed operation, it is common to pump a Ti:sapph laser with a doubled Nd:YAG laser (532 nm). For cw operation, a diode pumped solid state (DPSS) laser such as a Nd:YVO_4 laser (532 nm) can be used to pump the Ti:sapph laser. While Ar ion lasers were previously widely used as the pump laser, today solid state lasers are usually used. The emission wavelengths of the argon ion lasers are slightly more favorable than those of DPSS lasers, but the advantages of high reliability and much lower power consumption of DPSS lasers are overwhelming.

Ti:sapph lasers can be operated as either standing wave lasers or as ring lasers. A standing wave laser is illustrated schematically in Figure 9.7. The optical cavity for a Ti:sapph must be designed such that the pump laser illuminates the entire length of the Ti:sapph crystal. The laser pump beam is brought into the crystal collinear with the Ti:sapph laser beam. This can be accomplished by using a dichroic mirror that has a high reflectivity for the Ti:sapph wavelength but which is transparent for pump laser wavelength. This is shown schematically in Figure 9.7. The dichroic mirror through which the pump beam enters the laser cavity is located directly before the Ti:sapph crystal.

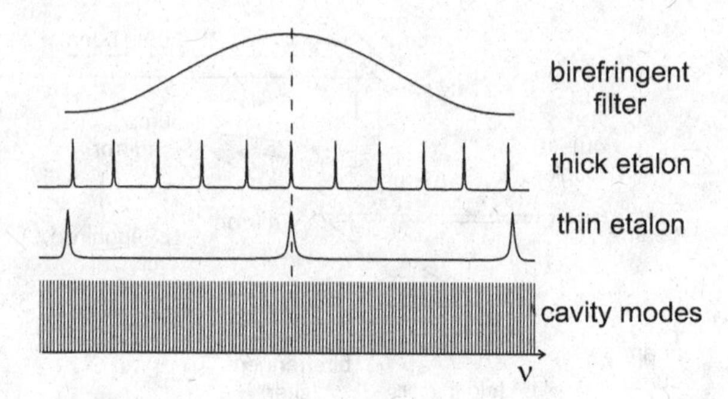

Fig. 9.8 Wavelength selection of a tunable laser using etalons and a birefringent filter.

A number of different optical elements are used to control the laser's output wavelength, only one of these, a birefringent filter, is shown in Figure 9.7. More typically, both a birefringent filter and a number of etalons are used (see Fig. 9.8). The etalons are not shown in the figure. The *birefringent filter* is a device that contains three optically active quartz plates, that rotate the polarization of a wave passing through the filter. The optical activity depends on the indices of refraction for opposite circular polarizations. These indices of refraction are functions of the wavelength. The thickness of the quartz plates is selected so that for some wavelength the net rotation is through an angle that is an integral multiple of 2π so that the polarization is left unchanged. This wavelength will have lower losses than other wavelengths as the wave passes through Brewster angle devices in the optical cavity.

With only the birefringent filter, the laser will operate multi-mode on many longitudinal modes. To select a single longitudinal mode Fabry Perot etalons are used. An etalon is characterized by two key parameters, the free spectral range (FSR) and the finesse $\mathcal{F}$. The free spectral range is the fringe spacing between adjacent modes. For an etalon with index of refraction n, thickness l and tilted at an angle θ, the free spectral range is

$$\mathrm{FSR} = \frac{c}{2\,n\,l\,\cos\theta}\,. \tag{9.6}$$

The finesse of the etalon defines how sharp the fringe transmission is. Assuming the same reflection coefficient on both sides of the etalon,

$$\mathcal{F} = \frac{\pi\sqrt{R}}{1-R}\,. \tag{9.7}$$

In Figure 9.8 two etalons, a thick and thin etalon, are used to select the wavelength. The tuning of an etalon can be controlled by either rotation (θ)

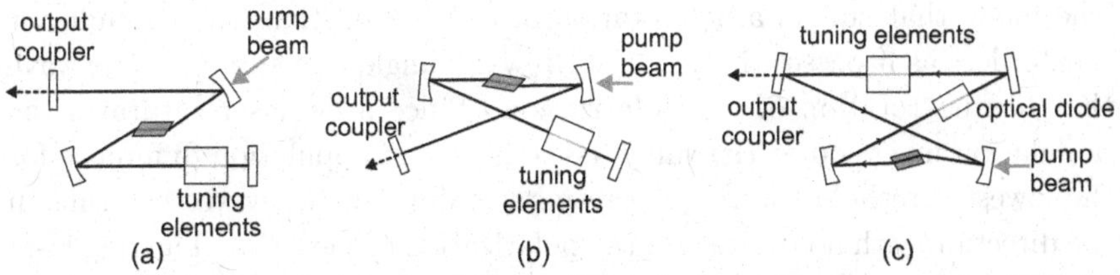

Fig. 9.9 Common optical cavity designs of Titanium:sapphire lasers: (a) z-fold, (b) x-fold, (c) ring laser.

or heating the etalon which causes the etalon to expand (changing l). The two etalons are adjusted to have only one wavelength that is passed by both etalons. The etalons in Fig. 9.8 both are shown as having a high finesse, however, a high finesse is not necessarily required. For a homogeneously broadened gain medium the losses of adjacent laser modes only needs to be reduced by a small amount to prevent lasing, and an un-coated etalon (R=0.04) with low finesse may suffice.

The operating wavelength of the laser is ultimately selected by narrowing the broad Ti:sapph gain envelope using the tunable birefringent filter and etalons. In a multi-mode laser more than one cavity mode still has positive gain. In a single-mode laser the gain window is narrowed down sufficiently to only permit a single cavity mode to lase. Further tuning can be achieved by varying the cavity length, and thus the exact frequency of the cavity modes.

The standing wave Titanium:sapphire laser illustrated in Figure 9.7 utilizes what is known as a z-cavity design, since the beam path between the two ends of the cavity trace out the shape of a 'z'. This is a bit more obvious in Figure 9.9 that includes a simplified version of the optical cavity. Figure 9.9 also includes two other common optical cavity designs for Titanium:sapphire lasers, an x-fold cavity and ring-laser configuration. the ring laser contains an optical diode to assure that the laser has a wave running in only one direction around the ring. The optical diode has two parts. The first part is a glass plate in the field of a permanent magnet. The Faraday effect in the glass produces a rotation of the polarization of a running wave going in either direction through the same angle. The second part is an optically active quartz crystal that that rotates the polarization of waves running in opposite directions through angles that are equal in magnitude but opposite in sign. The result is that for a wave going in one direction there is no net rotation of the polarization whereas the wave running in the opposite direction has a net rotation of the polarization.

The mode that suffers a net rotation of the polarization has a significant greater loss as it passes through the Brewster angle elements than the wave that has no rotation of the polarization. Since a cw laser saturates the population inversion at the value of the inversion population difference for the lowest threshold the ring laser operates only for the wave that runs in the direction with no rotation of the polarization of the wave. The ring laser also contains a galvo Brewster plate (not shown) that is used to change the path length of the cavity. The ring laser can be stabilized using an external cavity as a reference.

Titanium:sapphire lasers are widely used for experiments where it is desirable to tune a laser to a particular atomic or molecular transition. Titanium:sapphire lasers are also used as wide-band amplifiers. For example, *frequency combs* (covered in Sec. 13.5), use a mode locked Ti:sapph ring laser to generate an output consisting of a large number of equally spaced optical frequencies. Ti:sapphire lasers are also used to amplify short (femtosecond) pulses. The Fourier spectrum of a pulse contains a frequency spread $\pm 1/\tau$ from the center frequency where τ is the pulse duration. For a very short pulse, this requires an optical amplifier with a wide gain curve. There is a limit to the maximum output power of such a pulsed laser since the Ti:sapph crystal is damaged at high power densities. Nevertheless, incredibly high power pulses ($\geq 10^{11}$ W) can be generated with table-top Ti:sapph lasers using *chirped pulse amplification.* Before the pulse enters the Ti:sapph amplifier, the pulse is bounced off a series of gratings such that the path length for the short wavelength components of the pulse is longer (or shorter) than the path length for the longer wavelengths leading to a dispersion in the arrival times for different frequencies. This streches the length of the pulse at the amplifer by a factor of 10^3 or more and lowers the instaneous power by a similar factor. After the amplifier, another set of gratings are used to compress the pulse by having the short wavelengths take a shorter path. Ti:sapph lasers are today one of the most widely used lasers in scientific research.

Summary of Key Ideas

- There are a number of solid state fixed frequency lasers, including the ruby, Nd:YAG, Nd:Glass, glass and silica lasers. In all of these lasers the light is genenerated by emissions from a host ion emedded in a crystal (or glass) substrate.
- The ruby laser (Sec. 9.1) is a three level laser using Cr^{3+} ions in a sapphire crystal. The laser wavelength is 694.3 nm.

- Neodynymium lasers (Sec. 9.2) use Nd^{3+} ions in a varety of different host materials including YAG, YVO_4, and glass. The laser wavelength is near 1064 nm, but this is often frequency doubled to produce a green 532 nm output. Output can be either pulsed (flash lamp pumped) or CW (pumped by diode lasers).
- Fiber lasers use ions (typically Er^{3+}, but also Nd^{3+}, Yb^{3+}, Tm^{3+}, Pr^{3+}) distributed in a long optical fiber (Sec. 9.3). A diode laser beam is injected into the fiber for pumping.
- Titanium:sapphire lasers use Ti^{3+} ions in a sapphire crystal as the active medium (Sec. 9.4). They are tunable over a wide range of frequencies and can be doubled to increase the tuning range. They are reliable and are widely used for applications that require a tunable laser. Titanium:sapphire crystals are also used as wideband amplifier for pulsed light.

Suggested Additional Reading

William T. Silfvast, *Laser Fundamentals*, Cambridge University Press (1996).

T. H. Maiman, "Stimulated Optical Radiation in Ruby" *Nature* **187**, 493 (1960).

F. J. McClung and R. W. Hellwarth, "Giant Optical Pulsations from Ruby" *J. Appl. Phys.* **33**, 828 (1962) [Q-switching].

J. E. Geusic, H. M. Marcos and L. G. Van Uitert, "Laser Oscillations in Nd-Doped Yttrium Aluminum, Yttrium Gallium and Gadolinium Garnets" *Appl. Phys. Lett.* **4**, 182 (1964).

Oleg G. Okhotnikov, *Fiber Lasers*, Wiley-VCH (2012).

Thomas Pearsall, *Photonics Essentials, 2nd ed.*, McGraw-Hill Professional (2009).

Problems

1. A glass fiber has and index of refraction of 1.64 and the cladding has an index of 1.48. (a) What is the largest angle for which there is total internal reflection in the fiber? (b) Estimate the radius of curvature for which a 10 μm diameter glass fiber can no longer have total internal reflection.

2. Make a block diagram of a fiber laser system for data transmission.

3. A Lidar system is used to detect rain at a distance of 10 miles. What is the delay in the reflected signal?

4. Describe in your own words how a birefringent filter works.

5. Sketch out the details of a possible optical cavity for a Ti:sapph laser. Include the various elements required (optical diode, birefringent filter, *etc...*). Describe how such a laser works.

6. A Q-switched Nd:YAG laser with a frequency doubler emits a 0.5 J pulse with a wavelength of 532 nm and with a repetition rate of 30 pulses/s^{-1}. The pulse duration is 5 ns. (a) What is the peak power in the pulse? (b) What is the average power?

7. For the laser of problem 6, (a) How many photons are emitted in each pulse? (b) If the laser is focussed to a spot diameter 2μm what is the electric field at peak power?

8. A Ti:sapph ring laser has a length around the ring of 3 m. What is the maximum possible time between output pulses?

9. Explain carefully Why the output of a Ti:sapph laser is not absorbed by the $^2T_2$ level.

10. Optical pulses with wavelengths 589 nm and 800 nm are launched into an optical fiber simultaneously. the fiber has an index of refraction of 1.460 at 589 nm and of 1.452 at 800 nm. What is the separation in the time the two pulses reach a detector 1000 km from where they were launched? Which pulse arrives at the detector first?

Chapter 10

The Helium-Neon Laser

In previous chapters the operation of various lasers was discussed. None of these lasers were analyzed in great detail. The helium-neon (He-Ne) gas laser is relatively easy to understand in detail. In this chapter it is analyzed carefully enough that the reader can understand almost all of the problems associated with this laser. Understanding one laser in great detail will enable the reader to appreciate the problems in understanding other lasers. With a bright visible emission at 632.8 nm, the He-Ne laser was at one time a very common laser, with many applications. More recently, diode lasers have replaced He-Ne lasers for almost all applications. Helium-neon lasers, however, retain some advantages over diode lasers such as having a narrow linewidth with a known, fixed wavelength.

10.1 Theory of Operation

The He-Ne laser is a gas laser, that utilizes excitation transfer to produce the population inversion. An electric current is run through a tube containing helium and neon gas. Electron collisions with helium atoms excite atoms into long-lived metastable levels. Excitation transfer collisions between the metastable helium atoms and ground state neon atoms populate the upper level of the neon laser transition. In order to understand the He-Ne laser it is necessary to understand the energy levels of both the He and Ne atoms.

10.1.1 *Atomic Structure of He and Ne*

A Grotrian diagram of the relevant energy levels of He and Ne is shown in Figure 10.1. The He atom has two electrons. The ground level electron configuration of He is $1s^2$, which means that both electrons are in the

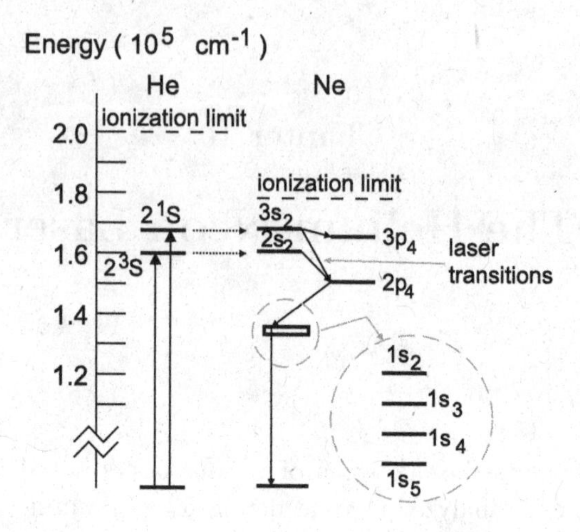

Fig. 10.1 The energy levels of the He and Ne. Note that the energies are measured in wave numbers above the ground level.

1*s* orbital. The orbital angular momentum for each electron is zero (i.e. $l_1 = l_2 = 0$ where the subscripts denote the two electrons). Hence the total orbital angular momentum ($\boldsymbol{L} = \boldsymbol{l}_1 + \boldsymbol{l}_2$) of the ground level is L=0. Levels with orbital angular momentum $L = 0, 1, 2, 3, \ldots$ are indicated as S,P,D,F,... levels. Thus the ground level of He is an S level. Since the two electrons have the same quantum numbers for n, l, and, m_l the Pauli exclusion principle requires that two electrons must have different values for, m_s, the z component of the spin angular momentum. The result is that the value of the z component of the spin must be zero and hence the total spin angular momentum of the ground level of He must be S=0. The Russell-Saunders LS spectroscopic notation for the ground level of He is 1^1S_0 where the 1 indicates the highest value of n for an electron in the atom (for most levels of He one of the electrons has n=1 so it is necessary only to indicate the highest value of n, which is the value of n for the other electron), where the superscript is the value of $2S + 1$ which for the ground level of He is 1 since S=0, where the notation S indicates the total orbital angular momentum of the ground level is L=0 (note that S is commonly and somewhat confusingly used to denote both the total spin angular momentum quantum number and to denote an atomic energy level with the zero orbital angular momentum), and where the subscript is the value of the total angular momentum, J, which is 0 for the ground level of He.

The lowest excited configuration of the He atom is $1s2s$, which means that one electron is in a $1s$ orbital and the other electron is in a $2s$ orbital. For this configuration both electrons have an orbital angular momentum of zero so that the total orbital angular momentum of the configuration is L=0. For this excited configuration the two electrons do not have the same quantum number for n so that the Pauli exclusion principle does not restrict the value of the spin angular momentum. Hence the $1s2s$ configuration can have both singlet and triplet S levels which are denoted by 2^1S_0 and 2^3S_1. Neither the 2^1S_0 nor the 2^3S_1 level of He can decay radiatively to the ground level by electric dipole emission, and hence both levels are *metastable.* Electric dipole emission is forbidden for both levels since the parity of the excited levels is the same as for the ground level and it is further forbidden for the singlet level since it would require a $0 \rightarrow 0$ transition and it is further forbidden for the triplet level since a triplet to singlet transition is forbidden. The higher $1snl$ configurations of He do not play a significant role in the He-Ne laser, and they are not discussed here.

In the discussion of the energy levels of He it has been assumed that Russell-Saunders coupling is applicable. In Russell-Saunders coupling the orbital angular momenta of the electrons in an atom are coupled together to form a total orbital angular momentum, L, and the spin angular momenta of the electrons in an atom are coupled together to form a total spin angular momentum, S. The total orbital and spin angular momenta are then coupled together to form the total angular momentum of the atom, J. For He Russell-Saunders coupling is an excellent approximation for the low lying energy levels. For He Russell-Saunders coupling first breaks down for the $1snf$ configurations. The lowest value of n for which there are f orbitals is n=4.

The energy levels of Ne are more complicated than for He and for some of the energy levels important in the He-Ne laser Russell-Saunders coupling fails. Neon has ten electrons, The ground configuration is $1s^2 2s^2 2p^6$. In the ground level the $1s$, $2s$, and $2p$ shells are full, and the total orbital and spin angular momenta are zero. The spectroscopic notation for the level is $^1S_0$. The excited configurations of Ne involve the excitation of one of the $2p$ electrons into a higher orbital. The $2p^5ns$ and $2p^5np$ configurations are important in the He-Ne laser. The $1s^2 2s^2$ electrons are ignored because they have zero orbital and spin angular momenta.

First consider the $2p^5ns$ configurations. The $2p^5$ core has orbital angular momentum l_1=1 and spin angular momentum s_1=1/2, and the outer ns

electron has l_2=0 and s_2=1/2. Russell-Saunders LS coupling gives energy levels denoted by $^1P_1$, $^3P_0$, $^3P_1$, and $^3P_2$ for this configuration. In order to describe the breakdown of Russell-Saunders coupling for the low lying levels of Ne, it is convenient to start with the LS-eigenlevels and allow for deviation from the LS-coupling by mixing the various LS-eigenlevels of the same total angular momenta. The total angular momentum, J, is always a good quantum number independent of whether LS-coupling is applicable or not. For the low lying configurations of Ne it is only important to consider mixing of levels within the same configuration; configuration mixing is negligible since there are no near lying configurations. In this situation the wave functions of the $2p^5ns$ levels can be expanded using LS-eigenfunctions as

$$\begin{array}{ll} \psi_1 = \alpha\,\phi(2p^5ns,\ ^1\mathrm{P}_1) + \beta\,\phi(2p^5ns,\ ^3\mathrm{P}_1) & 1s_2 \\ \psi_2 = \beta\,\phi(2p^5ns,\ ^1\mathrm{P}_1) - \alpha\,\phi(2p^5ns,\ ^3\mathrm{P}_1) & 1s_4 \\ \psi_3 = \phi(2p^5ns,\ ^3\mathrm{P}_0) & 1s_3 \\ \psi_4 = \phi(2p^5ns,\ ^3\mathrm{P}_2) & 1s_5 \end{array} \tag{10.1}$$

The $^3P_2$ and the $^3P_0$ levels are good LS-eigenfunctions since they have a unique value of J and therefore do not mix with any other member of the $2p^5ns$ configuration. On the other hand the wave functions ψ_1 and ψ_2 are mixed wave functions each with J=1. Thus two of the eigenfunctions for the $2p^5ns$ configuration are good LS-wave functions and two are not. Because the wave functions are not all good LS-wave functions it is necessary to adopt a notation for these levels. A commonly used notation is that due to Paschen where the levels corresponding to ψ_1, ψ_2, ψ_3, and ψ_4 wave functions for the $2p^53s$ configuration are called respectively the $1s_2$, $1s_4$, $1s_3$, and $1s_5$ levels. For the $2p^54s$ and $2p^55s$ configurations the breakdown of LS coupling is similar and the levels are denoted as $2s_2$ to $2s_5$ and $3s_2$ to $3s_5$ respectively. Electric dipole emission from the $1s_3$ and $1s_5$ levels to the ground level is forbidden since emission from the $1s_3$ level to the ground level is a $J=0 \rightarrow J=0$ transition and the emission from the $1s_5$ level to the ground level is also a forbidden transition since ΔJ=2 . These levels are therefore metastable. Both the $1s_2$ and the $1s_4$ levels are mixed levels with a combined singlet and triplet character, and they both can radiate by electric dipole emission to the ground level. The lifetime of the $1s_2$ level is about 14 times less than the lifetime of the $1s_4$ level from which one can conclude that the value of α^2 and β^2 are $\alpha^2 = 0.93$ and $\beta^2 = 0.07$.

The $2p^5np$ configurations can be analyzed in a similar manner to the $2p^5ns$ configuration. The $2p^5$ core has l_1=1 and s_1=1/2, and the outer electron has l_2=1 and s_2=1/2. This leads to the ten LS-wave functions corresponding to $^1S_0$, $^3S_1$, $^1P_1$, $^3P_0$, $^3P_1$, $^3P_2$, $^1D_2$, $^3D_1$, $^3D_2$, and $^3D_3$. The true wave functions are linear superpositions of the LS-wave functions within the configuration with mixing only among the wave functions with the same J. Only the wave function with J=3 is unmixed and hence a pure LS-wave function. The other nine levels are mixed with both singlet and triplet character. The levels resulting from the $2p^5 3p$ and $2p^5 4p$ configuration are called respectively the $2p_1$ to $2p_{10}$ and $3p_1$ to $3p_{10}$ levels in Paschen's notation.

The common red He-Ne laser operates at a wavelength of 632.8 nm which corresponds to emission from the $3s_2$ upper level (with J=1) into the $2p_4$ lower level (with J=2). Electric-dipole selection rules ($\Delta J = 0, \pm 1$, with $J = 0 \not\leftrightarrow J = 0$) also allow the decay of the J=1 $3s_2$ upper level into eight of the other $2p_x$ levels (all but the J=3 $2p_9$ level). By suppressing the gain on the 632.8 nm line it is possible to have a He-Ne laser operate on one of these other transitions with emission wavelengths in the range of 543-731 nm. In addition, the $3s_2$ upper level can also decay into levels of the $2p^5 4p$ configuration with J=0,1,2 (i.e., the $3p_1$-$3p_8$ and the $3p_{10}$ levels). These transitions all lie in the mid-infrared wavelength range. For example, one type of infrared He-Ne laser produces output at 3392.2 nm (3.39 μm) using the $3s_2 \rightarrow 3p_4$ transition. The $2p_4$ and $3p_4$ lower levels of these transitions each have J=2 and have a mixed singlet-triplet character. They can decay radiatively by electric dipole emission to the $1s_2$, $1s_4$, and $1s_5$ levels.

10.1.2 *Collision Processes*

The He-Ne laser is a gas discharge laser. Excited helium atoms are produced by electron impact collisions in a discharge. Excitation transfer collisions between helium atoms in long-lived metastable levels and ground state neon atoms create excited neon atoms in the upper level of the neon laser transition. Thus, there are three distinct steps in the operation of a He-Ne laser: (i) excitation of metastable He atoms, (ii) collisional transfer into excited Ne levels, (iii) stimulated emission on the Ne transition of interest.

First let us consider the collision processes involving the metastable helium atoms. The two metastable levels in He are populated by electron impact collisions in the discharge. Electrons with a minimum energy of 19.8 eV are required to excite a ground state helium atom into the $2\,^3S_1$ level and 20.6 eV for excitation into the $2\,^1S_0$ level. These are the highest minimum excitation thresholds for any neutral atom. The number density of electrons at each energy in the discharge is known as the electron energy distribution function. In general, it can be characterized by two parameters, the electron density n_e and the electron temperature T_e. The electron number density is the total number of electrons summed over all energies, whereas the electron temperature characterizes the relative number of electrons at each energy. A distribution with a 'hot' electron temperature has more high energy electrons than one with a 'cold' temperature. The requirement for high energy electrons ($E > 20$ eV) places a premium on operating the discharge with a high electron temperature as discussed in Sec. 10.1.3. Typically, a He-Ne laser operates with an electron temperature around 7 eV. Combining this temperature with the cross sections for electron-impact excitation into the metastable levels from the ground state of helium yields excitation rates of about $K_e^{gm} \approx 2 \times 10^{-10}$ cm^3 s^{-1} for both the He($2\,^3S_1$) and He($2\,^3S_1$) levels. The production rate of metastable helium atoms is found by multiplying these excitation rates by the number densities of electrons and ground state helium atoms, $K_e^{gm}\, n_e\, n_{He}$.

The gas pressure in the discharge is relatively high and hence the time between atom-atom collisions is very short. As is indicated in Figure 10.1 the energy above the He ground level of the He(2^1S_0) metastable level is nearly the same as the energy above the Ne ground level of the Ne($3s_2$) level, and the energy of the He(2^3S_1) metastable level is nearly the same as the energy of the Ne($2s_2$). Excitation transfer collisions between He and Ne result in an excited He atom in the 2^1S_0 or 2^3S_1 level returning to the ground level and a ground level Ne atom being excited to the $3s_2$ or $2s_2$ level. The small energy difference between the energies of the He and the Ne energy levels is made up by a change in the kinetic energy of the He and Ne atoms. In order for the energy difference between the He and Ne atoms to be taken up by a change in the kinetic energy of the atoms it is necessary that the energy difference be comparable to or smaller than a few times kT where k is Boltzmann's constant and T is the absolute gas temperature. The energy differences between the 2^1S_0 level of He and the $3s_2$ level of Ne is about 380 cm^{-1}, and the energy difference between the 2^3S_1 level of He and the $2s_2$ level of Ne is about 310 cm^{-1}. Room temperature is about 200 cm^{-1}. These small energy differences and other details of the collision process result in relatively large excitation cross sections. Nevertheless, the small energy barrier requires atoms moving with a velocity in the upper end of the Boltzmann distribution and thus the excitation transfer rate exhibits some sensitivity to the gas temperature. Neon atoms can also be excited into the $3s_2$ and $2s_2$ levels by electron-impact excitation of neon atoms. This mechanism, however, is relatively unimportant, contributing less than 5% to the total excitation rate for the $2s_2$ level and less than 2% for the $3s_2$ so that excitation transfer collisions are the primary mechanism for populating the upper laser level in Ne.

A Ne atom in the $3s_2$ level (a mixed J=1 level with both singlet and triplet character) can decay radiatively to any of the $2p_x$ or $3p_x$ levels with J=0, 1, or 2 by electric dipole emission. Thus an atom in the $3s_2$ level can decay to any of the $2p_x$ or $3p_x$ levels except the pure triplet level with J=3. As far as the operation of He-Ne lasers is concerned, the most interesting decays are into the $3p_4$ and the $2p_4$ levels, both of which have J=2. Decays from the $3s_2$ level into the $2p_4$ level result in the emission of a photon with a wavelength of 632.8 nm, while transitions into the $3p_4$ level have a wavelength of 3.39 μm. The $2s_2$ level decays to the $2p_4$ level with an emission wavelength of 1.15 μm. The $3p_4$ and $2p_4$ levels both decay rapidly to the $1s_x$ levels so that the population of these levels remains small. The lifetime of the $2p_4$ level is about 10^{-8} s, and the lifetime of the $3s_2$ level is about 10^{-7} s. The fact that the lifetime of the upper level is longer than the lifetime of the lower level aids in the formation of cw population inversion large enough that laser action can occur. A mixture of He and Ne gas can be made to lase easily on each of the three transitions discussed and can be made to lase on a number other transitions with the use of very high Q cavities (see Table 10.1 on p. 247).

10.1.3 *Gas Discharges*

Most commercial He-Ne lasers use a dc discharge although some early version designed for laser development operated with an rf discharge. Here only the dc discharge version is discussed. In a He-Ne laser, a potential difference of about $700eV$ is applied between the anode and cathode electrodes placed at opposite ends of a glass tube. Ions bombard the negatively biased cathode releasing electrons that are accelerated towards the anode. The voltage drop across the discharge is non-linear. There is a large electric field and hence a large potential drop in the immediate vicinity of the cathode (called the cathode fall), and then a weaker but uniform field in the *positive column* along most of the length of the discharge. Electrons are accelerated by the field and gain energy that is then lost in collisions with atoms. Two types of collisions are important. Electron-impact excitation produces excited atoms and ultimately light, whereas ionization produces

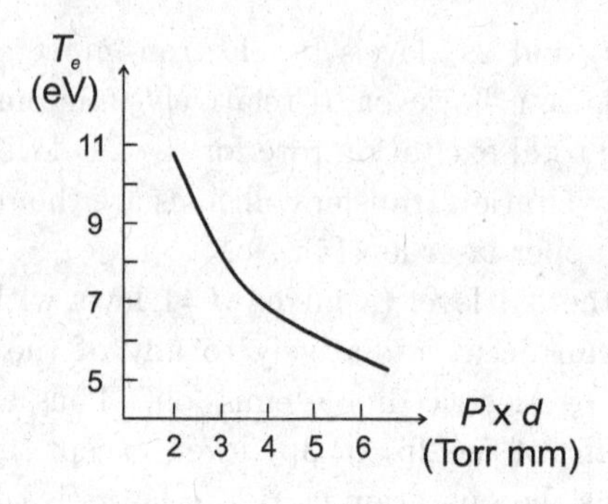

Fig. 10.2 Variation in electron temperature (T_e) with the product of the pressure and tube diameter ($P \times d$) for a DC discharge in a 5:1 He:Ne gas mixture. A temperature of 1 eV corresponds to 11,605 K. For a Boltzmann distribution, the average energy $\langle E \rangle = \frac{3}{2} T_e$.

new electron-ion pairs which sustain the discharge. Laser action occurs in the positive column of the discharge.

In the positive column portion of the discharge (where the electric field is uniform) the production rate of electron-ion pairs balances the loss rate of electrons and ions to the walls. The electron-ion production rate is equal to the product of the electron density, the atom density, and the ionization rate which is a function of both the ionization cross section and the energy distribution of the electrons. Since it requires a minimum energy of 24.6 eV to ionize a He atom and 21.6 eV for Ne, the ionization rates strictly depend upon the fraction of electrons with high energies ($E \gg 20$ eV). The loss rate of electron-ion pairs to the walls depends upon the distance to the walls (i.e. the radius of the discharge tube) and the ambipolar diffusion rate, which decreases as the gas pressure is increased. A full explanation of how dc discharges operate is beyond this text, but the basic idea is that the electron temperature (which characterizes the average energy of the electrons) varies inversely with the product of the pressure and tube diameter as shown in Fig. 10.2. The electron density is linearly dependent on the current density run through the discharge.

Next the relationship between discharge parameters and the production of laser light is calculated. The first step towards producing laser light is creating metastable helium atoms. In Sec. 10.1.2 the metastable production rate per unit volume was stated as being equal to $K_e^{gm}(T_e)\, n_e\, n_{\mathrm{He}}$. From this, it might appear that one can increase the laser output simply by increasing the number density/pressure of helium; but as shown by Fig. 10.2

T_e decreases (and thus $K_e^{gm}(T_e)$ decreases) as the pressure is increased for a fixed tube diameter. Decreasing the discharge tube diameter would appear beneficial for rasing T_e and thus the production rate of metastable helium atoms per unit volume, *but* this reduces total volume of the discharge, which lowers the total laser output. Due to these competing affects, almost all He-Ne lasers operate within a narrow range of $P \times d \approx (3.6$ to $4)$ Torr mm for a 5:1 He:Ne ratio. Since the ionization potentials differ between Ne and He, the T_e dependence curve varies with the gas composition. For a higher He:Ne ratio of 10:1 the optimum $P \times d$ shifts closer to 5 Torr mm, and decreases to near 3 Torr mm for a 3:1 ratio.

It still might appear that one can increase the laser's output by increasing the electron density n_e by cranking up the discharge current. This is not the case. In equilibrium the production rate of metastable helium atoms is equal to their loss rate. In addition to excitation transfer collisions to neon atoms, metastable helium atoms can also be lost/quenched by collisions with electrons. In the steady state equating the two rates yields

$$K_e^{gm}(T_e)\, n_e\, n_{\mathrm{He}} = K_e^{mq}(T_e)\, n_e\, n_{\mathrm{He}*} + K^{\mathrm{tr}}\, n_{\mathrm{He}*}\, n_{\mathrm{Ne}}\,, \tag{10.2}$$

where K_e^{gm} is the electron-impact excitation rate into the $2\,^1\mathrm{S}_0$ He metastable level, K_e^{mq} is the electron-induced quenching rate out of the metastable level, and K^{tr} is the excitation transfer rate from the $2\,^1\mathrm{S}_0$ level into the Ne $2\mathrm{s}_2$ level. Solving for He metastable density, $n_{\mathrm{He}*}$ gives,

$$n_{\mathrm{He}*} = \frac{K_e^{gm}(T_e)\, n_e\, n_{\mathrm{He}}}{K_e^{mq}(T_e)\, n_e + K^{\mathrm{tr}}\, n_{\mathrm{Ne}}}\,. \tag{10.3}$$

As long as $K_e^{mq}(T_e)\, n_e \ll K^{\mathrm{tr}}\, n_{\mathrm{Ne}}$ the metastable He density rises with the electron density, but when $K_e^{mq}(T_e)\, n_e \gg K^{\mathrm{tr}}\, n_{\mathrm{Ne}}$ the metastable He density becomes independent of the electron density. Once the metastable helium density starts to saturate, further increases in the electron density are detrimental to the laser's gain due to an increased electron-impact excitation rate into the lower level of the laser transition (i.e., the neon $2\mathrm{p}_4$ in the case of the 632.8 nm laser). Typically, the peak gain in He-Ne lasers occur when $n_e \approx (1-2) \times 10^{11}$ cm^{-3}.

While the production rate of metastable helium atoms increases with the electron density, the production rate of Ne($3\mathrm{s}_2$) eventually saturates. The excitation transfer rate for the $2\,^1\mathrm{S}_0$ level is given by $n_{\mathrm{He}*}\, \langle \sigma_{tr}(v)\, v_{\mathrm{He-Ne}} \rangle\, n_{\mathrm{Ne}}$ where $n_{\mathrm{He}*}$ is the number density of He atoms in the $2\,^1\mathrm{S}_0$ level, n_{Ne} is the number density of the ground level Ne atoms, $\sigma_{tr}(v)$ is the excitation transfer cross section as a function of velocity v

measured in the center-of-mass frame, and $v_{\mathrm{He-Ne}}$ is the relative velocity between the He and Ne atoms. The other major loss mechanism for metastable helium atoms is electron-atom collisions. Electron collision cross sections for atoms in metastable levels are generally orders of magnitude larger than cross sections out of the ground state. Further, unlike excitation from the ground state which requires high energy electrons, excitation into higher levels from a metastable level can be accomplished by low energy electrons. De-excitation to lower levels (so called super-elastic collisions) can be accomplished by electrons of *any* energy.

The brackets in the transfer rate indicate an average over the thermal velocity distribution. For simplicity the following approximations are used, (i) $v_{\mathrm{He-Ne}} \approx v_{\mathrm{He}}$ since helium is much lighter than Ne, and (ii) average numbers for the cross section and the velocity are used rather than averaging over the velocity distribution. For a $\sigma_{tr} \approx 4\times10^{-16}$ cm^2 and $v_{\mathrm{He}} \approx 1.5\times10^5$ cm/s at 400 K, so $\langle\sigma_{tr}(v)\, v_{\mathrm{He-Ne}}\rangle \approx 6\times10^{-11}$ cm^3 s^{-1}.

The discussion to this point has covered the key processes in how the gas discharge produces light at the laser emission wavelength, but there are a number of additional atomic physics processes at work that also contribute to explaining *why* a He-Ne laser works. For example, the $3s_2$ upper level of the 632.8 nm laser transition is a resonance level that can decay to the neon ground state. If one were to include this decay channel in calculating the radiative lifetime of the $3s_2$ level, it would no longer be true that the lifetime of the upper $3s_2$ level is longer than the lifetime of the $2p_4$ lower level, which would hamper laser operation. The reason it *can* be omitted is *radiation trapping*. When an atom does decay to the ground state, it emits a photon that is promptly reabsorbed by another neon atom before it escapes the discharge (i.e., the photon is trapped within the plasma). Due to this 'recycling', the total number of excited atoms in the discharge remains unchanged, and the system acts as if this decay route is much slower. While radiation trapping normally only occurs on transitions to the ground state, it can also affect transitions that terminate on metastable levels since the number density of these levels can also be quite high. Indeed, radiation trapping of the Ne($2p_4 \rightarrow 1s_5$) transition ultimately limits the population inversion by decreasing the decay rate of the lower level of the laser transition.

10.2 He-Ne Lasers in Practice

10.2.1 *Gas Discharge Components*

A schematic diagram of a He-Ne laser is shown in Figure 10.3. Typically a dc discharge is produced between the anode and the cathode electrodes. It is interesting to understand why He-Ne lasers have the construction shown in Figure 10.3. The mirrors are commonly glued on to the laser as indicated. The cavity is aligned before the glue sets. The anode is simply a tungsten feed-through. The cathode is made of ultra pure aluminum that has been treated to have an oxide coating. Aluminum outgasses very little which helps keep the laser tube free of impurities. Collisions between the metastable He and impurities result in ionization of the impurity (a process that is called Penning ionization). This reduces the pumping rate of the Ne. The build up of impurities is a common cause of failure for He-Ne lasers. The impurities when ionized are accelerated by the electric field between the anode and cathode. They are driven into the cathode or walls of the laser thereby cleaning the tube. In order to keep the tube clean it is desirable to run the laser regularly, and not let it sit idle for long periods of time. The oxide layer on the cathode enhances the electron emission of the cathode when it is bombarded by ions or uv photons. The electrons emitted by the cathode are necessary in order to sustain the discharge.

The narrow diameter of the capillary tube serves two functions. First, the product of the diameter and fill pressure ($P \times d$) determines the electron temperature of the discharge and thus the rates of electron-driven reaction rates. Second, the narrow diameter also shortens the travel time of atoms to the walls. The $2p_4$ level decays rapidly to the $1s_2$, $1s_4$, or $1s_5$ levels. The $1s_2$, and $1s_4$ levels decay to the ground level, but the $1s_5$ level is metastable. If

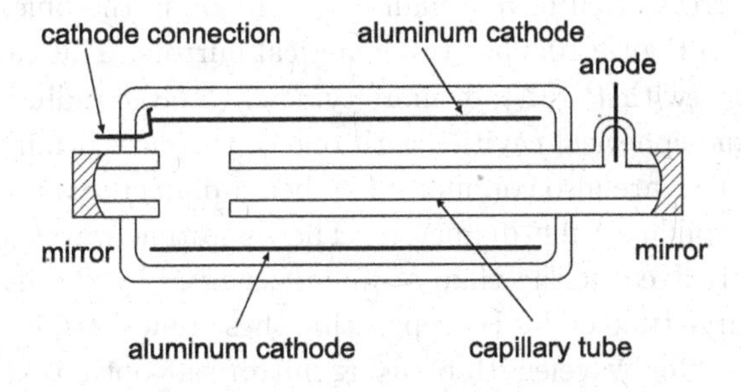

Fig. 10.3 A schematic diagram of a He-Ne laser.

the population of the $1s_5$ level becomes too large then the radiation emitted by the $2p_4$ level when it decays to the $1s_5$ level is absorbed by the atoms in the long-lived $1s_5$ level. This results in radiation trapping that keeps the population of atoms in the $2p_4$ level high, reducing the population difference between the upper and lower laser levels. Metastable atoms are non-radiatively de-excited to the ground level by collisions with the walls of the capillary tube. The capillary tube should have a small diameter so that the metastable Ne atoms reach the wall quickly thus keeping the population of atoms in the metastable level low.

A typical bore diameter for the capillary is 2 mm. Typical partial pressures in the He-Ne laser are 1.6 Torr of He and 0.3 Torr of Ne. It is interesting that almost all He-Ne lasers use ^{3}He in the mixture rather than the more common isotope ^{4}He. At a given temperature the lighter isotope has a higher velocity and hence the collisional excitation transfer rate (i.e. the pumping rate of the upper laser level) is higher for a laser filled with the lighter isotope. Finally the discharge in a He-Ne laser has typically an anode to cathode voltage of about 700 V. The discharge current scales with the cross sectional area of the capillary tube. For a 2 mm bore diameter, the discharge current is typically around 10 mA.

10.2.2 *Optical Components*

In Figure 10.3 the mirrors of the He-Ne laser optical cavity are an integral part of the discharge tube, forming the ends. The mirrors are commonly glued/bonded on to the capillary tube as indicated. The cavity is aligned before the glue sets. For operation at a fixed wavelength, thin-film dielectric coated mirrors that have a high reflection coefficient at the laser wavelength are used. Due to their high reflectivity at red wavelengths, for a 632.8 nm laser these mirrors often have a high transmission in the blue. The mirror configuration in Figure 10.3 has two spherical mirrors. This can be either a confocal cavity (with $R = L$), or more typically a large radius design (with $R \gg L$). Plano-spherical cavities with one of the curved mirrors replaced by a plane mirror are also common. The beam diameter is typically about a third the capillary tube diameter. The operating wavelength of most He-Ne lasers is fixed at the time of manufacturing by the use of integral mirror/discharge tube ends. By separating these functions, it is possible to select the operating wavelength by using mirror sets optimized for different wavelengths or by using a prism as shown in Figure 10.4.

It is straight forward to limit the laser so that it lases only on a single

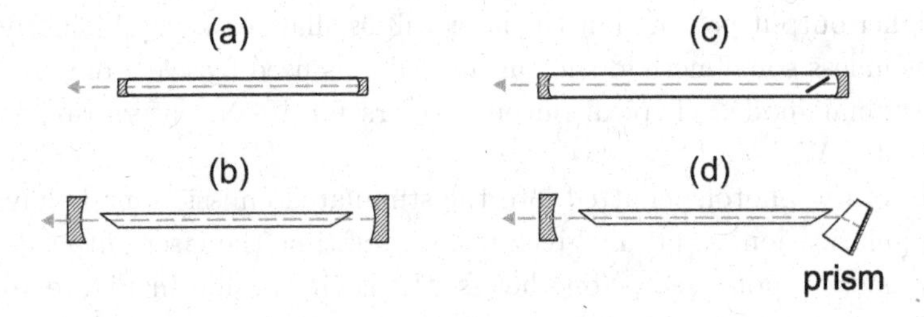

Fig. 10.4 Optical cavity designs for He-Ne lasers. (a) Mirrors as end windows: random polarization, fixed wavelength. (b) External mirrors, Brewster windows on discharge tube: linear polarization, mirror sets to select λ. (c) Mirrors as end windows with internal Brewster plate: linear polarization, fixed λ. (d) External cavity, rotatable prism to select wavelength, linear polarization.

transverse mode, the 0-0 mode. Typically, a He-Ne laser will lase on several longitudinal modes. The 632.8 nm transition in a He-Ne laser has a Gaussian line shape with a Doppler width of about 10^9 Hz. The small signal gain of the laser has a Gaussian line shape and is above the threshold value for some fraction of the Doppler line shape. The longitudinal cavity modes of the laser are separated by $\Delta\nu_c = c/2nL$. The laser can oscillate on all of the modes that are within that part of the Doppler line shape that has a small signal gain above threshold. This is possible because the different modes interact with different segments of the Doppler line shape and hence different Ne atoms. When the laser is oscillating the small signal gain curve has holes burned in it at the frequencies corresponding to each of the cavity modes that are within that part of the Doppler line shape with a small signal gain above the threshold value. The population difference at velocities corresponding to the frequencies of the oscillating modes is clamped at the threshold value.

It is possible to operate a He-Ne laser in a single longitudinal mode if the laser is short enough that only one cavity mode occurs within that part of the Doppler line shape with a small signal gain above threshold. If that part of the Doppler line shape with a small signal gain above threshold is about 10^9 Hz in width then one concludes that the maximum length for a single mode laser cavity is about L=15 cm. Such a short length for the cavity limits the output power of the laser to a rather low value. Thermal fluctuation induced changes in the cavity length can cause the output power of the laser to fluctuate as the exact frequency of the cavity varies with respect to the Doppler profile of the gain medium. Longer cavities allow for higher output powers, but the laser will oscillate on several longitudinal modes unless some method such as an etalon is used to select only a single longitudinal mode. Typical output powers for He-Ne lasers range from 0.5 to 50 mW.

All of the photons emitted into the stimulated emission mode have the same polarization, so at any given instance of time the laser output is generally linearly polarized. Nonetheless, the cavity design in Figure 10.4(a) is cylindrically symmetric, with no external element to define the direction of polarization. As a result, the laser is 'randomly polarized'. This direction will also fluctuate with time due to spontaneous emission similar to the discussion about the minimum bandwith of a laser in Sec. 6.3. With an external element, it is possible to fix the direction of the polarization. For other types of gas lasers (Chapter 9), this polarization control is done by attaching the end widows of the discharge tube at Brewster's angle as illustrated in Fig. 10.4(b). Light incident on a transparent material at Brewster's angle and with a polarization in the plane of incidence has a reflection coefficient of zero and hence there is no loss due to reflection, whereas other polarizations will have some losses. Since the population difference clamps at the population difference corresponding to the lowest threshold population difference, only the polarization direction with the lowest losses will lase. In a He-Ne laser with integral windows/mirrors, the same effect can be achieved by adding an internal window/element oriented at Brewster's angle within the discharge tube as indicated in Fig. 10.4(c). Polarized emissions can also be achieved by selectively populating select m_J states of the excited level. By applying a *uniform* magnetic field along the discharge tube, it is possible to use the Zeeman effect to shift the energies of the Ne(2s$_2$) $+m_J$ states closer into resonance with that of the He(2^1S_0) level and the Ne(2s$_2$) $-m_J$ states further from resonance, resulting in polarized emissions with a fixed polarization orientation.

10.2.3 *Operating Wavelength*

While the vast majority of He-Ne lasers are designed to operate on the red 632.8 nm $3s_2 \rightarrow 2p_4$ transition, they are also can operate at a number of other wavelengths as listed in Table 10.1. Both the oscillation at 3.39 μm and 632.8 nm originate with the same upper level, the 3s$_2$ level of Ne. The population difference clamps at the lowest threshold population difference for all transitions with the same upper level. From Eq. (3.14), this difference

Table 10.1 Most popular operating wavelengths of He-Ne lasers.

λ_{air}	Ne transition	color	A_{ij} (10^6 s^{-1})	'popularity'
543.4 nm	$3s_2 \to 2p_{10}$	green	0.28	●●
593.9 nm	$3s_2 \to 2p_8$	yellow	0.20	●●
604.6 nm	$3s_2 \to 2p_7$	orange	0.23	○
611.8 nm	$3s_2 \to 2p_6$	orange	0.61	●
632.8 nm	$3s_2 \to 2p_4$	red	3.39	●●●●●●
1.152 μm	$2s_2 \to 2p_4$	near-infrared	10.4	○
1.523 μm	$2s_2 \to 2p_1$	near-infrared	0.85	●
3.392 μm	$3s_2 \to 2p_4$	infrared	2.9	○

is equal to

$$\Delta n_{\rm th} = \frac{8\pi\,\nu_0^2\,\Delta\nu}{(c/n)^3}\,\frac{1}{A_{ij}\,\tau_{\rm ph}} = -\frac{8\pi\,\Delta\lambda}{\lambda_0^4}\,\frac{1}{A_{ij}\,\tau_{\rm ph}}\,. \tag{10.4}$$

Based upon their wavelengths, the threshold population difference for the 3.39 μm transition will be much lower than that of the 632.8 nm line. Lasing can nonetheless be obtained on the 632.8 nm transition by controlling the effective photon lifetime, $\tau_{\rm ph}$, for the two wavelengths by making the cavity have a high Q for one wavelength and a low Q for the other. One must select cavity Q's such that the threshold population difference is lower for the transition with the high Q cavity than for the transition with the low Q cavity, i.e. by controlling the mirror reflectivity at the two wavelengths. The population difference clamps at the lower threshold population difference and hence oscillation occurs on only the selected transition.

Having a lossy reflection coefficient at 3.39 μm is generally sufficient for short laser tubes, but for a long discharge tube the stimulated emissions from even the single pass gain of the 3.39 μm line may be a problem. Another means of altering the effective $\Delta n_{\rm th}$ is varying the linewidth $\Delta\nu$. Normally, the linewidth is dominated by the Doppler width. The Doppler width, Eq. (4.30), scales inversely with the wavelength, so the linewidth of the 3.39 μm transition is much narrower than that of the 632.8 nm line, yielding a lower threshold population difference. Adding an *inhomogeneous* magnetic field along the discharge tube will induce a broadening in both lines as the energy of each Zeeman component is shifted proportional to the magnetic field strength. The amount of Zeeman broadening is the same for both lines, but since the Doppler width of the 632.8 nm line is about five times larger than the 3.39 μm line, for $(0-100)$ Gauss magnetic fields the resulting 632.8 nm linewidth is only marginally increased from its

Doppler width while that of the 3.39 μm transition is increased substantially. This increases the required threshold population difference for the infrared 3.39 μm transition without much effect on the Δn_{th} required for the 632.8 nm line.

Unlike the laser wavelengths that all arise from pumping of the Ne(3s$_2$) level, the 1.15 μm and 1.523 μm lines arise from pumping of the Ne(2s$_2$) level by excitation transfer from He(2^3S). Oscillation can also be obtained on these transitions by using a high Q cavity optimized for one of these wavelengths.

10.3 Example: Analysis of a Low-Power He-Ne Laser

For illustration, consider the analysis of the sample He-Ne laser shown in Fig. 10.5. An estimate of the threshold population difference required for laser action on the 632.8 nm transition is given by $\Delta n_{\text{th}} = (8\pi\,\nu_0^2\,\Delta\nu)/(c/n)^3 \times (\tau_u/\tau_{\text{ph}})$. The lifetime of the upper level is $\tau_u = 1.7 \times 10^{-7}$ s. The line has a Doppler line shape with a width $\Delta\nu = 7.16 \times 10^{-7}(c/\lambda)(T/A)^{1/2} = 1.5 \times 10^9$ Hz where the temperature is taken as T=400 K. The photon lifetime, ignoring other losses, can be obtained from Eq. (3.6) using the mirror separation of L=25 cm and the reflectivity of the output coupler $\tau_{\text{ph}} = 2L/(cT_2) = 3.3 \times 10^{-8}$ s. These values lead to a threshold population difference of $\Delta n_{\text{th}} = 1.6 \times 10^{15}$ excited Ne atoms/m^3 = 1.6×10^9 excited Ne atoms/cm^3. In our sample case, the He and Ne pressures are 1.6 Torr and 0.2 Torr respectively. These pressures correspond to He and Ne number densities of 5.7×10^{16} cm^{-3} and 7.1×10^{15} cm^{-3} respectively. Thus about 2 Ne atoms in every 10^7 atoms must be in the upper laser level in order to reach threshold.

How threshold is reached in the He-Ne laser is examined next. Electron excitation of helium atoms produces atoms in the metastable 2^1S_0 level. The rate at which these excited He atoms are produced is equal to

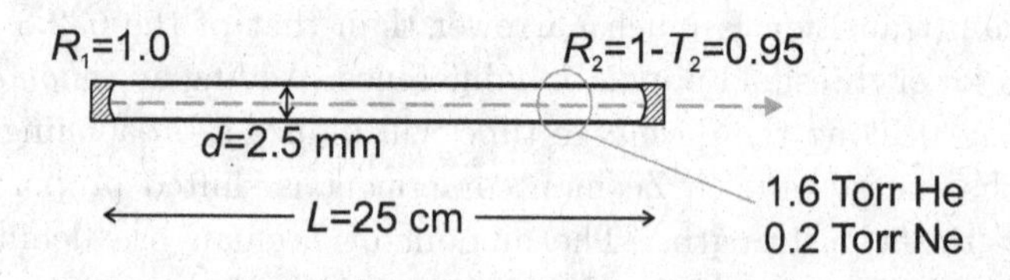

Fig. 10.5 Parameters for analysis of a sample He-Ne laser. For a discharge current of ~10 mA, the electron density is about 2×10^{10} cm^{-3} with a 400 K gas temperature.

$n_e \langle \sigma_{ex} v_e \rangle n_{He}$ where n_e is the electron number density, σ_{ex} is the electron excitation cross section from the ground level of He into the 2^1S_0 level, v_e is the electron velocity, which is essentially equal to the relative velocity between the electron and the He atoms, and n_{He} is the number density of ground level He atoms. The brackets indicate an average over the energy distribution. For simplicity average values are used. In steady state, the rate of production of excited He atoms must be equal to the loss rate. Neglecting the loss rates due to diffusion to the walls, the tiny radiative decay rate of metastable 2^1S_0 atoms and the loss rate due to other collision processes, the primary loss rate is assumed to be excitation transfer to Ne atoms. The excitation transfer rate is given by $n_{He*} \langle \sigma_{tr} v_{He-Ne} \rangle n_{Ne}$ where n_{He*} is the number density of He atoms in the metastable 2^1S_0 level, n_{Ne} is the number density of the ground level Ne atoms, σ_{tr} is the excitation transfer cross section, and v_{He-Ne} is the relative velocity between the He and Ne atoms, which is approximated as being equal to the velocity of the He atoms. In the absence of lasing the rate at which the excited Ne atoms are produced is nearly equal to the rate at which they decay radiatively. This leads to the result that in the absence of laser action $n_e \sigma_{ex} v n_{He} = n_{He*} \sigma_{tr} v_{He-Ne} n_{Ne} = n_{Ne*}/\tau_u$ where n_{Ne*} is the number density of Ne atoms in the $3s_2$ upper laser level and where $\tau_u = 1.7 \times 10^{-7}$ s is the radiative lifetime for $Ne(3s_2)$ atoms.

The electrons in the gas discharge have a range of electron energies, characterized by the electron temperature. Only a small fraction have high enough energy to excite the 2^1S_0 level in He. The electron temperature can be estimated from the product of the gas pressure and tube diameter. In this case, $P \times d = 4.5$ Torr mm. For a 8:1 He:Ne mix, this corresponds to about a 6.5 eV electron temperature. For a Maxwell-Botzmann energy distribution with T_e=6.5 eV, about 10% of all electrons have an energy greater than the 20 eV necessary to excite the $He(2^1S_0)$ level. Assuming the laser is operating close to the maximum optimal electron density of 2×10^{10} cm^{-3}, the effective electron density is approximately $n_e = 2 \times 10^9$ electrons/cm^3.

In this situation He atoms in the 2^1S_0 level are produced at a rate of $n_e \sigma_{ex} v_e n_{He} = 3 \times 10^{16}$ cm^{-3}/s where $v_e = 2.6 \times 10^8$ cm/s corresponding to an electron with 20 eV energy and $\sigma_{ex} = 1 \times 10^{-18}$ cm^2. Assuming that excitation transfer is the sole loss mechanism for the metastable helium atoms, this is also the production rate of excited Ne atoms in the upper laser level, and correspondingly the volume production rate of 632.8 nm photons. The number density of excited He* atoms can be obtained by equating the excitation transfer loss rate with the production rate, $n_{\mathrm{He*}}\,\sigma_{\mathrm{tr}}\,v_{\mathrm{He-Ne}}\,n_{\mathrm{Ne}} = 3 \times 10^{16}\ \mathrm{cm}^{-3}/\mathrm{s}$. Using $\sigma_{\mathrm{tr}} = 5 \times 10^{-16}\ \mathrm{cm}^2$ and $v_{\mathrm{He-Ne}} = 2.5 \times 10^5$ cm/s (corresponding to a temperature of 400 K) one obtains $n_{\mathrm{He*}} = 1.5 \times 10^{11}\ \mathrm{cm}^{-3}$ (or about 4 in 10^5 of the He atoms in the discharge). The number of excited Ne atoms in the $3s_2$ upper laser level in the absence of laser action can be found by dividing the volume production rate by the decay rate (equal to τ_u^{-1}), $n_{\mathrm{Ne*}} = 3 \times 10^{16}\,\mathrm{cm}^{-3}/\mathrm{s} \times \tau_u = 5 \times 10^9\ \mathrm{cm}^{-3}$. This simplified analysis indicates that in the absence of laser action the discharge can produce enough excited Ne atoms in the upper laser level to be about three times above threshold. A more complete analysis would average over the velocity distributions of the electrons and atoms and include additional collision processes that contribute to the loss and production rates. Nevertheless the ideas and the average numbers for the He-Ne laser are basically correct.

The power output of a He-Ne laser can be found by multiplying the production rate of Ne($2s_2$) atoms for the entire laser tube by the energy per emitted 632.8 nm photon. The production rate for the entire laser tube is found by multiplying the previously calculated production rate per cm^3 by the volume of the laser $V = \pi(d/2)^2\,L$. For our sample case with d=2.5 mm and L=25 cm, V= 1.2 cm^3, and the output power is calculated to be about $P_L = 3 \times 10^{16}\,\mathrm{cm}^{-3}/\mathrm{s} \times h\nu \times V = 10$ mW. This assumes the laser beam completely fills the discharge volume, more typically the laser beam diameter is about a third the capillary tube diameter, in which case $P_L \approx 1$ mW.

Summary of Key Ideas

- He-Ne lasers are a gas discharge lasers that use excitation transfer between metastable helium atoms and ground state neon atoms to create a population inversion.
- Typically, He-Ne lasers have a 5:1 He:Ne mix. The optimum total pressure (P) is a function of the capillary tube diameter d with $P \times d \approx$ 4 Torr mm.
- Most He-Ne lasers operate at 632.8 nm on the Ne($2s_2 \rightarrow 2p_4$) transition. These are typically cw lasers with mW power levels.

Suggested Additional Reading

D. C. Sinclair and W. E. Bell, *Gas Laser Technology*, Holt, Rinehart, and Winston (1969).

A. Javan, W. R. Bennett, Jr., and D. R. Herriott, "Population Inversion and Continuous Optical Maser Oscillation in a Gas Discharge Containing a He-Ne Mixture" *Phys. Rev. Lett.* **6**, 106-110 (1961).

William T. Silfvast, *Laser Fundamentals*, Cambridge University Press (1996).

Arnold L. Bloom, *Gas Lasers*, John Wiley & Sons (1968).

B. E. Cherrington, *Gaseous Electronics and Gas Lasers*, Pergamon Press (1979).

Problems

1. For a He-Ne laser tube with a capillary bore of 1 mm and gas density of 6×10^{16} cm^{-3} *estimate* the diffusion time to the wall for a Ne atom that starts at the center of the tube. Use a gas temperature of 400 K. You can assume the neon atom atom makes a random walk, with a mean free path given by $\ell_{\mathrm{MFP}} = 1/n\sigma$ where $\sigma \approx 5 \times 10^{-15}$ cm^2 is the gas kinetic cross section.

2. If a He-Ne laser is to oscillate on the 632.8 nm line rather than the 3.39 μm line, estimate the minimum ratio of the photon lifetimes for the 632.8 nm and 3.39 μm photons.

3. (a) Calculate the Doppler widths (in MHz) of the 632.8 nm and 3.39 μm transitions assuming a gas temperature of 400 K. (b) The Zeeman shift of an m_J state in an external magnetic field B_0 is equal to $\Delta E = m_J\, g_J\, \mu_B\, B_0$, where g_J is the Landé g-factor and μ_B is the Bohr magneton. Calculate the energy difference (in MHz) for the $m_J = \pm 1$ states of the Ne(2s$_2$) level for B_0=50 Gauss (typical of a small permanent magnet). The Landé g-factor for this level is 1.295. (c) Express the splitting as a percentage of the Doppler widths of both lines calculated in (a).

4. Is it possible to create a multi-line He-Ne laser that simultaneously lases on multiple lines? If so, are all combinations of lines possible? If not, what prevents simultaneous lasing? Explain.

5. For radiation trapping to suppress the resonance transition from Ne(2s$_2$) to the ground state, the mean free path for an emitted photon should be much less than the capillary tube radius. (a) Calculate the photon absorption cross section at line center (Chapter 2) for the 2s$_2$ ($J = 1$) $\rightarrow$ ground state ($J = 0$) transition, $\lambda = 60$ nm, $A_{ij} = 2.7 \times 10^7$ s^{-1} assuming a Doppler broadened line with T=400 K. (b) Estimate the mean free path (mfp $= 1/(\sigma n)$) for a resonance transition photon in a He-Ne laser filled with 0.3 Torr neon (T=400 K). How does this compare to a typical tube diameter of 2 mm?

6. A 632.8 nm He-Ne laser has a longitudinal mode spacing of 685 MHz, with a beam divergence (2θ full angle) of 1.34 mrad. The TEM$_{00}$ beam diameter at the output coupler is 0.60 mm. Determine the distance from the output coupler to the beam waist. Use this to determine the cavity

type. Make a sketch of the optical cavity, include the cavity length and indicate if each end mirror is curved or planar.

7. Design a He-Ne laser. You have at your disposal a long length of glass tubing with inner diameter of 5 mm that can be cut to your desired length and fitted with the required electrodes and end windows. You also have two spherical mirrors with R=2 m. The reflectivity of one mirror is 1.0 and the other is 0.99 at $\lambda = 632.8$ nm. Sketch out your laser and describe its design and operating characteristics in as much detail as possible. For example, provide the approximate fill pressures for the He and Ne gases. Is the laser single mode or multi-mode? Is it stable? Is the output linearly polarized?

Chapter 11

Gas Lasers

Historically, gas discharge lasers were among the first lasers produced, but they have been generally replaced by solid state lasers in most applications. Nonetheless, gas discharge lasers have some advantages and certain varieties are still widely used. For example, carbon dioxide lasers (Section 11.1) are used widely in applications such as the machining of metal where high power is required, and excimer lasers (Section 11.2) are used extensively in the photolithography used for the manufacturing of integrated circuits.

One general feature of gas lasers that enhances their utility is the fact that a gas heals itself if a damaging accident occurs in the gas. For example if a spark occurs in a gas when one desires a glow discharge it does not permanently damage the laser medium. This is in contrast to a solid state laser where an accident that damages the laser crystal may be irreparable.

There are an enormous number of different types of gas lasers and no attempt is made in this chapter to discuss all of them. Nevertheless, an attempt *is* made to sample the wide variety of different mechanisms used for producing the population inversion necessary for laser action. The population inversion can be produced by excitation transfer (as in the He-Ne laser), direct electron impact excitation, photodissociation excitation, chemical reaction excitation, and excimer excitation. In addition to the method used to produce the population inversion and hence the gain of laser, there are a number of other interesting features needed to understand each type of laser including the wavelength of the laser, the typical output power of the laser, and other characteristics of the laser.

11.1 CO_2 Laser

The CO_2 laser is a high power gas discharge laser. It lases on a vibrational transition in the ground electronic state. The CO_2 laser uses both excitation transfer and direct electron impact to produce the population inversion. There are a large number of other lasers that also lase on vibrational transitions in various molecules. Vibrational transitions usually have wavelengths in the infrared. The lasing wavelength for the CO_2 laser is 10.6 μm. The vibrational energy levels of CO_2 are shown in Fig. 11.1.

In order to understand the notation used in the CO_2 energy level diagram it is necessary to understand the vibrations of the CO_2 molecule. The CO_2 molecule is a linear molecule with the carbon atom between the two oxygen atoms so that the molecule is represented as O-C-O. This molecule has three different normal modes of vibration. They are the following: (i) the symmetric stretch mode, (ii) the asymmetric stretch mode, and (iii) the bending mode. The symmetric stretch mode is a vibration at a frequency ν_1, in which the two O atoms move in opposite directions along the internuclear axis of the molecule while the C atom remains at rest midway between the two O atoms. The asymmetric stretch mode is a vibration at a frequency ν_3, in which the two O atoms move along the internuclear axis in a direction opposite to the direction of motion of the C atom. The bending mode is a vibration at a frequency ν_2, in which the two O atoms move in a direction perpendicular to the internuclear axis while the C atom moves in the opposite direction. The bending mode can always be described as a linear combination of bending motions along two axes that are both normal

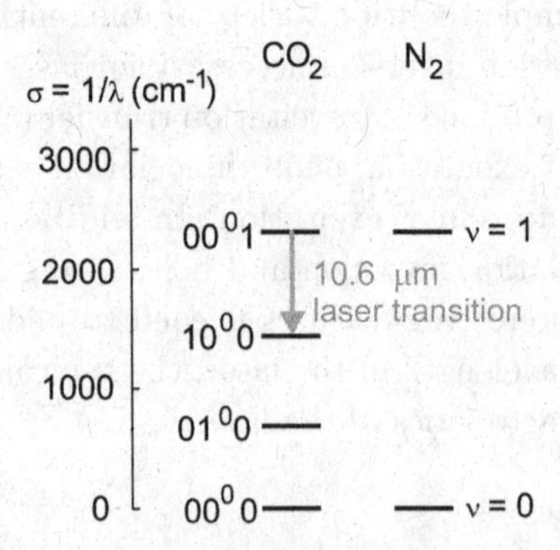

Fig. 11.1 The ground level vibrational energy levels of CO_2 and N_2 molecules. The energies are given in terms of the wave number in cm^{-1} above the ground vibrational level.

to the internuclear axis and are at right angles to each other. It is possible for this bending mode to have a component of angular momentum either parallel to or antiparallel to the internuclear axis if the linear combination of modes is phased so that the atoms rotate about the internuclear axis. The vibrational energy levels of a CO_2 molecule are described by specifying the frequencies ν_1, ν_2, and ν_3 and the angular momentum of the bending mode. The energy of a particular vibrational level is given by

$$W_n = n_1 h\left(\nu_1 + \frac{1}{2}\right) + n_2 h\left(\nu_2 + \frac{1}{2}\right) + n_3 h\left(\nu_3 + \frac{1}{2}\right) \tag{11.1}$$

where n_1, n_2, and n_3 are the vibrational quantum numbers corresponding to the three vibrational modes, ν_1, ν_2, and ν_3, of the CO_2 molecule. A given energy level is determined by the quantum numbers n_1, n_2, and n_3. The vibrational angular momentum quantum number is equal to the component of the vibrational angular momentum parallel to the internuclear axis divided by $h/2\pi$. The vibrational angular momentum quantum number of the bending mode is denoted as a superscript to the bending mode quantum number. Thus the CO_2 energy level denoted as 02^00 has n_1=0, n_2=2, n_3=0 and has zero vibrational angular momentum along the internuclear axis. The CO_2 laser operates between the 00^01 level (asymmetric stretch) and the 10^00 level (symmetric stretch) in the CO_2 molecule.

In addition to the energy level diagram for the vibrational levels of the CO_2 molecule Figure 11.1 also shows the energy levels for the ground level and the first excited vibrational level of the N_2 molecule. As seen in the Figure 11.1 the first excited vibrational level of the N_2 molecule has nearly the same excitation energy as the upper laser level in the CO_2 molecule. The excitation of the CO_2 molecule is achieved through electron collisions in a discharge either by exciting the CO_2 molecules to the upper laser level directly or after cascade from a higher level or by exciting a N_2 molecule to its first excited vibrational level followed by an excitation transfer collision between a vibrationally excited N_2 molecule and a ground level CO_2 molecule. A CO_2 laser usually contains N_2 and He gases in addition to the CO_2 gas. The ratios of the gases is typically CO_2:N_2:He=1:1.2:9. The He gas in the laser serves two purposes. The first purpose is to assure that the gas discharge is a uniform glow discharge. Helium gas readily supports a glow discharge. The second purpose of the He gas is to collisionally depopulate the lower laser level so that the loss rate out of the lower laser level is more rapid than would be the case if the level population decreased only by radiative decay.

The CO_2 molecule has a large number of rotational levels for each vibrational level. Because the CO_2 laser is usually operated at high pressures the rotational energy levels are normally equilibrated to a temperature that is nearly equal to the gas kinetic temperature. The possible laser transitions are between the various rotational levels of the 00^01 vibrational level and the various rotational levels of the 10^00 vibrational level. Thus there are a number of transitions that can lase.

The natural radiative lifetime for the upper laser level is 3 s. The Doppler width for the CO_2 laser is between 5 and 100 MHz depending on the gas kinetic temperature. The collision rate for a CO_2 molecule is such that for pressures of 10-20 Torr or higher the primary broadening mechanism for the laser transition is collisional broadening so that the laser transition is homogenously broadened. At pressures below 10 Torr the CO_2 laser transition is Doppler broadened and the transition is inhomogeneously broadened.

The CO_2 laser can operate either cw or pulsed with very high output powers. The CO_2 laser is very efficient. Lasers that emit powers of a few watts up to many kilowatts have been constructed. Due to their combination of high output powers, ease of construction, and reliablity, CO_2 lasers are widely used for laser machining of various materials and for other industrial applications. The use of a laser for machining permits one to cut metals and other materials in delicate and intricate patterns. This type of machining is very difficult to carry out without the use of high power lasers.

11.2 Excimer Lasers

An *excimer* molecule is a molecule with an unbound ground level (i.e. a molecule whose ground level rapidly disassociates into individual atoms). A potential energy curve for an excimer laser (KrF) was presented in Fig. 3.2 on page 46. Excimer lasers are important pulsed gas lasers that have very high output powers. The most commonly used excimer lasers are the XeF, KrF, and XeCl lasers although F_2, ArF, and KrCl lasers are used to obtain laser action at wavelengths that can not be obtained from XeF, KrF or XeCl. Rare gas dimmers such as Xe_2, Kr_2, or Ar_2 are also excimer molecules that can serve as the active medium for a laser. Chemicals containing F atoms are often very dangerous and must be handled with extreme care. All the laser surfaces need to be passivated to avoid chemical reactions with the excimer gases. The wavelengths of the XeF and KrF lasers are centered at 351 nm and 248 nm respectively, see Table 11.1 for other excimer laser

Table 11.1 Wavelengths of excimer lasers.

λ (nm)	Excimer	λ (nm)	Excimer
126	Ar_2	222	KrCl
146	Kr_2	248	KrF
157	F_2	282	XeBr
172	Xe_2	308	XeCl
193	ArF	351	XeF

wavelengths. Typical output powers for commercial XeF and KrF lasers are 90 and 200 mJ per pulse respectively. These lasers can operate with repetition rates of about 100 Hz. Thus the average output powers for these lasers are generally in the 5-10 W range.

Excimer lasers can be excited in two different ways. The first method of exciting an excimer laser is by electron collisions in a glow discharge. The commercial excimer lasers use this technique and are usually constructed in a manner similar to a N_2 laser. A fast pulsed glow discharge is struck between two parallel electrodes that are about 0.5-1.0 m in length. Often there is a method of providing some *preionization* for a discharge type excimer laser. The function of the preionization is to provide seed ionization in the gas so that the laser breaks down into a glow discharge along the entire length of the electrodes and not into one or a few arcs at localized places along the electrodes. The preionization can be accomplished in a number of ways. For example there might be a set of corona discharges near the region where the glow discharge is to be formed. The corona sparks are shielded from the electric fields of the electrodes by a conducting screen, but the ultraviolet light from the corona produces a weak preionization in the region where the glow discharge is to form. The second method of exciting the excimer gas is by a high energy electron beam that bombards the excimer gas thereby producing intense ionization in the gas.

The population inversion in the excimer gas is produced by reactions that leave the excimer molecule in a bound low lying excited level that decays radiatively to the unbound ground level. The ground level is unbound because it has a repulsive potential. This results in the lower laser level always being unpopulated.

The reaction pathways that lead to the population inversion in excimer lasers are very complicated. In this paragraph some of the important reaction pathways are discussed, but there are numerous other reactions that occur. A typical KrF excimer laser that is driven by glow discharge might have input gases which are Ar, Kr, and F_2. A typical gas mixture might be in the ratios Ar:Kr:F_2=0.95:0.048:0.002. The high high energy tail of the electron energy distribution ionizes the Ar gas. The process of ionizing the Ar also produces a large number of secondary electrons of lower energy. The principal gas phase reactions that lead to KrF^* are the following:

$$\begin{aligned}
&\mathrm{Ar} + \mathrm{e}^- \rightarrow \mathrm{Ar}^+ + 2\mathrm{e}^- \\
&\mathrm{F_2} + \mathrm{e}^- \rightarrow \mathrm{F}^- + \mathrm{F} \\
&\mathrm{Ar}^+ + 2\mathrm{Ar} \rightarrow \mathrm{Ar}_2^+ + \mathrm{Ar} \\
&\mathrm{Ar}_2^+ + \mathrm{Kr} \rightarrow \mathrm{Kr}^+ + 2\mathrm{Ar} \\
&\mathrm{Kr}^+ + \mathrm{F}^- + \mathrm{Ar} \rightarrow \mathrm{KrF}^* + \mathrm{Ar}
\end{aligned}$$

or

$$\begin{aligned}
&2\mathrm{Ar}^+ + \mathrm{F}^- + \mathrm{Ar} \rightarrow \mathrm{ArF}^* + \mathrm{Ar}^+ \\
&\mathrm{ArF}^* + \mathrm{Kr} \rightarrow \mathrm{KrF}^* + \mathrm{Ar}
\end{aligned}$$

or

$$\begin{aligned}
&\mathrm{Ar}_2^+ + \mathrm{F}^- \rightarrow \mathrm{ArF}^* + \mathrm{Ar} \\
&\mathrm{ArF}^* + \mathrm{Kr} \rightarrow \mathrm{KrF}^* + \mathrm{Ar}
\end{aligned}$$

or

$$\begin{aligned}
&\mathrm{Kr}^+ + \mathrm{Kr} + \mathrm{Ar} \rightarrow \mathrm{Kr}_2^+ + \mathrm{Ar} \\
&\mathrm{Kr}_2^+ + \mathrm{e}^- \rightarrow \mathrm{Kr}^* + \mathrm{Kr} \\
&\mathrm{Kr}^* + \mathrm{F_2} \rightarrow \mathrm{KrF}^* + \mathrm{F} \quad [\text{“harpoon reaction”}]
\end{aligned}$$

The extra Ar that appears on both sides the third reaction in the first pathway (as well as a number of other reactions) is needed for conservation of energy and momentum. Although a large number of reactions have been listed there are yet many other reaction pathways that can lead to the excited excimer molecules that form the population inversion between the bound upper level of the KrF^* molecule and the unbound ground level. Excimer lasers have high gain and typically operate with an unstable resonator.

The high power and the ultraviolet radiation make excimer lasers very useful. The average output energy from a typical commercial excimer laser using KrF^* as the excimer molecule is 0.2 J per pulse. The pulse duration is about 10 ns so that the peak output power is 20 MW. Commercial excimer lasers can operate at a repetition rate of 100 pulses per second. The primary use of excimer lasers (KrF and ArF lasers in particular) is in the exposure of the photo-resists used in the manufacture of semiconductor chips. Using a variety of advanced techniques, the semicoductor manufacturing industry has pushed ArF lasers with $\lambda = 193$ nm to pattern chip features with dimensions smaller than 50 nm.

11.3 Argon and Krypton Ion Lasers

Argon and Krypton lasers are gas lasers that produce high energy outputs on particular transitions in the visible. These lasers are examples of gas lasers for which the population inversion is produced by direct electron impact excitation. Argon ion lasers typically have output powers of 4-30 W. For most applications Ar and Kr ion lasers have been replaced by solid state lasers. Argon and Krypton lasers can be used to pump either dye lasers or titanium-sapphire (Ti-Sapph) lasers and they are used as very bright light sources for Raman or Rayleigh scattering experiments or for medical applications. One might not realize how bright an Ar ion laser is, thinking that 4-30 W is much less than a 100 W incandescent light bulb. The incandescent light source, however, has its output spread over 4π steradians and over a very wide bandwidth in wavelength. On the other hand the Ar ion laser typically operates at a few wavelengths in a single transverse mode and is extraordinarily bright.

The Ar atom has a ground configuration $1s^22s^22p^63s^23p^6$ and the Ar ion has a ground configuration $1s^22s^22p^63s^23p^5$. In our discussions we ignore the $1s^22s^22p^63s^2$ core since the inner shells are filled for the levels of importance in the Ar ion laser. The laser action in an Ar ion laser is between excited levels of the ion. The Ar ion laser can operate on a number of transitions. The most intense laser action occurs on the two transitions, the $3p^44p\ ^2\mathrm{D}^o_{5/2} \rightarrow 3p^44s\ ^2\mathrm{P}_{3/2}$ transition at 488.0 nm and the $3p^44p\ ^4\mathrm{D}^o_{5/2} \rightarrow 3p^44s\ ^2\mathrm{P}_{3/2}$ transition at 514.5 nm. The two lower laser levels decay radiatively to the ground level of the ion with very short lifetimes so that these levels are usually almost completely unpopulated. The population of the ground level ions is normally low so that radiation trapping does not keep the lower laser levels populated and hence radiation trapping is not a problem as it is for the He-Ne laser. The relevant energy levels of the Ar ion are shown in Figure 11.2. In addition to the two transitions shown in Figure 11.2, argon ion lasers can also operate at lower output levels on a number of other $3p^44p \rightarrow 3p^44s$ transitions. These are listed in Table 11.2 on page 266.

The population inversion necessary for laser action is produced by electron-impact excitation in an intense gas discharge. The lower levels of the laser transitions are about 17 eV above the ground level of the ion so that there is no significant thermal Boltzmann population of the lower laser levels even though the temperature in the gas discharge is very high. The excitation of the upper laser levels is a multi-step electron-impact process.

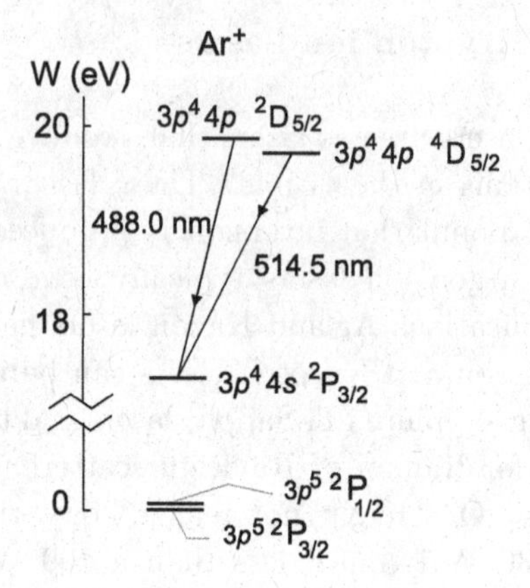

Fig. 11.2 The energy levels of the Ar^+ ion. The energies are in eV above the ground level of the Ar^+ ion.

Most of the excitation is accomplished via two steps:

$$(i)\quad \mathrm{e}^- + \mathrm{Ar}(3p^6) \rightarrow \mathrm{Ar}^+(3p^5) + 2\mathrm{e}^- \qquad (E > 15.8\ \mathrm{eV})$$

$$(ii)\quad \mathrm{e}^- + \mathrm{Ar}^+(3p^5) \rightarrow \mathrm{Ar}(3p^4 4p)^+ + \mathrm{e}^- \qquad (E > 20\ \mathrm{eV})\ .$$

In the first step an Ar atom is ionized by electron impact, and in the second step the ion is excited to one of the upper laser levels by electron impact. Excitation can also proceed through a single step:

$$\mathrm{e}^- + \mathrm{Ar}(3p^6) \rightarrow \mathrm{Ar}^+(3p^5 4p) + 2\mathrm{e}^- \qquad (E > 35.5\ \mathrm{eV})\ .$$

This second excitation method requires electrons with much higher incident energy than the two step process and is generally much less important than the two-step processes except in pulsed Ar ion lasers. For the two step process, the excitation rate for the laser levels scales with the density of electrons times the density of argon ions. Plasmas are quasi-neutral, so $n(\mathrm{Ar}^+) \approx n_\mathrm{e}$, and the output power is expected to scale with n_e^2.

A schematic diagram of an Ar ion laser is shown in Figure 11.3. The Ar ion laser operates with a dc discharge between the anode and cathode. The gas density in the Ar ion laser tube is low so that the discharge operates with a very high electron temperature. At low gas density a high electron temperature is produced when the laser operates with a high current density in the small bore of the BeO plasma tube. That a high electron temperature is produced at low gas density and a high current density follows from simple models of the gas discharge, but the demonstration is beyond the scope of

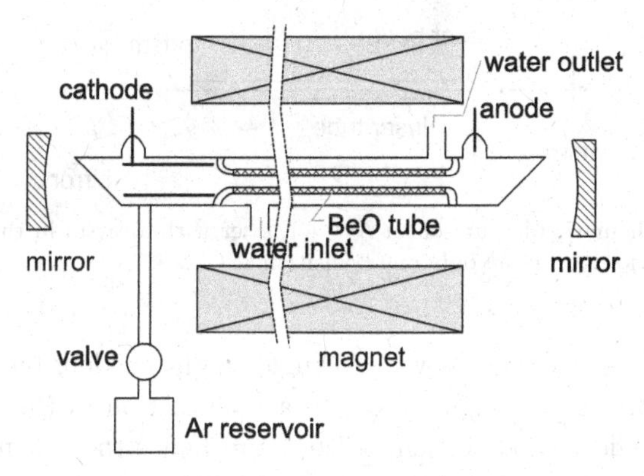

Fig. 11.3 A schematic diagram of an Ar^+ ion laser. Typically the BeO plasma tube is 1-2 m in length.

this book. Because two steps are required for the electron-impact excitation process one might expect that the gain and the output power would vary approximately as the square of the current density, and this is found to be nearly correct. The current density and the power density in the laser are closely related so that it is essential that the laser be operated with a very high power density in the discharge in the laser plasma tube for maximum output laser power. Typical power densities in the laser tube are greater than 5000 W/cm^3. The total input power to an Ar ion laser is typically 3×10^4 W for an output power of about 15 W on all lines. A power of 15 W requires a laser tube about 2 m in length. It is necessary to have the laser plasma tube constructed so that it can tolerate the heating produced by the power dissipated in it. The laser tubes are constructed using refractory materials such as BeO, which has a relatively high thermal conductivity and good strength at elevated temperatures. Because of the large power dissipation in the plasma tube, water cooling is essential. The anode to cathode voltage is typically 500 V, and the current drawn through the discharge is typically 30-60 A. With such large currents, collisions between the ions and the neutral Ar atoms transfer substantial momentum to the Ar gas and the Ar gas tends to flow toward the cathode. In order to prevent a piling up of the gas near the cathode it is necessary to provide a return path for the gas. Low power ion lasers are also available that use only air-cooling, these lasers produce a maximum output of a couple hundred milliwatts.

Typically the Ar ion laser plasma tubes have output windows that are

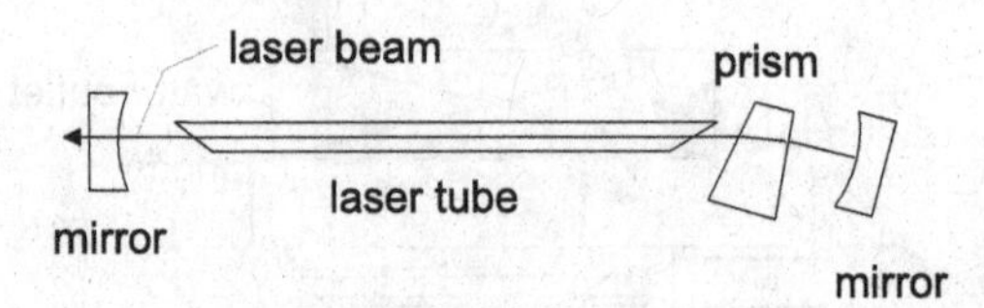

Fig. 11.4 A schematic diagram of an Ar^+ ion laser with a prism in the optical cavity so that laser action occurs on only one transition.

attached onto the tube at Brewster's angle. Light incident on a transparent material at Brewster's angle and with a polarization in the plane of incidence has a reflection coefficient of zero and hence there is no loss due to reflection. Because the reflection loss for a light beam polarized in the plane of incidence is less than for a beam polarized perpendicular to the plane of incidence the threshold population difference is less for the one polarization than the other. For a cw laser the population difference clamps at the population difference corresponding to the lowest threshold population difference. Thus the output beam from an Ar ion laser is polarized in the plane of incidence for the Brewster's angle windows.

An axial magnetic field is applied to the plasma laser tube. The magnetic field helps prevent the electrons in the discharge from reaching the walls of the plasma tube. This enables the electrons to gain more energy before making a wall collision. The electrons lose only a small fraction of their energy in elastic collisions in the gas. Thus the electron energy distribution is shifted toward higher energies with increasing magnetic field. This is important because the upper laser levels are so high in energy above the ground level of the ion. Only electrons in the high energy tail of the electron energy distribution have sufficient energy to ionize the atomic Ar or to excite ground level Ar^+ ions into the upper laser levels. The number of electrons at a given energy in the high energy tail of the distribution is increased when the electron energy distribution is pushed toward higher energies, and hence the excitation rate for the production of Ar ions in the upper laser levels is increased as the electron energy distribution is pushed to higher energies.

If an Ar ion laser is set up so that it has a cavity that uses mirrors with a high reflectance over a broad spectral range in the visible and near ultra-violet then the laser will lase simultaneously on several transitions (multi-line operation). Alternatively, it is possible to use a prism inside the optical cavity as shown in Figure 11.4 so that the cavity is aligned for only for the wavelength of one transition (single-line operation). In this situation the Ar ion laser lases on only one transition rather than on several transitions. Almost all of Ar ion lasing transitions terminate on the same two lower levels ($3p^4 4s\ ^2P_{1/2}$ and $^2P_{3/2}$). One might think that there would be strong competition between the various laser transitions that terminate on the same *lower* level. It is, however, found that switching from a high reflector to a prism does not change the power on a given line by more than about 10%.

In the same way there seems to be very little competition between lines with a common *upper* level. This is the case, even though one might expect that the system would lase only on the transition that has the lowest threshold population difference. Nevertheless because of a phenomenon called *spatial hole burning* lasing occurs on several lines with the same upper level. Spatial hole burning occurs as follows. In a standing wave cavity there are nodes and anti-nodes of the electric field of the standing electromagnetic field of the laser beam in the cavity. Stimulated emission does not occur at the nodes of the electric field. Different laser transitions emit different wavelengths. Modes with different wavelengths have their nodes and anti-nodes located at different positions in the laser cavity and hence interact with atoms at different locations in the cavity. Since modes with different wavelengths are interacting with physically different atoms it is possible for the laser to operate on different transitions with the same upper level. Because of the relative insensitivity to competition the highest output powers are obtained by running an ion laser multi-line. This is useful when the ion laser is used as the pump for another laser. On the other hand, single-line operation using a prism to select a particular line is useful for eliminating lines that are not used in a given application because the power on the unused lines produces excess heating or other undesirable effects.

The output of the laser is typically in a single transverse TEM_{00} mode, but oscillates on several longitudinal modes that fall within that part of the Doppler line shape, for which the gain is above threshold. The output power of an Ar ion laser is very dependent on the pressure and purity of the gas in the tube. Most Ar ion lasers have a gas handling system that allows the easy replacement of gas that has been ionized and driven into the cathode or walls of the laser. The most common failure method of Ar ion lasers is a failure of the laser tube. The typical lifetime of an Ar ion laser plasma tube is in the range of thousands of hours of operation.

Although the Ar ion laser operates with the highest intensity on two lines at 514 and 488 nm, there are about eight other transitions in the

Table 11.2 Laser transitions of argon ion and krypton ion lasers. Normalized output powers (P_{out}) are only approximate.

Ion	λ(nm)	Transition	P_{out}	Ion	λ(nm)	Transition	P_{out}
Ar^{2+}	351.1	$4p\ ^3P_2 \rightarrow 4s\ ^3S_1^o$	25	Kr^{2+}	406.7	$5p'\ ^1F_3 \rightarrow 5s'\ ^1D_2^o$	25
Ar^{2+}	363.8	$4p\ ^1F_3 \rightarrow 4s\ ^1D_2^o$	40	Kr^{2+}	413.1	$5p\ ^5P_2 \rightarrow 5s\ ^3S_1^o$	50
Ar^+	454.5	$4p\ ^2P_{3/2}^o \rightarrow 4s\ ^2P_{3/2}$	5	Kr^+	468.0	$5p\ ^2S_{1/2}^o \rightarrow 5s\ ^2P_{1/2}$	15
Ar^+	457.9	$4p\ ^2S_{1/2}^o \rightarrow 4s\ ^2P_{1/2}$	20	Kr^+	476.2	$5p\ ^2D_{3/2}^o \rightarrow 5s\ ^2P_{1/2}$	10
Ar^+	465.8	$4p\ ^2P_{1/2}^o \rightarrow 4s\ ^2P_{3/2}$	15	Kr^+	482.5	$5p\ ^4S_{3/2}^o \rightarrow 5s\ ^2P_{1/2}$	10
Ar^+	472.7	$4p\ ^2D_{3/2}^o \rightarrow 4s\ ^2P_{3/2}$	10	Kr^+	530.9	$5p\ ^4P_{5/2}^o \rightarrow 5s\ ^4P_{3/2}$	50
Ar^+	476.5	$4p\ ^2P_{3/2}^o \rightarrow 4s\ ^2P_{1/2}$	25	Kr^+	568.2	$5p\ ^4S_{3/2}^o \rightarrow 5s\ ^2P_{1/2}$	25
Ar^+	488.0	$4p\ ^2D_{5/2}^o \rightarrow 4s\ ^2P_{3/2}$	80	Kr^+	647.1	$5p\ ^4P_{5/2}^o \rightarrow 5s\ ^2P_{3/2}$	100
Ar^+	496.5	$4p\ ^2D_{3/2}^o \rightarrow 4s\ ^2P_{1/2}$	25	Kr^+	676.4	$5p\ ^4P_{1/2}^o \rightarrow 4d\ ^4D_{3/2}$	30
Ar^+	501.7	$4p'\ ^2F_{5/2}^o \rightarrow 3d\ ^2D_{3/2}$	10	Kr^+	752.4	$5p\ ^4P_{3/2}^o \rightarrow 4d\ ^4D_{3/2}$	40
Ar^+	514.5	$4p\ ^4D_{5/2}^o \rightarrow 4s\ ^2P_{3/2}$	100	Kr^+	799.3	$5p\ ^4P_{3/2}^o \rightarrow 4d\ ^4D_{1/2}$	10
Ar^+	528.7	$4p\ ^4D_{3/2}^o \rightarrow 4s\ ^2P_{1/2}$	20				

450-530 nm wavelength range that can also lase (see Table 11.2). Output on a number of lines in the 275-385 nm ultraviolet range is also possible, the two most intense of these transitions are included in Table 11.2. Lasers operating at these UV wavelengths require different optics than the standard visible-range lasers. The emitting levels are also due to doubly-ionized argon, so that the discharge conditions are also a bit different. Nonetheless, the Ar ion laser lines in the near ultra-violet region of the spectrum are useful for pumping dye lasers that operate with blue emitting dyes.

A Kr ion laser typically produces less power than an Ar ion laser, but it has laser transitions that are spread throughout the visible portion of the spectrum and slightly beyond (see Table 11.2). The Kr ion laser has some lines in the red that are very useful for pumping dye lasers operating with dyes that emit in the red, and the Kr ion laser also has a substantial emission in the near ultra-violet. Thus Kr ion lasers have a number of applications where Ar ion lasers can not be used or where Ar ion lasers are inefficient. Mixed gas Ar-Kr ion lasers produce an approximately white-light output from the mix of red (Kr^+), blue(Ar^+), and green (Ar^+) emissions.

11.4 Nitrogen Laser

The N_2 laser is a laser that produces a high power output in the near uv. Nitrogen gas is excited by a fast-pulsed glow discharge that acts as the active medium for the laser. For this laser, the differences in the direct electron impact excitation cross sections between the C $^3\Pi$ $v = 0$ vibrational

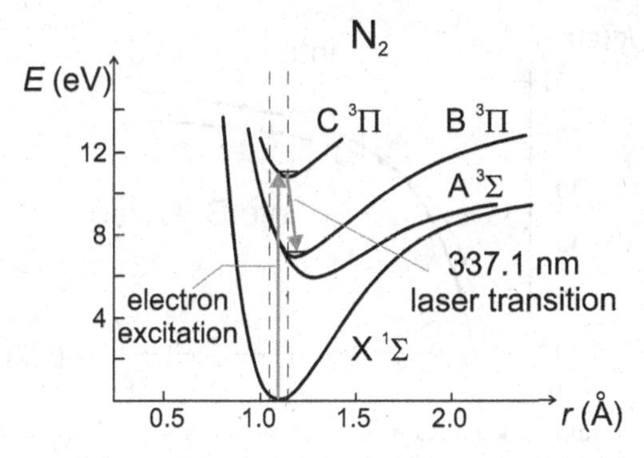

Fig. 11.5 Potential energy level diagram for the N_2 molecule with the energy levels relevant to the N_2 laser.

level and the B $^3\Pi$ $v = 0$ vibrational level produce a short-lived population inversion. The output laser pulse from a N_2 laser is at a wavelength of 337.1 nm. An energy level diagram for the N_2 molecule showing the energy levels relevant to N_2 laser is shown in Figure 11.5.

It is often difficult to achieve a population inversion in highly excited atomic levels by direct electron impact excitation. The size of excitation cross section depends upon the overlap of the initial and final state wave functions. For progressively higher levels the wave functions becomes more diffuse, so cross sections into higher levels tend to progressively decrease due to the poorer overlap with the compact ground state wave function of the initial state. In molecules, however, the Frank-Condon principle dominates the overlap of initial and final states. An electron-molecule collision occurs over a timescale much shorter than the vibrational frequency, so the radial separation of the two N atoms is essentially fixed. As seen in Figure 11.5, the overlap of the X $^1\Sigma$ ground level is much higher for the lowest vibrational level of the C $^3\Pi$ excited level than the B $^3\Pi$ level. The electron excitation rates into the B $^3\Pi$ and C $^3\Pi$ levels from the X $^1\Sigma$ level are shown in Figure 11.6 as a function of the electron temperature. At high electron temperatures, the rate into the C $^3\Pi$ level is greater due to the better Frank-Condon overlap. At low electron temperatures, the rate into the B $^3\Pi$ level dominates since it requires less energy to excite the B $^3\Pi$ level. For a discharge with an electron temperature higher than a few eV, electron collisions in the discharge pump the population inversion.

The radiative lifetime of the B $^3\Pi$ lower level (τ_B =6 μs) is much longer than that of the C $^3\Pi$ upper level (τ_C = 40 ns). As a result, once laser action starts to occur the number of molecules in the lower level rises rapidly

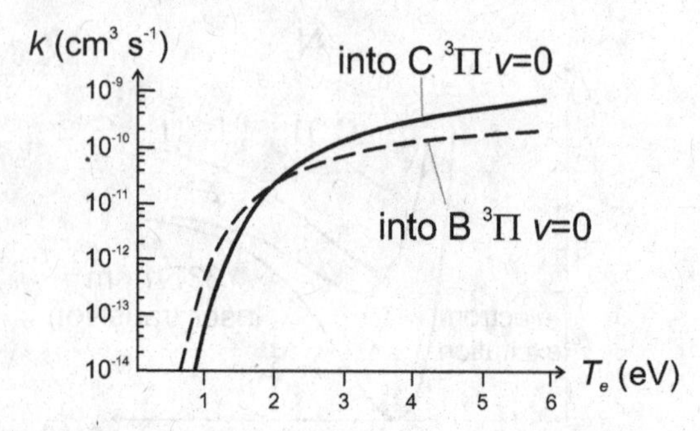

Fig. 11.6 Electron-impact excitation rates into the $N_2(B\ ^3\Pi)$ $(v = 0)$ and $N_2(C\ ^3\Pi)$ $(v = 0)$ levels as a function of the electron temperature.

and kills the population inversion. Thus N_2 lasers of necessity operate in a pulsed mode. The laser action can involve several rotational levels in the lowest vibrational levels.

The output of a N_2 laser can be modeled by calculating how the number densities of the upper and lower laser levels vary over the course of a pulse. Neglecting excitation into other levels,

$$\frac{dN_C}{dt} = n_e\, N_{\text{gas}}\, k_{X\to C}(T_e) - \sigma_{\text{se}}\, n_{\text{ph}}\, c\,(N_c - N_B) - \frac{N_C}{\tau_C} \tag{11.2}$$

$$\frac{dN_B}{dt} = n_e\, N_{\text{gas}}\, k_{X\to B}(T_e) + \sigma_{\text{se}}\, n_{\text{ph}}\, c\,(N_c - N_B) + \frac{N_C}{\tau_C} - \frac{N_B}{\tau_B} \tag{11.3}$$

$$\frac{dn_{\text{ph}}}{dt} = \sigma_{\text{se}}\, n_{\text{ph}}\, c\,(N_c - N_B) - \frac{n_{\text{ph}}}{\tau_{\text{ph}}}\,, \tag{11.4}$$

where n_e and T_e are the instantaneous electron density and electron temperature, N_{gas} is the ground state N_2 number density, and σ_{se} is the stimulated emission cross section. The resulting time evolution of the discharge parameters and laser output are illustrated in Figure 11.7 for a particular laser system. At the start of a pulse, there are a few electrons and ions remaining from the previous discharge pulse or from a glow discharge run to produce weak ionization in the N_2 gas, so $n_e(t = 0)$ is a small but finite value. The voltage between the two electrodes on either side of the discharge rapidly increases from zero volts to about 20 kV. The electrons left from the previous pulse are accelerated and begin to make ionizing collisions. As the ionizing collisions occur, the electron and ion densities in the N_2 gas increase, the resistance of the gas falls, and more current is drawn from the voltage source. The electron temperature in an N_2 discharge is

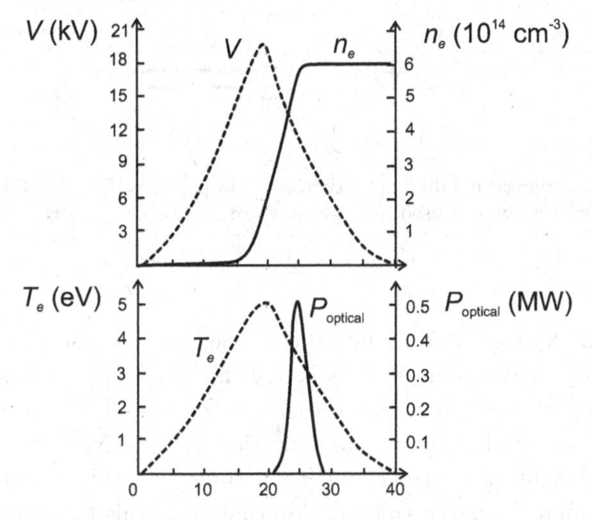

Fig. 11.7 Simulation of the time dependence of the voltage, the electron temperature, and the optical output for a pulsed N_2 laser at 65 Torr with a 2.54 cm electrode separation.

a monotonically increasing function of E/N where E is the magnitude of the electric field in the gas and N is the number density of the gas. Thus, the electron temperature T_e can be determined directly from the measured voltage. While the electric field is increasing the electron temperature increases, typically reaching a maximum value in the range of 4-5 eV. As the discharge develops and the electron and ion densities increase the current drawn from the pulsed voltage source increases rapidly. As the current increases the voltage across the N_2 laser tube decreases due to the internal resistance of the voltage source. The population inversion occurs during the short time when the electron density n_e is high enough that the electron impact excitation rate is relatively high, *and* the electron temperature is is high enough that the excitation rate into the C $^3\Pi$ level is greater than the excitation rate into the B $^3\Pi$ level. At later times, the combination of the higher production rate for the $B^3\Pi$ level over that for the C $^3\Pi$ level and the accumulation of molecules in the lower level kills the population inversion so that the laser action ceases. As a result the N_2 laser pulse lasts for a very short time, typically 4-10 ns. For a laser pulsed at 60 Hz, there is a 16.7 ms delay before the next pulse. During this period, the excited molecules are quenched back to the X $^1\Sigma$ ground level by a combination of radiative decays and collision processes and some excited molecules have been replaced by flowing the N_2.

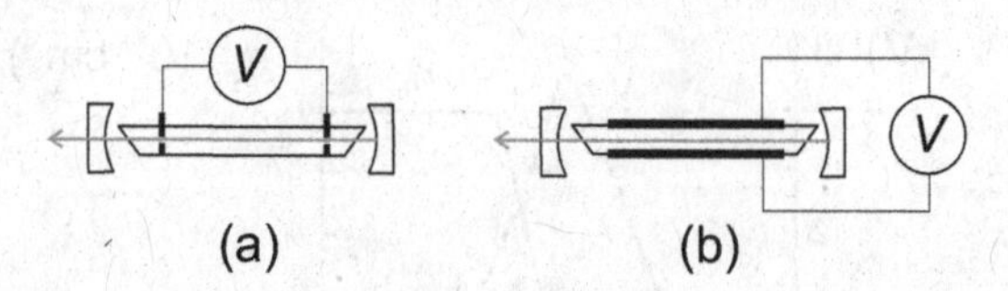

Fig. 11.8 (a) Longitudinal discharge, typical of low pressure He-Ne and Ar^+ lasers; (b) transverse discharge typical used for high pressure N_2 and CO_2 lasers.

A typical N_2 laser is is about 50 cm in length. The breakdown voltage of a discharge scales with $P \times l$ where P is the gas pressure and l is the electrode separation. If the electrodes were located at the ends of the discharge tube, similar to the configuration in a He-Ne Laser, l = 50 cm and only a low fill pressure ($P \approx 1$ Torr) could be used without having an excessively high discharge voltage. Alternative to this longitudinal design, in a transverse design two long electrodes parallel to each other that run the length of the discharge (see Figure 11.8). In this case the discharge separation is ~2.5 cm and a much higher fill pressure (~ 70 Torr) can be used. The higher fill density is beneficial for producing a higher output power. It is not possible to sustain a cw discharge with a sufficiently high electron temperature at this pressure, so that N_2 lasers must operate in a pulsed mode. The laser action is produced by pulsing the voltage and thereby striking a pulsed glow discharge between the two electrodes.

A N_2 laser can have a very high peak output power, typically 0.5-1.0 MW. The N_2 laser is usually operated with a repetition rate of 20-100 pulses per second. If the pulse rate is 60 Hz, and has a peak output power of 1 MW with a pulse duration of 5 ns, then the average output power is about 0.3 W. The N_2 laser is a very high gain laser and is normally operated using a mirror at one end of the discharge and a quartz window at the other end. Since there is only a very low Q optical cavity the N_2 laser usually operates on many modes and the output laser beam is very divergent.

The N_2 laser can be used as the pump laser for a pulsed dye laser (Sec. 12.2.1). The 337.1 nm ultraviolet wavelength of the N_2 laser permits the use of dyes with output wavelengths throughout the visible range ($\lambda >$ 380 nm). For most uses the N_2 laser has been replaced by either pulsed Nd:YAG or excimer lasers both of which can operate with higher peak and average output powers than the N_2 laser.

As seen in Figure 11.5, the overlap of the X $^1\Sigma$ ground level is also higher for the lower vibrational levels of the B $^3\Pi$ excited levels than the A $^3\Sigma$ level. As a result, a population inversion can also occur for a number of B $^3\Pi \rightarrow$A $^3\Sigma$ transitions in the near-infrared portion of the spectrum ($\lambda \approx$ 750-1000 nm). The A $^3\Sigma$ lower level is metastable, so as with C $^3\Pi \rightarrow$B $^3\Pi$ laser transition the output is pulsed as the population inversion is self-terminating. Since it requires less energy to excite the B $^3\Pi$ upper level, a lower peak electron temperature can be used. The maximum B $^3\Pi \rightarrow$A $^3\Sigma$ output is obtained with a fill pressure about three times higher than in a N_2 laser operating on the C $^3\Pi \rightarrow$B $^3\Pi$ transition. The peak output powers for the near-infrared lines are typically much less than those of the 337.1 nm line. The wavelengths are also not nearly as useful as the C $^3\Pi \rightarrow$B $^3\Pi$ ultraviolet output.

11.5 Copper Vapor Laser

The Cu vapor laser is another laser that operates by the use of direct electron impact excitation to produce a population inversion. The ground level of Cu is $1s^2 2s^2 2p^6 3s^2 3p^6 3d^{10} 4s\ ^2$S. The lowest excited level is $1s^2 2s^2 2p^6 3s^2$ $3p^6 3d^9 4s^2\ ^2$D and the next higher level is $1s^2 2s^2 2p^6 3s^2 3p^6 3d^{10} 4p\ ^2$P. The ^{2}D level is metastable. The cross section for the electron impact excitation of the optically allowed ^{2}P level from the ground level is much larger than the cross section for the electron impact excitation of the ^{2}D level from the ground level. Thus a fast pulsed discharge in an atmosphere of Ne and Cu vapor produces a population inversion between the ^{2}P level and the ^{2}D level. Neon is used as the carrier gas for the discharge because some excitation transfer from the Ne metastable levels enhances the population inversion. Intense laser action occurs at wavelengths of 510.5 nm and 578.2 nm. The operation of this laser is pulsed in order to allow time for the depopulation of the lower metastable laser level. Note that the ^{2}P level is dipole-allowed to decay back into the ^{2}S ground state. This decay channel reduces the population of the excited state. Nevertheless, the typical vapor pressure of copper atoms in the laser of around 1 Torr is high enough that radiation trapping of the resonance transition substantially blocks off this radiative decay route. The Cu vapor laser can operate at pulse repetition rates of 10^4 s^{-1} or higher and with peak powers of about 5×10^4 W and pulse durations of 10^{-8} s. Thus the average power can be several Watts.

A temperature of $\sim$1500° C is required to produce a copper vapor pressure of 1 Torr. The discharge tube is heated to this temperature by the discharge in the neon carrier gas. This requires a well insulated discharge tube and extremely high discharge currents. Copper-chloride lasers work much the same as copper vapor lasers, but rely upon dissociation of CuCl to produce the copper atoms. This permits the use of a much lower operating temperature.

The copper vapor laser is one member of the class of metal vapor lasers. Two others in this class include the lead vapor laser ($\lambda = 722.9$ nm) and the gold vapor laser. The gold vapor laser has two laser lines with a weak output at 312.2 nm and stronger output at 627.8 nm. The operating temperature of a lead vapor laser is a bit lower than that of copper vapor laser, while the temperature of a gold vapor laser is a bit higher. Of the three, copper is the most widely used. Copper vapor lasers can be used for pumping dye lasers, but for the most part they have been replaced by solid state for most uses.

11.6 Helium-Cadmium Laser

Helium-cadmium lasers are an interesting mix of many of the previously discussed gas discharge lasers. Cadmium is a metal, so the He-Cd laser is somewhat similar to the copper/metal vapor lasers of Sec. 11.5. The actual laser transitions, however, are between Cd^+ levels, so the He-Cd laser is somewhat similar to the argon ion lasers discussed in Sec. 11.3. Finally, the excited Cd^+ ions are created primarily by collision transfer from helium metastable atoms, similar to the He-Ne laser described in Chapter 10.

Almost all He-Cd lasers work at one of two wavelengths, 325.0 nm or 441.6 nm. The Cd^+ energy levels involved in each transition are indicated

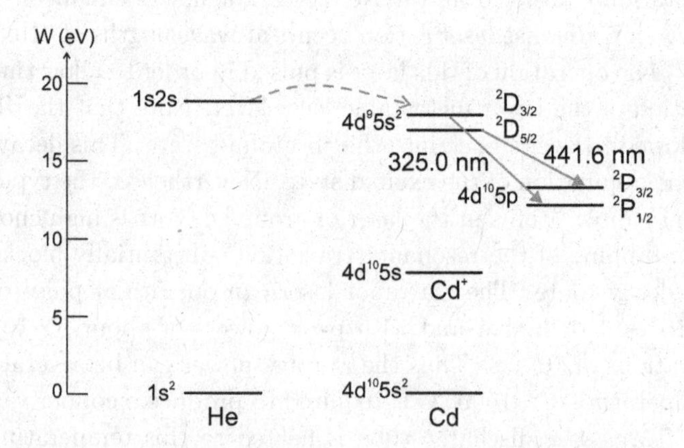

Fig. 11.9 The energy levels of He and Cd of interest in a He-Cd laser.

in Figure 11.9. The $4d^95s^2\ ^2D_{3/2} \rightarrow 4d^{10}5p\ ^2P_{3/2}$ transition at 353.6 nm can also be made to lase but a lower output power than the $4d^95s^2\ ^2D_{3/2} \rightarrow 4d^{10}5p\ ^2P_{1/2}$ transition at 325.0 nm and is thus rarely if ever used. The general design of a He-Cd laser is similar to that of a He-Ne laser, with two important additions. A heated Cd reservoir ($\sim 300°$ C) is used to supply the Cd vapor at a pressure of approximately 0.1 Torr. Eventually, the Cd vapor condenses on colder parts of the laser tube. As it does so, it buries some of the He gas (nominally at a pressure of 10 Torr). As a result, a He-Cd laser also requires a reservoir of helium to replenish the losses. The Cd reservoir is located on the anode end of the discharge tube. Once the cadmium is ionized, the Cd^+ ions are drawn towards the cathode end of the discharge tube, a process called *cataphoresis*, thus distributing the Cd along the length of the discharge.

Helium-cadmium lasers typically have cw output powers in the 10-200 mW range depending on the length of the discharge tube and discharge current. Low power lasers have a high quality TEM_{00} output beam, higher power lasers are more typically muti-mode. The short, blue/ultra-violet output wavelengths are useful in many photochemical active applications such as photolithography and holography. He-Cd lasers are also used in confocal microscopy, where the laser excites a fluorescent probe attached to the cellular specimen.

11.7 Iodine Lasers

The iodine laser is an example of a laser for which the population inversion is produce by photodissociation of a complicated polyatomic molecule such as CF_3I. The CF_3I molecule has a broad bandwidth absorption near 270 nm. The excited level of the molecule dissociates into CF_3 and I^*. The excited I^* is in the metastable $^2P_{1/2}$ level of the ground configuration. This metastable level decays by magnetic dipole radiation to the $^2P_{3/2}$ level of the ground configuration. Laser action occurs on this transition. The radiative lifetime of the metastable level is long. If the diffusion time of the metastable level to the walls is also long one can store the excited atoms for many excitation pulses of the 270 nm light that dissociates the CF_3I molecule thereby producing a relatively high density of excited I^* atoms. This storage laser can dump the stored energy into a pulse with a duration of a nanosecond or less. This results in a very intense laser pulse. The wavelength of the I^* laser is 1.315 μm.

In addition to the photodissociation method, iodine lasers have also been built using chemical reactions to produce the excited I* atoms. In a *chemical oxygen iodine laser* (COIL), the metastable iodine is produced by near-resonant collisional transfer reactions between molecular iodine and metastable oxygen molecules,

$$I_2 + O_2(a\,^1\Delta) \rightarrow I^* + I + O_2 \,. \qquad (11.5)$$

Various chemical reactions can be used to produce the singlet oxygen. This method can produce very high cw output levels (kW to MW class).

11.8 HF Laser

The HF laser is an example of a laser that utilizes chemical reactions that leave the HF molecule in a excited vibration-rotation level. The HF laser typically uses a mixture of either SF_6 or CF_4 together with H_2. A dc or microwave discharge dissociates the fluorine containing molecule so that atomic F is formed. The atomic F then reacts with H_2 to produce the population inversion in the HF laser by the reaction

$$F + H_2 \rightarrow HF^* + H \,. \qquad (11.6)$$

The HF* is formed in an excited vibration-rotation level so that a population inversion occurs as a result of the chemical reaction that produces the excited HF* molecule. The HF* decays by the emission of radiation, and laser action can occur on a large number of vibration-rotation transitions. The initial vibration-rotation distributions that result from the chemical reactions relax toward an equilibrium distribution via vibration-vibration($V-V$) excitation transfer collisions, vibrational relaxation collisions, and rotational relaxation collisions. Although one might think that for near atmospheric pressures rotational relaxation is so rapid that rotational equilibrium would exist, experiments have shown that this is not the situation, but that the rotational relaxation rate is usually comparable to the stimulated emission rate. The properties of the HF laser are determined by a competition between the stimulated emission rate and the $V-V$ changing collision rate, the vibrational relaxation rate, and the rotational relaxation rate.

Laser action is easily obtained between the v=2 and v=1 and between v=3 and v=2 vibrational levels of HF where v is the vibrational quantum number. A number of the rotational transitions can be made to lase for each of these vibrational transitions. It is also possible to obtain laser action on the $v=1 \rightarrow v=0$ vibrational levels with some difficulty. The HF laser can operate with wavelengths between 2.5 and 3.4 μm. The DF molecule, where D is the deuterium atom, also can be produced in excited levels by chemical reactions and can be made to operate as a laser. The DF laser operates at wavelengths between 3.5 and 4.1 μm. The HF or DF lasers can be operated at very high powers either pulsed or cw. It is possible to operate the HF laser with cw output powers between 1 and 10^3 W. There are numerous other chemical lasers such as the chemical iodine oxygen laser mentioned in the last section, but the HF is one of the most important chemical lasers and illustrates the principles of chemical lasers well.

Summary of Key Ideas

- The CO_2 laser is a gas discharge laser that operates in the infrared ($\lambda = 10.6\,\mu$m) on transitions between vibrational levels (Section 11.1). It can be operated at very high power levels and has many industrial applications.
- Excimer lasers operate in the uv on transitions between a bound excited level and an unbound ground level (Section 11.2). Excimer lasers can produce high power levels and are used extensively in the production of semiconductor chips.
- Argon ion and Krypton ion lasers are high power gas lasers the operate in the visible between excited levels of their respective ions (Section 11.3).
- The N_2 laser operates in the near uv on transitions between two bound molecular levels (Section 11.4). The excitation occurs in a pulsed gas discharge.
- Copper vapor lasers operate in the visible with high repetition rates (Section 11.5).
- The population inversion in iodine lasers (Section 11.7) is produced by photo dissociation, while the population inversion in HF lasers (Section 11.8) is produced by chemical reactions.

Suggested Additional Reading

D. C. Sinclair and W. E. Bell, *Gas Laser Technology*, Holt, Rinehart, and Winston (1969).

William T. Silfvast, *Laser Fundamentals*, Cambridge University Press (1996).

William B. Bridges, "Laser Oscillation in Singly Ionized Argon in the Visible Spectrum" *Appl. Phys. Lett.* **4**, 128 (1964).

R. Paschotta, *Encyclopedia of Laser Physics and Technology*, Wiley-VCH (2008).

C. K. N. Patel, "Continuous-Wave Laser Action on Vibrational-Rotational Transitions of CO_2" *Phys. Rev.* **136**, A1187 (1964).

W. A. Fitzsimmons, L. W. Anderson, C. E. Riedhauser, and J. M. Vrtilek, "Experimental and theoretical investigation of the nitrogen laser" *IEEE Journal of Quantum Electronics* **QE-12**, 624 (1976).

Jerome V. V. Kasper and George C. Pimentel, "HCl Chemical Laser" *Phys. Rev. Lett.* **14**, 352 (1965).

D. J. Spencer, T. A. Jacobs, H. Mirels, R. W. F. Gross, "Continuous wave Chemical Laser" *Int. J. Chem. Kinetics* **1**, 493 (1969) [HF and DF lasers].

Dirk Basting, Gerd Marowsky, Eds., *Excimer Laser Technology*, Springer (2005).

M. L. Bhaumik, R. S. Bradford Jr. and E. R. Ault, "High-efficiency KrF excimer laser" *Appl. Phys. Lett.* **28**, 23 (1976).

A. E. Siegman, *Lasers*, University Science Books (1986).

Problems

1. For a CO_2 laser operating at 10.6 μm and with an output power of 10^3 W calculate the number of photons/sec emitted by the laser.

2. In order for a HF laser to operate at 3 μm for one hour with a cw output power of 10 W how many moles of H_2 must be consumed?

3. If a KrF excimer laser has an output energy of 0.25 J per pulse how many photons are emitted per pulse?

4. The gas temperature of a He-Ne laser is about 400 K, and that of a He-Cd laser is about 600 K. (a) Calculate the Doppler widths of the He-Ne laser transition at 632.8 nm and the He-Cd laser transition at 441.6 nm. (b) Due to the finite size of the nucleus, energy levels of *s*- orbitals vary with the atomic mass (*isotope shifts*). There are two primary isotopes of Ne, ~90% ^{20}Ne, and ~10% ^{22}Ne. If the frequency of the 632.8 nm line for the ^{22}Ne isotope is shifted up by 1 GHz relative to that for the ^{20}Ne isotope, make a sketch the composite He-Ne laser line shape. (c) Make a sketch the composite He-Cd laser line shape assuming the cadmium sample used in the laser is composed of equal amounts of ^{112}Cd and ^{114}Cd and the isotope shift between the two is 1.5 GHz. (d) Is the line broadening caused by isotope shifts a form of homogeneous or inhomogeneous broadening? Explain.

5. The Doppler width for the Ar ion laser transition at 514 nm is 3×10^9 Hz. If the Ar ion laser cavity is 2 m long calculate the number of longitudinal modes within the Doppler width. Can you think of any way to obtain laser action on only a single longitudinal mode?

6. Explain in your own words the concept of spatial hole burning. Does spatial hole burning permit laser action on several longitudinal modes for a homogeneously broadened line or is laser action on several longitudinal modes only possible for an inhomogeneously broadened line?

7. Explain why the Cu vapor laser must operate in a pulsed mode. Why is the lower level in the Cu vapor laser metastable?

Chapter 12

Tunable Lasers

Some laser applications only require a bright spot or pulse of light, with only limited requirements on the wavelength of the emissions, as long as the light is reflected or absorbed by the target the exact laser wavelength does not matter. Other applications, however, require a much finer control over the wavelength of the laser emissions. For example, one might want to excite a single atomic transition at the exclusion of all others, by having the laser wavelength be within a linewidth of the transition. When laser cooling atoms (Chapter 14), even finer control the wavelength of the laser is required, the laser must be tuned a few MHz to the red of the transition. *Tunable lasers* are used in these types of applications since their operating wavelength can be varied until their output corresponds to the desired wavelength.

The operating wavelength of a laser is set by a combination of the gain envelope of the active medium and the feedback provided by the optical cavity. Fixed frequency lasers typically have an active medium which only has gain over a narrow range of wavelengths. These types of lasers can generally operate with mirrors that have broad reflectance. The optical cavity controls which logitudinal modes are permitted, but the location of the maximum in gain curve generally controls the laser's operating wavelength. Some lasers such as He-Ne and Ar ion lasers have gain curves which permit lasing at multiple distinct wavelengths. For these types of lasers the reflectivty of the cavity mirrors play a larger role in selecting the operating wavelength. By using a prism it is possible to have the user select which of the fixed frequencies the laser operates at. Nevertheless, these lasers are still generally considered fixed wavelength lasers. Tunable lasers generally have a medium which supports gain over a much broader wavelength range than than the fixed frequency varieties. The feedback provided by the optical cavity can be controlled by the user to select any operating wavelength from within this gain window.

There are many types of tunable lasers. Previous chapters have already covered two of these types. The Titanium:sapphire laser (Section 9.4) is a broadly tunable laser that can operate in the 670-1150 nm wavelength range. By controlling the temperature, injection current, and external cavity feedback it is possible to tune diode lasers over a few to tens of nanometers (Section 8.2.4). This chapter adds two more varities of tunable lasers: the free electron laser and the dye laser. Both of these lasers are broadly tunable over a wide range of wavelengths. Optical parametric oscillators (OPOs) are also commonly used to generate tunable beams of coherent light. In a parametric oscillator the output arises from nonlinear optical effects in a crystal. As such, the discussion of OPOs is deferred until Chapter 13 on nonlinear optics.

- free electron lasers (Section 12.1)
- dye lasers (Section 12.2)
- Ti:Sapphire lasers (Section 9.4)
- diode lasers (Section 8.2.4)
- optical parametric oscillators (Chapter 13)

12.1 Free Electron Lasers

The free electron laser (FEL) operates by stimulating radiation in a completely classical process. A charged particle radiates when it is accelerated. In a free electron laser a beam of relativistic energy electrons passes through a region of space where there is a series of transverse magnetic fields. The electron radiates spontaneously as it passes through the spatially oscillatory transverse magnetic field. Any process that can radiate spontaneously must be capable of undergoing stimulated radiation. Therefore radiation at the correct frequency passing through an electron beam undergoing transverse acceleration can stimulate radiation and thereby produce laser action. The spontaneous radiation of an electron undergoing an oscillatory transverse acceleration is a process that is explained by purely classical physics. There is no essential quantum mechanical feature to the radiation from an accelerated electron. The free electron laser is a classical laser.

12.1.1 *Operating Wavelength*

Both free electron lasers and synchrotrons produce emissions from the bending of electrons by a magnetic field. In a *synchrotron*, electrons with energy E travel is a circle with a constant radius R. The constant acclereation produces a broadband emissions. These emissions can be calculated from first principles and as a result, synchrotron emissions are used as a standard to calibrate other sources and detectors. Half of the output occurs at a wavelength less than the 'critical wavelength', $\lambda_C = 0.56R/E^3 = 1.86/(BE^2)$ where λ_C is measured in nm, R in meters, E in GeV and B in the constant magnetic field in Telsa. In the rest frame of the circling electrons, the emissions occur in all directions, but in the lab frame the relativisitic motion of the electrons converts this into a collimated beam that sweeps around the ring. This is illustrated in Figure 12.1(a).

The output of a synchrotron into a particular direction can be boosted by replacing part of the beam line by an *undulator* or *wiggler*. These are devices are a straight segment of beamline which have a spatially oscillatory transverse magnetic field as indicated in Figure 12.1(b). As an electron passes through the spatially oscillatory transverse magnetic field it experiences a transverse force that causes the electron to undergo an oscillatory transverse acceleration and emit light. For an undulator or wiggler these emissions all occur in the same direction, thus producing a more intense ouput. The difference between a wiggler and an undulator is that an undulator produces relatively small transverse accelerations of the electron beam whereas an wiggler produces relatively large transverse accelerations. In a synchrotron, the emissions from the wiggler (or undulator) are due to spontaneous emission. A free electron laser adds an optical cavity around

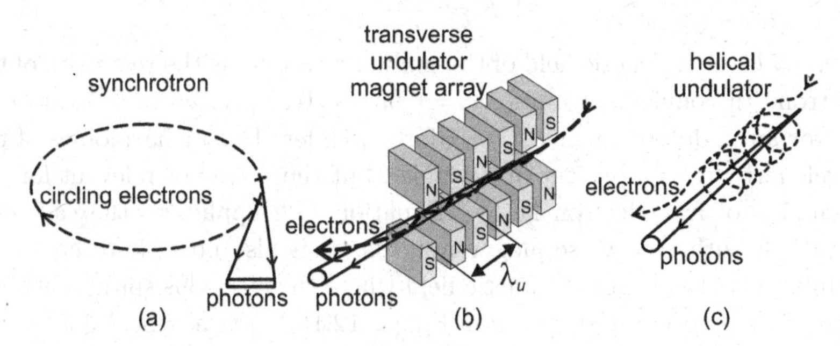

Fig. 12.1 Devices to create light from accelerated electrons: (a) synchrotron, (b) transverse undulator (or wiggler), (c) helical undulator.

the undulator. Stimulated emission by the light in the cavity affects the emission properties of the undulator's output.

The peak emission wavelength of a linearly polarized undulator can be calculated as follows. The undulator is composed of a number of transverse magnetic fields that repeat with a length λ_u. In the lab frame the electron is moving with a velocity near the speed of light as it passes through a transverse oscillatory magnetic field. In the electron's frame the electron sees the transverse field as a pulsed electromagnetic wave. The wavelength of this radiation is Lorentz contracted relative the length in the lab frame by a factor $\gamma = (1 - v^2/c^2)^{-1/2}$ where v is z-component of the velocity in the undulator. This is the wavelength of the emitted light in the rest frame of the electrons. The motion of the electrons Doppler shifts this to a shorter wavelength when the light is measured in the lab frame. This introduces another factor of γ. Hence, the radiated wavelength is proportional to the undulator spacing doubly shifted. The B-field produces a transverse motion of the electron in the undulator. The energy of the electron and hence the velocity of the electron is unchanged, but the z component of the velocity is reduced in the undulator. The γ referes to the z-component of the in the undulator and is given by

$$\gamma = \frac{E}{mc^2}\left(1 + \frac{K^2}{2}\right)^{-1/2} \tag{12.1}$$

where the quantity K is called the *undulator parameter*. Fully accounting for the transverse motion in the undulator,

$$\lambda^* = \frac{\lambda_u}{2\gamma^2} = \frac{\lambda_u (mc^2)^2}{2E^2}\left(1 + \frac{K^2}{2}\right), \tag{12.2}$$

The undulator parameter has a value of

$$K = \frac{\lambda_u \, eB}{2\pi \, m \, c}, \tag{12.3}$$

where B is the magnetic field of the undulator and m is the rest mass of the electron. In convienient units, $K = 0.934\, B\,[\text{Tesla}]\ \lambda_u\,[\text{cm}]$. K is of order one for an undulator, while $K \gg 1$ for a wiggler. Higher harmonics of the wavelength are also present in a wiggler, but these are not relevant for our discussion of free electron lasers. Equation 12.2 applies to the case of a undulator with transverse magnetic fields. It is also possible to create an undulator with a helical magnetic field where the electrons spiral down the path of the undulator as shown in Figure 12.1(c). For a helical undulator,

$$\lambda^*_{\text{helical}} = \frac{\lambda_u^{\text{helical}} (mc^2)^2}{2E^2}\left(1 + K^2\right). \tag{12.4}$$

Note that the method of selecting the operating wavelength is to vary the energy of the electron beam. The linewidth of the undulator's emissions varies inversely with the number of periods in the undulator, N_u,

$$\frac{\Delta\lambda^*}{\lambda^*} \approx \frac{1}{N_u} \,. \tag{12.5}$$

12.1.2 *FEL Gain*

The emission wavelength for an undulator (Equations 12.2 and 12.4) correspond to the peak wavelength for spontaneous emission. To understand how a free electron laser operates, one needs to fold in the stimulated emission and resulting gain. This section presents a simplified treatment of the gain of a free electron laser with a helical undulator. In order to do this, the problem is treated in the rest frame of the electron and then shifted back to the lab frame. The magnetic field is taken as helical. A light beam, that is propagating parallel to the electron beam is amplified. In the electron's rest frame the transverse helical magnetic field essentially appears as a circularly polarized plane electromagnetic wave coming toward the electron. It is called the pump beam. The light wave coming toward the electron from the opposite direction (i.e. traveling in the same direction as the electrons) is called the probe beam. The frequency of the probe beam in the electron rest frame is Doppler shifted from its frequency in the lab frame. In the electron's rest frame the angular frequencies of the pump and probe beams are called ω_0 and ω respectively and their intensities are called I_0 and I.

Thomson scattering is the scattering of light by a free charge. In the electron's rest frame, the pump beam undergoes spontaneous Thomson scattering in the backward direction (i.e. in the diection the electrons are travelling in the lab frame) at a rate per unit solid angle and per unit volume that is given by

$$R = \frac{2\pi \, n_e \, r_e^2 \, I_0}{h\omega_0} \tag{12.6}$$

where n_e is the density of electrons and where the cross section for Thomson scattering in the backward direction is given by r_e^2 where r_e is the classical radius of the electron. If the probe beam has the same polarization as the spontaneous back scattered pump beam then the probe beam will stimulate the Thomson scattering of the pump beam. The total rate of Thomson scattering into a mode including both spontaneous and stimulated emission is given by $(n+1)$ times the spontaneous Thomson scattering rate into a given mode where n is the number of photons in the mode. The spontaneous scattering rate into a given mode, γ, is equal to the spontaneous scattering rate per unit solid angle and per unit volume divided by the number of modes per unit solid angle and per unit volume. The gain coefficient for stimulated Thomson scattering in the backward direction is equal to the rate of stimulated scattering into a given mode divided by n and divided by the speed of light, c, so that the gain coefficient is given by γ/c. The number of modes per unit volume per unit solid angle for photons of the same polarization as the pump beam is given by $(\omega_0^2\,\Delta\omega_0)/(2\pi\,c)^3$ where $\Delta\omega_0$ is the bandwidth of the scattered light. Thus for light of the proper polarization

$$\frac{\gamma}{c} = \frac{(2\pi\,c)^3\,R}{c\,\omega_0^2\,\Delta\omega_0} = \frac{16\pi^4\,c^2\,n_e\,r_e^2\,I_0\,g\left(\omega - \left[\omega_0 - \frac{h\omega_0^2}{\pi\,mc^2}\right]\right)}{h\,\omega_0^3} \tag{12.7}$$

where $\Delta\omega_0$ is replaced by the inverse of the normalized line shape at the center of the recoil shifted frequency.

The net gain is equal to the gain minus the loss. The primary loss mechanism is the inverse process where the pump beam stimulates the Thomson scattering of the probe wave in the back direction. The net gain is given by

$$\alpha = \frac{16\pi^4\,c^2\,n_e\,r_e^2\,I_0}{h\,\omega_0^3}\left[g\left(\omega - \omega_0 + \frac{h\omega_0^2}{\pi\,mc^2}\right) - g\left(\omega - \omega_0 - \frac{h\omega_0^2}{\pi\,mc^2}\right)\right]. \tag{12.8}$$

Since the recoil is small the net gain is approximately given by

$$\alpha = \frac{32\pi^3\,n_e\,r_e^2\,I_0}{m\,\omega_0}\,\frac{d}{d\omega}g(\omega - \omega_0)\,. \tag{12.9}$$

The above expression is very interesting for several reasons. First the gain does not depend on Planck's constant, h. This is because the process is essentially classical even though the derivation given for the gain involved quantum ideas. It is possible to derive the identical expression for α using only classical physics, but the classical derivation is somewhat more complicated than the one given above. Second the gain coefficient, α, is directly proportional to the derivative of the line shape. For a symmetrical line shape this means that the gain is zero at line center. The system has gain at frequencies above the line center frequency and absorption at frequencies below the line center frequency.

An electron in the lab frame passes through the spatial region of the transverse magnetic field which has a length L. In the electron's rest frame the electron sees a pulsed electromagnetic wave. If the duration of the pulsed magnetic field is τ then the normalized line shape is given by the absolute square of the Fourier transform of a pulsed sine wave of duration τ, which is

$$g(\omega - \omega_0) = \frac{\tau \sin^2 \frac{(\omega-\omega_0)\,\tau}{2}}{2\pi \left[\frac{(\omega-\omega_0)\,\tau}{2}\right]^2} \tag{12.10}$$

so that the net gain is given by

$$\alpha = \frac{8\pi^3\, n_e\, r_e^2\, I_0}{m\,\omega_0}\, \frac{d}{d\eta}\left(\frac{\sin^2 \eta}{\eta^2}\right) \tag{12.11}$$

where $\eta = (\omega - \omega_0)\,\tau/2$.

Note that the gain α calculated in Equation 12.11 is the gain in the electron's rest frame. We are more interested in the free electron laser's gain in the lab frame. This is done by the use of the Lorentz transformation and by making the approximation that the velocity of the electrons is very close to the speed of light so that $v \approx c$. In this situation one obtains the following quantities:

(i) The coherence time, τ, is given by

$$\tau = \frac{L}{\gamma v} \approx \frac{L}{\gamma c} \tag{12.12}$$

where L is length of the undulator in the laboratory frame, where $\gamma = (1 - \beta^2)^{-1/2}$, and where $\beta = v/c$.

(ii) The probe frequency in the electron rest frame is given by

$$\omega = \omega^* \left(\frac{1 - v/c}{1 + v/c}\right)^{1/2} \approx \frac{\omega^*}{2\gamma} \tag{12.13}$$

where ω^* is the probe wave frequency in the lab frame.

(iii) If l is the period length of the undulator's transverse helical magnetic field in the lab frame ($l = L/N_u = \lambda_u$) the pump wave frequency in the electron rest frame is given by

$$\omega_0 = \frac{2\pi\,\gamma\, v}{l} \approx \frac{2\pi\,\gamma\, c}{l}\,. \tag{12.14}$$

(iv) If the magnitude of the transverse component of the undulator's helical magnetic field in the laboratory frame is B then the intensity of each linearly polarized component of the circularly polarized wave in the electron rest frame is given by

$$I_0 = \frac{c\,\gamma^2\, B^2}{2\,\mu_0}\,. \tag{12.15}$$

(v) The electron density in the electron rest frame, n_e, is given by $n_e = n_e^*/\gamma$ where n_e^* is the electron density in the laboratory frame.

(vi) The homogeneous line shape variable η is given by

$$\eta = -\frac{(\omega - \omega_0)\,\tau}{2} = \frac{1}{2}\,L\left(\frac{\omega^*}{2\gamma^2\,c} - \frac{2\pi}{l}\right). \tag{12.16}$$

For a free electron laser the small signal gain means that the pump beam is not too strong so that the motion of the electrons is relatively small and the change in the signal beam is small. The small signal gain per pass of the laser amplifier can be obtained by finding the separation L' in the electron rest frame between the following two events: (a) the front of the probe wave enters the helical field, and (b) the front of the probe leaves the helical field. Use of the Lorentz transformation with $v \approx c$ yields the result that $L' = L/2\gamma$. Therefore the gain per pass in the electron rest frame is $G = L\alpha/2\gamma$. This is the logarithm of the ratio of the intensity at (a) to that at (b) so that G is also the gain per pass in the laboratory. Thus the small signal gain α^* in the laboratory is given by

$$\alpha^* = \frac{G}{L} = \frac{3}{2}\,(n_e^*\,\sigma_T)\,\frac{B^2}{\mu_0}\,\frac{1}{2mc^2}\,\frac{\lambda_0\,L^2}{\gamma^3}\,F\,\frac{d}{d\eta}\left(\frac{\sin^2\eta}{\eta^2}\right) \tag{12.17}$$

where the final result is expressed in terms of the Thomson cross section which is given by $\sigma_T = 8\pi\,r_e^2/3$ and where F is a filling factor determined by the fraction of the cross section of the electromagnetic wave that interacts with the electron beam.

Gain occurs for values of $\eta > 0$ and loss occurs for values $\eta < 0$. The gain is a maximum at approximately $\eta = 1.3$. It is interesting to note that loss in the optical beam corresponds to a gain in kinetic energy of the electrons. There are proposals to construct particle accelerators using what are essentially free electron lasers run in the loss mode. In evaluating the expression for the small signal gain of the free electron laser one must take care to understand that γ is evaluated for the z component of the velocity of the electron beam inside the helical magnetic field of the undulator.

Amazingly, the preceeding discussion of the free electron laser has been very simplified. Only single electrons were taken as interacting with the optical beam. In fact, due to the interaction of the electrons with the optical field the electrons are bunched and radiate coherently as bunches. Saturation of a free electron laser occurs when the electrons are well bunched and the bunches achieve their maximum modulation. Also omitted from the discussion, was the number of period lengths in the wiggler. For too few period lengths the radiation is broadened too much and for too many the electrons are slowed too much by the incident wave. This is particularly a problem since a spread in energies of the electrons in the beam is detrimental to the laser performance (i.e. the output wavelength is a function of the electron energy). A more complete analysis of FELs can be found in the review article by C. Pelleggrini included in the additional readings. An interesting conculsion of this more complete analysis is the existence of a high gain region where the gain is maximum at the synchronism condition ($\eta = 0$) where the gain is zero in the small signal gain derivation.

12.1.3 *FEL Operation*

The free electron laser has been discussed as an amplifier, but of course the free electron laser can also be used as an oscillator provided the appropriate positive feed back is employed. Figure 12.2 shows the set up for deflecting an electron beam through an optical cavity in order to use the free electron laser as an oscillator. While Section 12.1.2 discussed a helical undulator that produces a circularized polarized wave, Figure 12.2 shows a schematic diagram of a linearly polarized undulator using a transverse undulator. In fact, this set up is the most commonly used one.

Via Equation 12.2 the operating wavelength of a free electron laser is seen to be a simple function of the electron energy. By varying the electron energy, one can tune the laser over a range of wavelengths. The wavelength range of a particular FEL is primarily determined by the type of electron accelerator used to create the high energy electrons. Table 12.1 lists the wavelength ranges possible for a few different types of electron sources.

Free electron lasers have two major advantages over other types of lasers. First, in principle, they can be operated at any wavelength. Since the active medium is free electrons, there is no issue with absorption within the gain medium. Further, the operating wavelength can be tuned by varying the

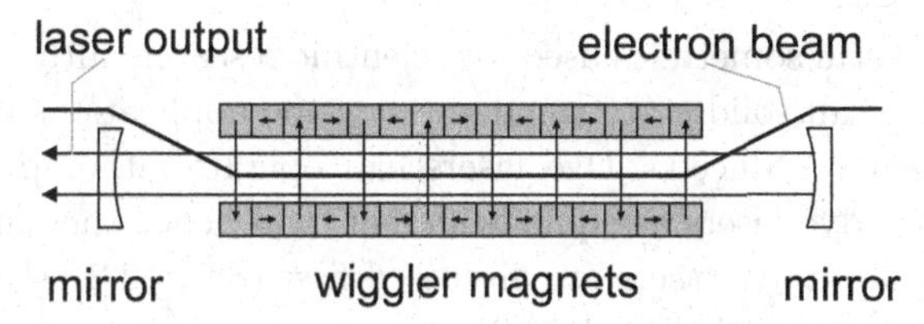

Fig. 12.2 A schematic diagram of a free electron laser. The electron beam is deflected through a undulator. The mirrors provide the optical cavity and the feedback necessary for laser oscillation.

Table 12.1 Approximate wavelength range for free electron lasers using different types of electron accelerators. Values are based upon Equation 12.2 assuming typical transverse undulator values of $\lambda_u \approx 3$ cm and $K \approx 3$.

Source	Energy	emission λ
electrostatic accelerator	1-10 MeV	20 mm - 200 μm
linear induction accelerator	1-50 MeV	20 mm - 10 μm
storage ring	0.1-10 GeV	2 μm - 0.2 nm
rf linear accelerator	1-25 GeV	20 nm - 0.03 nm

electron beam energy. Thus, the wavelength can be adjusted to maximize absorption in the target and minimize absorption in the beam path outside the laser. Second, FELs are relatively efficient and can produce relatively high output powers. The major disadvantage to FELs is that they are very complex and therefore very costly. The complexity stems from the need to build high quality accelerators and undulators. As a result, the major niche for FELs is for operation in the far infrared and the far ultraviolet and x-ray regions of the spectrum where other lasers (tunable or fixed frequency) are not available. For operation at x-ray wavelengths where the construction of mirrors are exceeding difficult, the free electron laser can be operated as a self amplified spontaneous emission (SASE) laser without any mirrors. The photons from the electrons are amplified by stimulated emission as they make a single pass through the undulator. The radiation is very forward directed (as is also the case in a non-lasing undulator), but is not in a single mode.

Free electron lasers can be expected to be important for the understanding and development of new solid state materials and to play a very important role in understanding significant biological problems.

12.2 Dye Lasers

Dye lasers are still sometimes used for scientific research, although with the development of all solid state tunable lasers the applications of dye lasers has been greatly reduced. Dye lasers can operate either pulsed or cw. Pulsed and cw dye lasers are quite different, and hence they are discussed separately. Pulsed dye lasers are discussed first (Sec. 12.2.1), and then cw dye lasers are discussed (Sec. 12.2.2).

The excitation of dye lasers is by the absorption of light. Therefore it is necessary to understand the absorption and the emission of light by dye

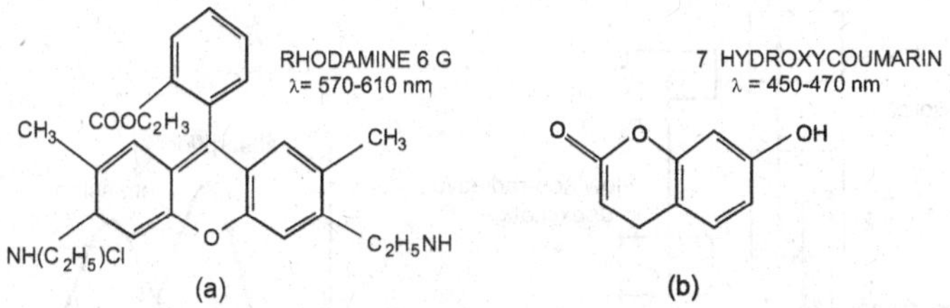

Fig. 12.3 The chemical structure of two dye molecules. The wavelengths for laser action are indicated.

molecules. Dye molecules are complicated organic molecules. Figure 12.3 shows the chemical structure of two different dye molecules that are used in dye lasers, Rhodamine 6G (R6G) and 7 Hydroxycoumarin. As can be seen in Figure 12.3 the R6G molecule is indeed very complicated having four carbon rings connected together and with various radicals attached to the rings. The dye molecules are typically dissolved in a solvent. For example R6G can be used as a 5×10^{-3} molar solution in ethyl alcohol (EtOH). One of the roles of the solvent is to broadened collisionally the molecular vibrational-rotational lines.

An energy level diagram for a typical dye molecule is shown in Fig 12.4(a). The energy levels are either singlet or triplet levels depending upon whether the spins of the electrons in the molecule are paired off or not. The singlet electronic levels are indicated as S_1, S_2, and $S_3 \ldots$ and the triplet electronic levels are indicated as T_1 and T_2. For each electronic level there are large number of vibrational levels and for each vibrational level there are a large number of rotational levels as is indicated in Fig 12.4(a). The collisional frequency in the liquid is high enough that each vibrational-rotational level is greatly broadened. Collisional non-radiative processes rapidly quench the vibrational and rotational excitation of a dye molecule so that dye molecules in the lowest electronic level are all in vibrational-rotational levels within about kT of the lowest level and the dye molecules in an excited electronic level are all within about kT of the lowest vibrational-rotational level in that excited electronic level. A dye molecule can absorb light at wavelengths corresponding to the energy difference between the lowest vibrational-rotational level of the ground electronic level and any vibrational-rotational level in the excited electronic level consistent with the normal selection rules. The dye molecules in excited vibrational-rotational levels are very rapidly de-excited collisionally to the lowest vibrational-rotational levels in that electronic level. The excited electronic levels are not rapidly de-excited collisionally.

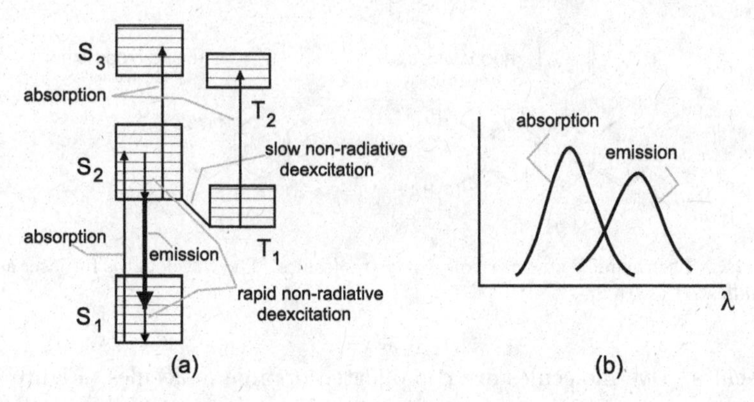

Fig. 12.4 (a) The energy levels of a typical dye molecule. The singlet levels are indicated by S and the triplet levels by T. The rapid non-radiative decays are also indicated. The vibrational-rotational energy levels within a particular electronic level are very closely spaced and are not indicated to scale. (b) Typical absorption and emission spectra of a dye molecule.

The emission is at wavelengths corresponding to the energy difference between the lowest vibrational-rotational energy of the excited electronic level and any vibrational-rotational level of the ground electronic level. This means that the absorption spectra typically occurs at shorter wavelengths than the emission spectra. This is similar to the absorption and emission in a Ti:sapph crystal. Figure 12.4(b) shows typical absorption and emission spectra of a dye molecule. Both the absorption and emission spectra have a very broad bandwidth. The broadening of a given level due to collisions is several times kT. This is indicated by the fact that it is possible to produce very short pulses using dye lasers. The dye laser absorption and emission has a homogeneous packet width that permits dye laser pulses as short as tens to hundreds of femtoseconds. The absorption spectrum of a dye is at shorter wavelengths than the emission spectrum. This means that the emission is not absorbed by the dye itself. This is true because the emission mostly occurs between the lowest vibrational-rotational levels in the excited electronic level and excited vibrational-rotational levels in the ground electronic level. This is of course advantageous for laser action since the lower level is usually unpopulated.

Some of the excited singlet dye molecules undergo non-radiative collisional transfer into triplet levels. Decay of excited triplet molecules to the ground singlet level is forbidden for radiative decay. This is a problem since molecules in the lowest triplet level can absorb the dye laser emissions. If the concentration of excited triplet molecules becomes large then laser action is adversely affected. In order to avoid this both pulsed and cw dye lasers flow the dye solution in order to remove the excited triplet molecules from the region of the active medium. The dye can be recirculated after the excited triplet molecules are slowly quenched by collisions back to the ground electronic level.

12.2.1 *Pulsed Dye Lasers*

12.2.1.1 *General Description*

Figure 12.5(a) shows the schematic layout of a pulsed dye laser. The pump light is usually supplied by another pulsed laser such as a doubled Nd:YAG laser ($\lambda_{\text{pump}} = 532$ nm). This type pump laser works well for red-orange emitting dyes, but since the absorption wavelength for the dye needs to be less than the emission wavelength, pump lasers with shorter wavelengths are required for emissions at shorter wavelengths. For example, blue emissions ($\lambda \approx 450$ nm) would require something like a tripled pulsed NdYAG pump laser operating with $\lambda_{\text{pump}} = 355$ nm or a N_2 laser operating at $\lambda_{\text{pump}} = 377$ nm. The pump light is focused into a line at the location of the dye cell by a cylindrical lens. Since the pump light is readily absorbed by the dye, the location of this line focus should be relatively close to the cell window. The dye solution flows past the line focus in a direction that is upward out of the plane of the diagram. In Fig. 12.5(a) the dye cell is shown as a parallelogram. The windows on the side of the dye cell are usually mounted so that they are not normal to the dye laser beam and hence prevent reflections from the windows acting as cavity mirrors.

At the start of the pump laser pulse some of dye fluorescence (spontaneous emission) occurs along the line focus from the pump laser. These

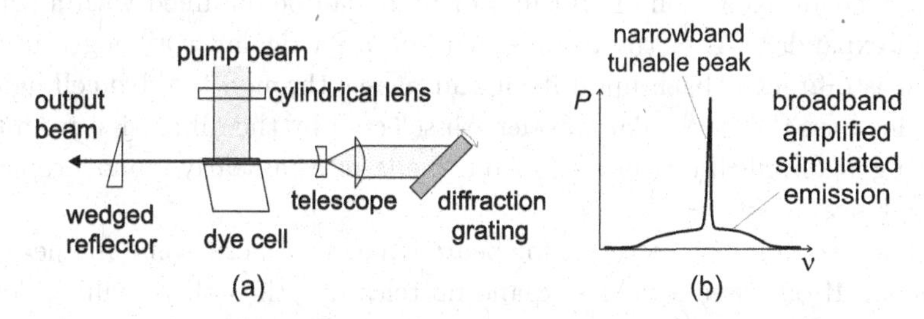

Fig. 12.5 (a) A schematic diagram of a pulsed dye laser. The beam expansion is accomplished with a Galilean telescope. (b) Spectral output of laser.

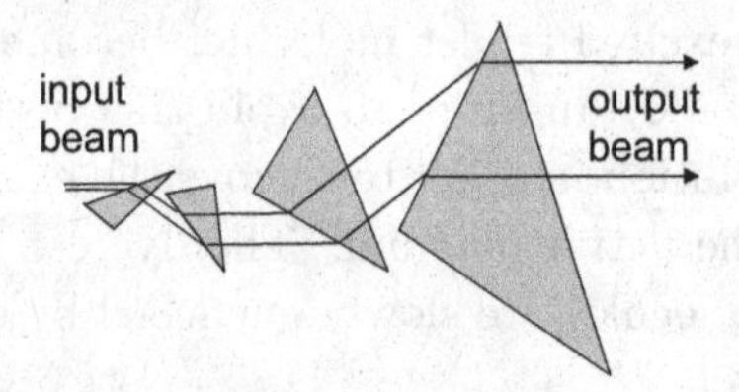

Fig. 12.6 A schematic diagram of a prism beam expander for use in a dye laser. Note that although the beam is expanded it is not deflected through an angle.

emissions will be amplified. This amplified spontaneous emission emerges through the side windows of the cell. Let us consider the light directed to the left. Some of this light (about 4%) is reflected back into the dye cell by the normal surface of the glass Herschel wedge. This light is further amplified as it passes back through the dye cell. A glass wedge is used as the reflector in order to avoid interference in the reflected light from the second surface. As the amplified light emerges from the dye cell on the right side, the beam is expanded before it reaches a diffraction grating. Since the resolving power of a grating is proportional to the number of grooves illuminated, the beam expansion must be large enough that the dye laser beam fills the diffraction grating. By utilizing the full resolution of the grating, one can obtain a relatively narrow output laser bandwidth.

Figure 12.5(a) indicates that the beam expansion is accomplished by the use of a Galilean telescope. The beam expansion can be also accomplished by a prism beam expander. Figure 12.6 show schematically an arrangement of four prisms that expands a laser beam without deflecting it. The surfaces of the prisms must be oriented so that they do not reflect the light back into the dye cell and so that no two of the surfaces are parallel and act as an etalon. The prisms in the beam expander should be made of Schlieren grade fused silica and the surfaces of the prisms must be interferometrically flat. A beam expansion of 80 times or more can be obtained with a prism beam expander. After the beam expansion a particular wavelength in the beam is diffracted back upon itself and passes through the dye cell again. On this pass the particular wavelength selected by the diffraction from the grating is amplified and emerges from the laser as a nearly monochromatic output beam.

The two different methods for beam expansion merit some further discussion. If one uses a beam expanding telescope then there will be some reflection off the center of the first lens that will be returned to pass through the dye cell where it is amplified. This light has not been diffracted by the grating and therefore contains broad bandwidth light. This leads to the output beam of the dye laser having an intense narrow bandwidth component plus a broad bandwidth component as indicated in Fig. 12.5(b). For many applications this is satisfactory, but there are many other applications where even a relatively weak broad bandwidth component of the output light is detrimental (although there will always be *some* weak broadband emissions at the start of the pulse before any light has had enough time for a round-trip journey to the grating). With the four prism beam expander there is no light reflected from the prisms that is reflected back through the dye cell so that the output laser beam contains 'only' narrow bandwidth light. The prism beam expander has, however, higher losses than the telescopic beam expander. For this reason the telescopic beam expander is useful for dye lasers for which the gain is only marginally high enough to assure laser action. If a telescopic beam expander is used in the dye laser then the telescope should be focused so that a parallel beam of light returning from the diffraction grating is focused at the center of the dye cell.

As with the tunable external cavity diode laser described in Section 8.2.4, the diffraction grating is used in a Littrow mounting so that the wavelength, λ, of the light that is diffracted back into the dye cell satisfies

$$n\,\lambda = d\,(\sin\theta + \sin\theta_d) = 2d\,\sin\theta \qquad (12.18)$$

where n is the order number, d is the grating spacing, θ is the angle of incidence which is also equal to the angle of diffraction, θ_d. Unlike the setup used in the diode laser, however, an Echelle grating can be used at a large angle of incidence. Such gratings are designed to work at high order number (i.e., $n \approx 10$). The shallow angle of incidence onto the grating also aids in filling the grating and thus obtaining a narrow bandwidth.

The overall length of a pulsed dye laser should be kept as short as possible. The reason for this is that the dye excitation is limited to the time of the pump laser pulse duration. For a typical Nd:YAG laser pump pulse this is about 10 ns. For a N_2 laser, the pulse duration may be around 5 ns. There is a time delay in the turn on of the dye laser equal to $2L/c$ where L is the overall length of the dye laser optical cavity and c is the speed of light. This means that the shorter the overall length of the dye laser the longer the pulse duration of the dye laser. On the other hand, for a short cavity length, the angular divergence of the beam tends to be greater which leads to an increase in the laser's bandwidth.

12.2.1.2 *Example Design Analysis*

To make this discussion more concrete, the use of a laser similar to the design of Fig. 12.5(a) pumped with 5 ns pulses from a pump laser is analyzed. The excited region of the dye is a tiny cylinder about 2 cm in length produced by focusing the pump laser into a line focus with a cylindrical lens. The transverse dimensions of the cylindrical focus might be about 0.02 cm. The distance of the wedge from the dye cell determines the angular divergence of the beam that is reflected back through the excited dye cell. With a separation of about 5 cm it is possible to obtain a divergence of about 2 milli-radians. The distance of the dye cell from the negative (diverging) lens of the Galliean telescope might be about 5 cm. The magnification of the Galliean telescope is the ratio of the focal length of the positive (converging) lens to the focal length of the negative lens. If the positive lens has a focal length of 3 cm and if the focal length of the negative lens is 0.5 cm then the magnification then the magnification of the telescope is $6\times$. The distance from the telescope to the diffraction grating can be made very short. The overall length of the dye laser from the glass wedge to the grating is typically about 20 cm. A handy number to remember is the speed of light is 30 cm/ns or just about one foot per nanosecond. Since the laser output beam has traversed the system two complete times (plus the initial 5 cm distance from the cell to the wedge), the laser beam has traveled a total distance of about 50 cm ($\sim$2 feet) before emerging from the laser, so the narrow bandwidth laser light does not turn on until about 2 ns after the pump laser first begins exciting the dye in the dye cell. If the pump laser has a duration of 5 ns then the narrow bandwidth laser pulse can only be about 3 ns in duration.

Using a high efficiency Echelle grating at $\theta \sim 80°$, this laser can obtain bandwidths of about $\Delta\lambda \approx 0.05$ nm. The short cavity dye laser just described is a variant of the original dye laser design by Hänsch, which used a beam expanding telescope with a magnification around $20\times$ and a grating oriented at a more modest $\theta \sim 60°$. With a longer telescope, the cavity length of this design is $L \approx 60$ cm and there is about a 4 ns delay before the start of the narrow band dye laser output.

If one inserts an etalon inside the dye laser cavity it is possible to narrow the bandwidth of the laser. An etalon inside the dye laser cavity results in a longer delay before the laser turns on and a smaller output power.

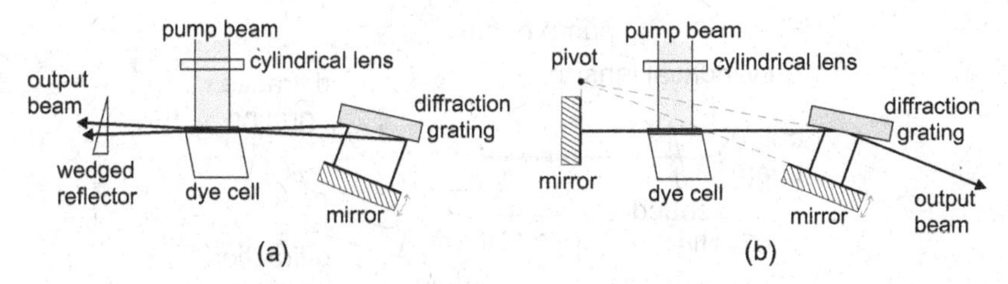

Fig. 12.7 Schematic diagrams for dye lasers: (a) uses a grating at grazing incidence followed by a mirror to eliminate the need for a beam expander; (b) with proper placement of pivot point for mirror rotation, it is possible scan output wavelength continuously (no mode hops) over a wide wavelength range.

12.2.1.3 *Alternative Cavity Designs*

In addition to the Littrow grating design of Fig. 12.5, a number of other cavity designs are also widely used. Three of these common variants are illustrated in Figures 12.7, 12.8, and 12.9. Figure 12.7 is a dye laser with no beam expanding telescope. This type cavity design is commonly known as a Littman or Littman-Metcalf configuration. The function of the beam expanding telescope was to enlarge the laser beam so that the diffraction grating is filled with light. This enables one to make use of the full resolution of the grating. In the set up shown in Figure 12.7 the diffraction grating is illuminated at an extremely grazing angle so that the light fills the grating. The mirror simply reflects one wavelength back onto the grating and then to the dye cell. This type of dye laser produces a narrow bandwidth and a quite high output power since the laser is very short. The output wavelength is selected by rotating the angle of mirror. With careful placement of the mirror's pivot point (Fig. 12.7b), it is possible to vary the length of the optical cavity along with wavelength selected by the grating so as to achieve continuous single-mode scanning of the output wavelength over a wide range

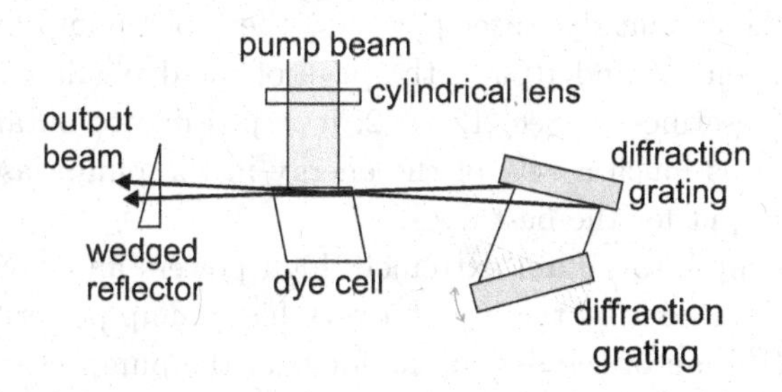

Fig. 12.8 A schematic diagram of a dye laser that uses a diffraction grating at grazing incidence followed by a second diffraction grating.

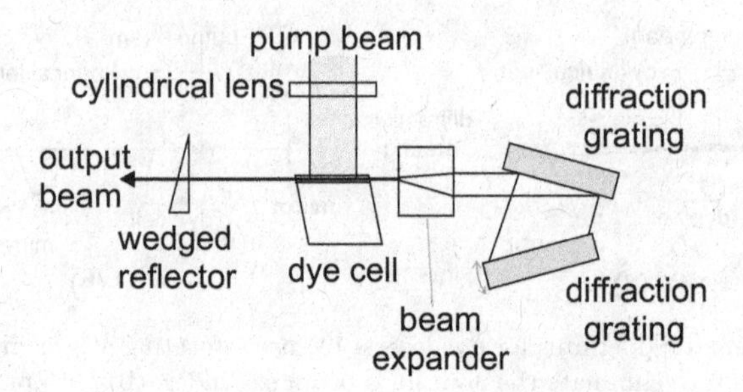

Fig. 12.9 A schematic diagram of a dye laser that uses a prism beam expander followed by a grazing incidence diffraction grating followed by a second diffraction grating.

with no mode hops. This type cavity design is also commonly used for tunable diode lasers.

Figures 12.8 and 12.9 show schematically two designs for lasers in which the mirror of Figure 12.7 is replaced by a second diffraction grating in a Littrow configuration so that the beam is reflected back onto itself just as it would be with the mirror. This laser has a narrower bandwidth than the laser of Figure 12.7 and is also relatively efficient. The version shown in Figure 12.9 adds a prism beam expander designed to produce a small beam expansion and two diffraction gratings. This laser produces an extremely narrow bandwidth (only one or two longitudinal modes) and a relatively high efficiency.

12.2.1.4 *Output Power and Amplifiers*

The dye laser output power depends upon a number of factors including the type of dye, the cavity design and the pump laser power. For an efficient dye such as R6G the output power of the dye laser is about 10% of the pump laser power near the peak of the gain curve. For other laser dyes the ratio of the output dye laser power to the input pump laser power is typically between 1% and 10% at the peak of the dye gain curve. For the laser system described in Sec. 12.2.1.2, it is possible to obtain an energy efficiency with as much as 4% of the energy in the pump laser converted into dye laser light for the best dyes.

If one attempts to obtain extremely high powers and a narrow bandwidth from a dye laser by the use of a very high pump power the resulting very high excitation of the dye at the focus of the pump laser results in a large output of broad bandwidth amplified spontaneous emission. In order to obtain a very high output power and a narrow bandwidth it is necessary

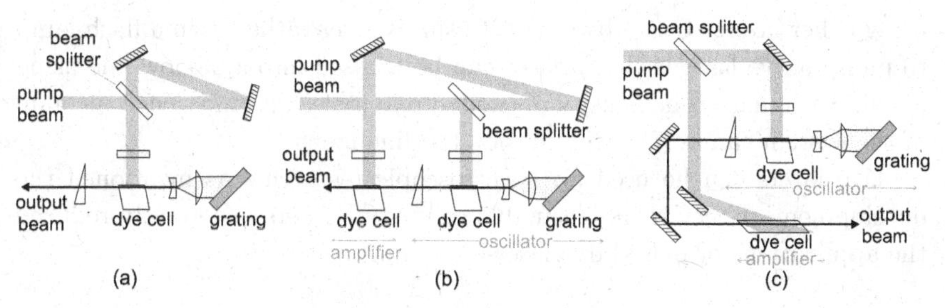

Fig. 12.10 Schematic diagrams for dye lasers amplifiers: (a) transversely-pumped single cell, (b) transversely-pumped amplifier, (c) longitudinally pumped amplifier. Typically $\sim 8\%$ of the pump beam is directed to the oscillator stage, with the rest directed to the amplifier.

to use a relatively low power narrow bandwidth dye laser as a low-power oscillator and then amplify the output of the dye laser with a separate amplifier stage. Three schemes for amplifier stages are shown in Fig. 12.10. For each scheme, the pump laser beam is split into two paths using a beam splitter. By varying the path lengths the two pulses take, the pulse to the amplifier is delayed so that the pump laser pulse arrives at the amplifier at the same time the dye laser pulse arrives at the amplifier. The amplifier stages in Fig. 12.10(a) and (b) use transversely pumped cells the same as those described previously for use in the oscillator stage. In Fig. 12.10(a) the same dye cell is used for the oscillator and amplifier. The first pulse is very weak, so there is very little broadband amplified stimulated emission. The stronger pump pulse arrives just as narrow-band feedback from the grating arrives at the dye cell. Figure 12.10(b) is similar, but uses separate cells for the oscillator and amplifier stages. If the pump laser has an excellent quality beam profile as can be the case for a Nd:YAG laser then one can use a longitudinally pumped cell as illustrated in Fig. 12.10(c). Only very high gain dyes such as R6G can be used in longitudinally pumped amplifiers. Multiple amplifier stages can also be used. Using a transversely pumped oscillator, followed by a transversely pumped amplifier and finally a longitudinally pumped amplifier it is possible to produce dye laser pulses with 0.1 J of energy per pulse and with a pulse duration of 5 ns. Such a pulse has a peak power of 20 MW. Amplifier stages are particularly useful for boosting the power output of oscillator cavity designs that traded off output power for a narrow bandwidth (i.e., those with intra-cavity etalons).

Finally it should be mentioned that in the past dye lasers were sometimes pumped using flash lamps. Today dye lasers are, however, pumped using other lasers. One advantage of using a laser rather than a flash lamp to pump a dye laser is that a laser can be focused into a narrow line using a cylindrical lens. The density of excited dye molecules and hence the gain of the system can be very high inside the line focus.

Dye lasers can be used for spectroscopic research. As mentioned the development of very reliable tunable solid state lasers has greatly reduced the applications of pulsed dye lasers.

12.2.2 *CW Dye Lasers*

Dye lasers can be operated in a cw mode as well as in a pulsed mode. A cw dye laser can have a high average output power and can have a very monochromatic character. Continuous wave dye lasers can operate with either a standing wave or ring configuration. The best cw dye lasers utilize the ring laser concept so that we focus our discussion on this type of cw dye laser. In a standing wave laser the mirrors are configured such that the laser beam is reflected back and forth between the mirrors thereby setting up a standing wave. This is illustrated in Figure 12.11. In a ring laser the mirrors are configured so that the laser beam is reflected in a ring thereby setting up a running wave. This is illustrated in Figure 12.12. A ring dye laser contains an optical diode so that there is a running wave in only one direction around the ring. For either a standing wave laser or a ring laser there must be some device for selecting a narrow bandwidth of wavelengths.

The basic advantage of a ring dye laser over a standing wave dye laser can be understood as follows. A standing wave laser will exhibit spatial hole burning effects so that the laser will oscillate on several longitudinal modes at the same time unless the device for selecting a narrow bandwidth of wavelengths selects a bandwidth so small that it encompasses only a single longitudinal mode. The selection of this narrow a bandwidth requires

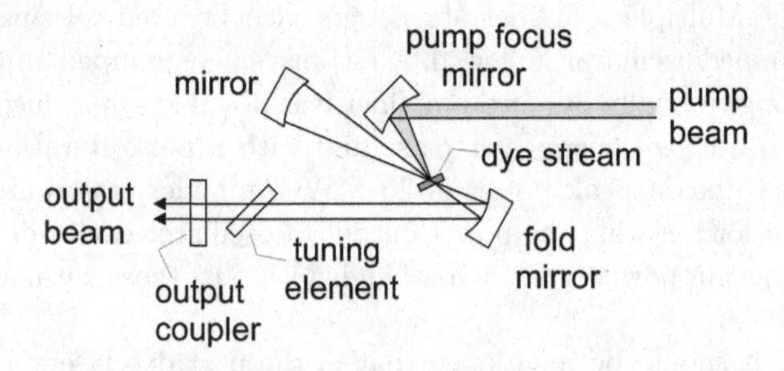

Fig. 12.11 A schematic diagram of a standing wave cw dye laser.

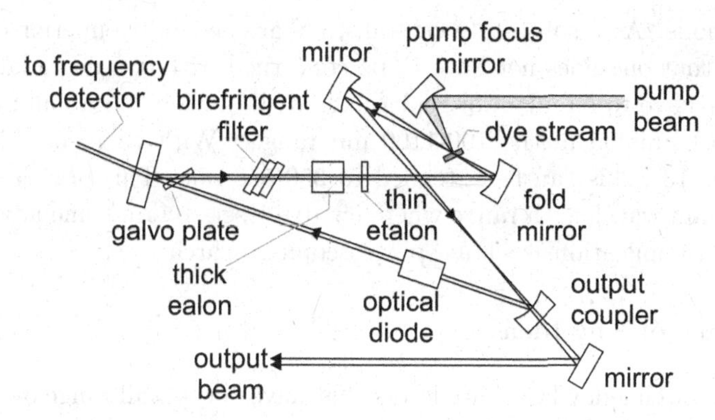

Fig. 12.12 Schematic diagram of a ring dye laser. The direction of circulation of the dye laser beam in the ring cavity is shown.

that the standing wave will not make use of a large fraction of the excited dye molecules due to the spatial hole burning. This results in a relatively low output power for a standing wave laser that is used for single longitudinal mode operation. For a ring dye laser the running wave does not permit spatial hole burning since it runs past all the excited dye molecules. Therefore it is possible to obtain single longitudinal mode operation with much higher output powers from a ring dye laser than from a standing wave dye laser. One might wonder why a ring laser has longitudinal modes at all since there is no standing wave. The existence of longitudinal modes can be understood when one remembers that the boundary condition for steady state cw operation of a laser is that the wave must have the same amplitude and phase after one complete round trip around the cavity. The second condition on the phase of the wave after a round trip in the cavity means that there must be an integral number of wavelengths in a complete round trip, and therefore there must be longitudinal modes.

The ring dye laser can be pumped by a reliable solid state laser. The pump laser is focussed by the pump mirror onto a dye jet. The use of a stream of dye eliminates the use of windows so that losses are minimized. The design of the jet is very important since the surface of the flowing liquid must be interferometrically flat over the active region of the laser. Other features of the ring dye laser such as the use of an optical diode to restrict the laser to run in one direction around the ring and the wavelength tuning by a birefringent filter and etalons are essentially the same as those described for a Ti:Sapph laser (Section 9.4).

Indeed, titanium:sapphire lasers have replaced cw dye lasers in most applications. As a solid-state crystal, they are generally superior to a dye laser in that one does not have to be concerned with dye degradation nor does one have to circulate or replace the dye solution. Titanium:sapphire lasers are tunable in the 700-1100 nm range. With frequency doubling (Chapter 13), this can be extended to 350-550 nm. This only leaves the 550-700 nm wavelength range where cw dye lasers retain some advantages for use in applications such as spectroscopic research.

Summary of Key Ideas

- Fixed frequency lasers are lasers that have very small range over which they can be tuned, and are used only to provide laser power at a single frequency. Tunable lasers have a wide range of frequencies over which they can be tuned.
- Free electron lasers use spatially varying magnetic fields to cause a fast electron beam to emit stimulated radiation (Sec. 12.1). The frequency of a free electron laser can be varied by changing the electron energy. For a transverse undulator,
$$\lambda^* = \frac{\lambda_u}{2\gamma^2} = \frac{\lambda_u (mc^2)^2}{2E^2}\left(1 + \frac{K^2}{2}\right) \quad (12.2)$$
where K is the undulator parameter,
$$K = \frac{\lambda_u \, eB}{2\pi \, m \, c} = 0.934 \, B\,[\mathrm{T}]\, \lambda_u\,[\mathrm{cm}]\,. \quad (12.3)$$
- Dye lasers use complex organic molecules with absorption frequencies that are higher than the emission frequencies to obtain a tunable output (Sec. 12.2).
- The operating wavelength of pulsed dye lasers is typically tuned by using a diffraction grating in a Littrow mount, where
$$n\,\lambda = 2d\,\sin\theta \quad (12.18)$$
where n is the order number ($n \approx 10$), d is the grating spacing, and θ is the angle of incidence.

Suggested Additional Reading

A. Yariv, *Quantum Electronics 3rd Ed.*, John Wiley and Sons (1989).

John M. J. Madey, "Stimulated Emission of Bremsstrahlung in a Periodic Magnetic Field " *J. Appl. Phys.* **42**, 1906 (1971).

L. R. Elias, W. M. Fairbank, J. M. J. Madey, H. A. Schwettman, and T. I. Smith, "Observation of Stimulated Emission of Radiation by Relativistic Electrons in a Spatially Periodic Transverse Magnetic Field" *Phys. Rev. Lett.* **36**, 717 (1976).

R. L. Spencer, J. E. Lawler, and L. W. Anderson, "Stimulated Thomson scattering" *Phys. Rev. A* **20**, 304 (1979).

P. Schmüser, M. Dohlus, J. Rossbach, and C. Behrens. *Free-Electron Lasers in the Ultraviolet and X-Ray Regime*, Springer (2014).

J. E. Lawler, J. Bisognano, R. A. Bosch, T. C. Chiang, M. A. Green, K. Jacobs, T. Miller, R. Wehlitz, D. Yavuz, and R. C. York, "Nearly co-propagating sheared laser pulse FEL undulator for soft x-rays" *J. Phys. D: Appl. Phys.* **46**, 325501 (2013). [An extremely original method for producing electromagnetic waves with tunable wavelengths.]

J. B. Murphy and C. Pellegrini, "Generation of high-intensity coherent radiation in the soft-x-ray and vacuum-ultraviolet region" *J. Opt. Soc. America B* **2**, 259 (1985).

C. Pellegrini, "The history of X-ray free-electron lasers" *Eur. Phys. J. H* **37**, 659 (2012).

William T. Silfvast, *Laser Fundamentals*, Cambridge University Press (1996).

A. E. Siegman, *Lasers*, University Science Books (1986).

Michael Stuke, *Dye Lasers: 25 Years*, Springer (2014).

T. W. Hänsch, "Repetitively Pulsed Tunable Dye Laser for High Resolution Spectroscopy" *Appl. Optics* **11**, 895 (1972).

J. E. Lawler, W. A. Fitzsimmons, and L. W. Anderson, "Narrow bandwidth dye laser suitable for pumping by a short pulse duration N_2 laser" *Appl. Optics* **15**, 1083 (1976).

I. Shoshan, N. N. Danon, and U. P. Oppenheim, "Narrowband operation of a pulsed dye laser without intracavity beam expansion" *J. Appl. Phys.* **48**, 4495 (1977); M. G. Littman and H. J. Metcalf, "Spectrally narrow pulsed dye laser without beam expander" *Appl. Optics* **17**, 2224 (1978). ["Littman-Metcalf" cavity design.]

Karen Liu and Michael G. Littman, "Novel geometry for single-mode scanning of tunable lasers" *Optics Letters* **6**, 117 (1981).

P. W. Milonni and J. H. Eberly, *Laser Physics 2nd Ed.*, John Wiley and Sons, (2010).

Problems

1. What is the output polarization for the free electron laser shown in Fig. 12.2?

2. The expression for the small signal gain in the laboratory (Equation 12.17) is zero at $\eta = 0$. Using Equation 12.16 find an expression for ω^* and λ^* when $\eta = 0$. Verify that this matches the value $\lambda^* = \lambda_u/(2\gamma^2)$ given by Equation 12.4 when $l = L/N_u = \lambda_u$.

3. The US Navy is interested in using $\sim$ 10 MW free electron lasers to shoot down missles. FELs are desirable in this application because they can produce high powers and their wavelength can be tuned to 'sweet spots' where atmospheric absorption is very low. (a) For an FEL with λ_u=4 cm and B=0.5 T, what energy electrons are required to generate emissions at two of the atmospheric sweet spots of λ =1.6 μm and λ =2.2 μm? (b) An infrared FEL at Jefferson National Lab has achieved $\sim$ 1.5% wall plug efficiency. What are the electrical power requirements for a 10 MW FEL?

4. Draw a schematic diagram for a vertical undualtor with 20 period lengths. Carefully label λ_u and L. The fractional bandwidth of a free electron laser is given approximately by $1/N_w$ where N_w is the number of periods in the undulator. Estimate the bandwidth of a free electron laser operating at 1 mm if the undulator has 20 period lengths.

5. What is the frequency separation of the longitudinal modes of a ring laser operating at 600 nm if the path length around the ring is 1.2 m?

6. Sketch a possible optical cavity for a standing wave dye laser that contains a birefringent filter, a thin etalon and a thick etalon. Describe how such a laser works.

7. An experiment wishes to excite Na atoms from the $3^2S_{1/2}$ ground state to the $4^2D_{3/2}$ excited level using the $3^2P_{1/2}$ as an intermediate level. The wavelength of the Na($3^2S_{1/2} \rightarrow 3^2P_{1/2}$) transition is 589.6 nm, and the wavelength of the Na($3^2P_{1/2} \rightarrow 4^2D_{3/2}$) transitions is 568.3 nm. Design a pulsed dye laser that will lase at both the wavelengths at the same time.

8. A pulsed dye laser operating at 500 nm has 0.01 J pulses with a duration of 5 ns. (a) If the repetition rate of the laser is 20 Hz what is the peak output power, and what is the average output power? (b) If the

number of photons per pulse remains constant as the wavelength is scanned, what is the energy per pulse when the dye laser operates at 600 nm?

9. What must be the bandwidth of a pulsed dye laser if it is to match the Doppler width of the Na $3^2S_{1/2} \rightarrow 3^2P_{1/2}$ absorption ($\lambda = 590$ nm) at T=500 K? If it is to match the natural radiative line width ($\tau_u = 16$ ns)?

10. If a pulsed dye laser has a pulse duration of 5 ns what is the minimum obtainable laser bandwidth in Hz?

Chapter 13

Nonlinear Optics Other Methods for Producing Coherent Beams of Light

Lasers provide a useful method of generating coherent beams of light. There are, however some wavelengths that are not easily produced using existing lasers. Coherent beams of light can in some situations be produced at wavelengths that are not easily produced by lasers using nonlinear optics. It is for this reason that a chapter on nonlinear optics is included in this book. The subject of nonlinear optics is enormous and complex. It includes such subjects as frequency doubling and tripling, frequency summing and differencing, optical parametric amplifiers and oscillators in addition to many other topics. The invention of the laser provided a bright enough light source that these techniques could be investigated. There are entire books on the subject of nonlinear optics. In a text on lasers only a few of the many significant nonlinear optics techniques can be discussed. This chapter focusses on frequency doubling (Sec. 13.1), frequency summing and differencing (Sec. 13.3), optical parametric amplifiers and oscillators (Sec. 13.4), and frequency combs (Sec. 13.5). These topics have been selected because they are among the most important applications of nonlinear optics. These techniques are especially important since they permit one to generate intense coherent beams of light at frequencies that are difficult to obtain using conventional lasers. In addition these topics provide a useful introduction to the field of nonlinear optics.

13.1 Frequency Doubling

Probably the most widely used nonlinear technique is frequency doubling. Frequency doubling was the first demonstrated by Franken *et al.* in 1961. In that experiment the beam from a ruby laser was focused onto a piece of quartz, and the output from the quartz was shown to contain not only light at the ruby laser wavelength, 694 nm, but also light at half that wavelength i.e. light at twice the frequency of the incident light. Although the fraction of light at half the wavelength was small it was easily separated from the light at the ruby wavelength since the wavelengths are very different. The technique of frequency doubling enables one to produce coherent light at higher frequencies than may be convenient with a laser directly. Frequency doubling of tunable lasers is especially valuable since it permits the generation of coherent tunable light at wavelengths with high efficiency where tunable lasers do not operate.

One might, correctly, expect that the incident light intensity must be high in order to permit frequency doubling since one must somehow combine two photons into a single photon. Indeed a high intensity light source is essential for all nonlinear processes. The invention of the laser provided a coherent light source with intensity high enough to enable one to observe and use nonlinear processes. A 1 MW laser pulse from a Nd:YAG laser can be focused tightly enough that the electric field in the laser pulse is on the order of 10^8 V/cm which is approaching the strength of the atomic electric fields. Under these conditions the polarization of the material is not a linear function of the electric field in the laser pulse and a wide variety of nonlinear processes occur.

A nonlinear process such a frequency doubling can be explained in terms of the generation of a polarization with a component at an angular frequency of 2ω when a wave with angular frequency of ω is incident on a crystalline solid with a nonlinear susceptibility. Typically the component of the polarization with angular frequency 2ω is small and varies as the square of the intensity in the incident wave. There are both coherent and incoherent nonlinear processes. It is much easier to observe a coherent processes than an incoherent processes because one can observe them in a particular direction and at a large distance from the nonlinear crystal so that the background light can be reduced to a low level so that the coherent emission is observable. The important applications of frequency doubling all involve the generation of coherent frequency doubled light in a crystalline solid.

At low electric fields the polarization of an isotropic crystalline solid is normally described as $\boldsymbol{P} = \epsilon_0 \chi \boldsymbol{E}$ where the electric susceptibility, χ, is a scalar constant. The susceptibility is related to the index of refraction by the equation $n^2 = \epsilon/\epsilon_0 = 1 + \chi$. For an isotropic solid the low field polarization is a linear function of the electric field, and the polarization is in the same direction as the electric field. If the crystalline solid is anisotropic then the susceptibility is not a scalar but is instead is represented by a 3×3 matrix. For the case of an anisotropic crystalline solid the low field polarization is still a linear function of the electric field, but the polarization is not necessarily in the same direction as the electric field. For reasons that will be explained frequency doubling is almost always done in anisotropic crystals.

Optical frequency doubling is usually accomplished with the use of a crystalline solid that does not have an inversion symmetry. For a crystalline solid that does not have an inversion symmetry the electrons in the solid move as if they are bound in an anharmonic potential well and in the presence of a time varying electric field their motion is that of a driven anharmonic oscillator. Under this condition the electrons in the solid move in response to an applied electric field at optical frequencies with an acceleration that is not a sinusoidal function at the optical driving frequency, but instead they move with an acceleration that contains higher harmonics of the optical driving frequency in addition to the fundamental optical driving frequency. The accelerated motion of the electrons in the solid generates radiation that contains harmonics of the optical driving frequency. This results in the generation of the light with a frequency at the second harmonic of the fundamental driving frequency i.e. light with angular frequency 2ω where the incident light has angular frequency ω. Thus crystalline solids that lack an inversion symmetry can be used for frequency doubling. In terms of the electric polarization this is expressed by the statement that the polarization is not a linear function of the electric field.

For an anisotropic crystal the ith component of the polarization of the solid is given in a Taylor series expansion by the expression

$$\boldsymbol{P}_i = \epsilon_0 \left(\sum_j \chi_{ij}\, \boldsymbol{E}_j + \sum_{jk} d_{ijk} \boldsymbol{E}_j\, \boldsymbol{E}_k + \dots \right) \tag{13.1}$$

where the first term is summed over j, where the second term is double summed over j and k, and where the coefficient of the nonlinear $\boldsymbol{E}_j\,\boldsymbol{E}_k$ term is denoted by d_{ijk}. If the electric field is small then the polarization is well approximated as a linear function of the electric field, but when the electric field is large then higher order nonlinear terms are important in the polarization. The coefficients d_{ijk} form a third rank tensor. The coefficients must be the same independent of the order of the electric field components $\boldsymbol{E}_j$ and $\boldsymbol{E}_k$ so that $d_{ijk} = d_{ikj}$. This means that the last two indices jk can be replace by a single index so that there are not as many independent coefficients as one might have initially thought. In fact there are only 18 such independent coefficients rather than the 27 one might have expected. The values of the coefficients d_{ijk} are different for different materials. Some materials have a larger nonlinear response to a driving electric field than other materials and for this reason some materials are better for frequency doubling than other materials. It can be shown that if the solid has an inversion symmetry then all the coefficients d_{ijk} must vanish. There are higher order terms in the polarization including those that vary as the third and fourth powers of the electric field. These higher order terms play a role in many important effects.

If the electric field components $\boldsymbol{E}_j$ and $\boldsymbol{E}_k$ are generated by the same laser beam then they have the same time-frequency dependence, $\sin \omega t$, where we have selected the zero of time to be such that the phase of the sinusoidal time dependence of the electric field is zero. In this situation the nonlinear term varies as $\sin^2 \omega t = (1 - \cos 2\omega t)/2$ so that the polarization has a term that varies as the second harmonic of the fundamental. The coherent radiation from this polarization will have a component at the second harmonic.

13.2 Phase Matching

From the discussion in section 13.1 of this chapter one might think that all that is required to produce a second harmonic beam of light is to obtain a nonlinear crystal and illuminate it with an intense laser beam. While this is essentially true there is a very important consideration, called phase matching, that arises if one wishes to generate a second harmonic wave with substantial intensity. In order to understand the need for phase matching the details of second harmonic generation in a crystal are analyzed.

For most crystals the index of refraction at the fundamental frequency ω is smaller than the index of refraction at the second harmonic frequency 2ω so that the phase velocity of a wave at the fundamental frequency is higher than the phase velocity of a wave at the second harmonic frequency. A wave at the fundamental frequency will at a given position in the crystal generate a wave at the second harmonic frequency with a particular phase relationship between the fundamental and the second harmonic waves generated from it. This relationship will be the same throughout the crystal. If the fundamental wave has a different velocity in the crystal than the second harmonic wave then the second harmonic wave generated at one position in the crystal will not be in phase with the second harmonic wave generated at a different position in the crystal. If the crystal is very thin then the electric field of the second harmonic waves from the different locations in the crystal are nearly in phase and the electric field generated at different locations in the crystal will add together nearly in phase to produce a net electric field. For a longer beam path in the crystal, however, the electric field of the second harmonic generated in one part of the crystal is out of phase with the electric field of the second harmonic generated in another part of the crystal. Thus as the length of the crystal increases from a very short length the second harmonic electric field will at first increase, but as the beam path in the crystal increases the second harmonic field emerging from the crystal will increase more slowly, will stop increasing and will begin to decrease. In fact if the crystal has a length such that the second harmonic electric field generated anywhere in the crystal is exactly out of phase with the second harmonic electric field generated at exactly one other location in the crystal then there will be no net second harmonic generation, and the second harmonic intensity will be zero.

The coherence length of a crystal is defined to be the one half of the shortest length for which there is no second harmonic generation in a crystal. The coherence length of a crystal is equal to the thickness of crystal that results in the maximum second harmonic generation. As the thickness of the nonlinear crystal is increased beyond the coherence length the second harmonic generation decreases and increases repeatedly with a period equal to twice coherence length. The physics of the processes is very interesting. As the thickness of the nonlinear crystal increases from a very thin crystal at first the fundamental wave is converted into a second harmonic wave. As the crystal thickness increases beyond the coherence length the second harmonic wave is converted back into the fundamental. When the thickness of the crystal is equal to two coherence lengths there is no second harmonic generation, and as was explained when the thickness increases beyond two coherence lengths the process is repeated with a repetition length of two coherence lengths.

The coherence length of a crystal can be calculated as follows. The coherence length is equal to one half the crystal thickness for which there are m wavelengths of the fundamental and $2m+1$ wavelengths of the second harmonic where m is an integer. The wavelength of the fundamental in the crystal is equal to $\lambda_1 = \lambda_0/n(\omega)$ where λ_0 is the vacuum wavelength of the fundamental and $n(\omega)$ is the index of refraction at the frequency of the fundamental. The wavelength of the second harmonic is equal to $\lambda_2 = \lambda_0/2n(2\omega)$ where $n(2\omega)$ is the index of refraction at the frequency of the second harmonic. Setting

$$m\,\lambda_1 = (2m+1)\lambda_2 \tag{13.2}$$

yields

$$m\left(\frac{1}{n(\omega)} - \frac{2}{n(2\omega)}\right) = \frac{1}{n(2\omega)}\,. \tag{13.3}$$

Solving for m and determining the coherence length from $l_{\text{coh}} = m\lambda_1/2$ yields

$$l_{\text{coh}} = \frac{\lambda_0}{4\,[n(2\omega) - n(\omega)]}\,. \tag{13.4}$$

For example, consider the use of the nonlinear crystal ammonium dihydrogen phosphate (ADP) for frequency doubling the wavelength of a Nd:YAG laser. The free space wavelength of the laser fundamental is $\lambda_1 = 1.06\ \mu$m, while the frequency doubled beam has a wavelength of $\lambda_2 = 0.53\ \mu$m. For a beam propagating along the optic axis of the ADP crystal, the values of the index of refraction at the fundamental and second harmonic frequency are $n(\omega_1)$=1.51 are $n(\omega_2)$=1.53. The resulting coherence length is $l_{\text{coh}} = 1.3 \times 10^{-3}$ cm. This means that the second harmonic generation reaches a maximum value for a thickness of only $1.3{\times}10^{-3}$ cm.

One expects that the short coherence length would result in the generation of only very small amounts of second harmonic light. In order to generate substantial amounts of second harmonic light it is necessary to have a coherence length that is as large as or larger than the macroscopic size of the crystal used for second harmonic generation. In order to do this it is necessary to have both the fundamental and second harmonic waves travel with almost the same phase velocity in the crystal. This is called *phase matching.* For example if one requires a coherence length of 1 cm for the wavelength of a Nd:YAG laser the difference in the indices of refraction at the fundamental and second harmonic frequencies can be only $2.6{\times}10^{-5}$. Since the second harmonic generation involves the coherent summation of the second harmonic waves from different parts of a nonlinear crystal it can be shown that the intensity of the second harmonic wave increases approximately as the square of the thickness of the crystal for crystal path lengths much less than the coherence length. For a material suitable for doubling the light from a Nd:YAG laser the fractional increase in the second harmonic wave for a crystal 1 cm in length divided by the fractional increase in the second harmonic wave for a crystal that is 1.3×10^{-3} cm in length is 6×10^5. Thus phase matching is essential if one is to generate a second harmonic wave with substantial intensity.

The process of phase matching is equivalent to the statement that the vector sum of the wave vectors of the two incident photons that are destroyed in the frequency doubling process is equal to the wave vector of the final photon so that

$$2\boldsymbol{k}_i(\omega) = \frac{4\pi\, n(\omega)}{\lambda_0} = \boldsymbol{k}_f(2\omega) = \frac{2\pi n(2\omega)}{\lambda_0/2} = \frac{4\pi\, n(2\omega)}{\lambda_0} \tag{13.5}$$

It is clear that the condition on the wave vectors is identical to the condition that the index of refraction of the fundamental at angular frequency ω be equal to the index of refraction of the second harmonic at 2ω so that both the fundamental and second harmonic waves travel with the same phase velocity.

It might seem that phase matching is an impossible dream since the index of refraction of any material varies as a function of frequency. There are, however, methods that enable one to obtain phase matching. Phase matching can be accomplished by the use of an anisotropic crystal (also called a birefringent crystal or a double refracting crystal). In an anisotropic crystal the speed of light varies for different polarizations of light. Hence, the basic idea of phase matching is to have one polarization at angular frequency ω have the same phase velocity as the other polarization at the angular frequency 2ω.

For an isotropic crystal, the unit cells have equal spacing in each direction while in an anisotropic crystal the spacing are not all the same. For a uniaxial crystal, the spacing is the same in two directions, while the spacings are unequal in all three directions in a biaxial crystal. Both uniaxial and biaxial crystals are used for second harmonic generation, but we limit this discussion to uniaxial crystals since the analysis is simpler and yet it still shows almost all the essential features. For a uniaxial crystal there is one unique axis, called the optic axis, while the other two axes are equivalent. One might think that since the length of the unit cells in a uniaxial crystal also have one axis with an unequal spacing from the other two axes that the crystal axis and optical axis are the same. This is not necessarily the case. As mentioned in the last paragraph, phase matching requires using two different polarizations of light. By having the optic axis rotated relative to polarization axis of the light entering the crystal, the electric field of electrons oscillating along the crystal's axes can produce polarization components ortohogonal to the input. The response of the electrons in the atoms and molecules to the optical electric field determintes the index of refraction and this is generally not parallel to the crystal axes. The phase velocity of a particular polarization of light depends upon the index of refraction along that axis. Linearly polarized light impinging of the crystal can be decomposed into ordinary and extraordinary wave components based upon their relative orientation with the optic axis. For the *ordinary ray* the polarization axis is perpendicular the optical axis, so the index of refraction is unchanged for that polarization as the crystal is rotated relative to the optical axis. The extraordinary wave has a polarization component parallel to the optic axis. It has a phase velocity that depends on its direction of propagation. A uniaxial crystal can be analyzed in terms of only two different indices of refraction, n_o and n_e. The index of refraction of the ordinary wave, denoted by n_o, is independent of the crystal rotation angle θ and the ordinary wave propagates with a phase velocity c/n_o. For the perpendicular polarization, the index, and thus the phase velocity varies as the crystal is rotated. As a result, the index of refraction for the *extraordinary wave* varies with the rotation angle θ. For a uniaxial crystal, the extraordinary wave propagates with a phase velocity c/n_o along the direction of the optic axis, but along either of the other two axes the extraordinary wave propagates with a phase velocity c/n_e.

It is possible for n_o to be either larger or smaller than n_e. A crystal for which n_o is greater than n_e is called a *negative uniaxial crystal*, and one for which n_o is less than n_e is called a *positive uniaxial crystal*. Some of the most important nonlinear crystals are negative uniaxial crystals. As was explained the ordinary wave has a phase velocity in the crystal equal to c/n_o independent of the direction of propagation in the crystal. The extraordinary wave has a phase velocity equal to c/n_o along the optic axis and equal to c/n_e along any direction in the plane defined by the other two axes. An extraordinary wave traveling at an angle θ between the optic axis and the $\boldsymbol{k}$ vector has a phase velocity equal to $c/n_e(\theta)$. The extraordinary index of refraction for propagation at an angle θ between the optic axis and the direction of propagation is given by

$$\frac{1}{n_e^2(\theta)} = \frac{\cos^2\theta}{n_o^2} + \frac{\sin^2\theta}{n_e^2}\,. \tag{13.6}$$

The above equation includes what has already been stated in words, namely that the extraordinary index of refraction is equal to n_o for θ=0 and n_e for $\theta = \pi/2$.

One common method of obtaining phase matching is to have the incident wave polarized such that it is purely an ordinary wave in a negative uniaxial crystal. This is called the Type I method of phase matching. For this case the phase matching condition is that $n_o(\omega) = n_e(\theta, 2\omega)$. Thus one uses

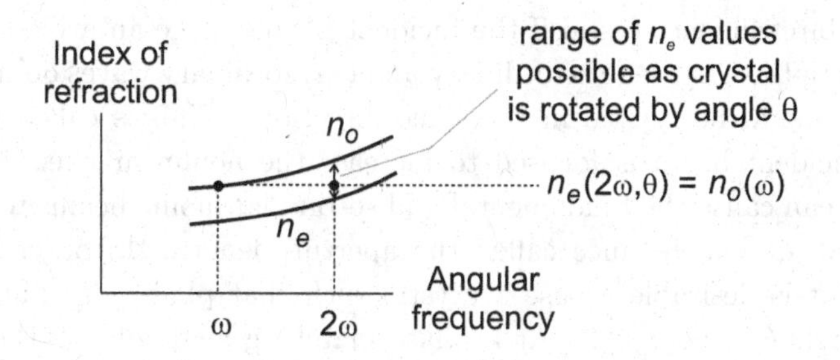

Fig. 13.1 The variation of the indices of refraction of a uniaxial crystal with angular frequency. For an extraordinary ray with frequency 2ω the index of refraction can vary between $n_e(2\omega)$ and $n_o(2\omega)$ as a function of the angle θ. For type-I phase matching, the angle θ is chosen such that $n_e(2\omega,\theta) = n_o(\omega)$.

"birefringence to offset dispersion". This is illustrated in Figure 13.1. For Type I phase matching the phase matching angle, θ_{pm}, is given by

$$\sin^2\theta_{\text{pm}} = \frac{n_o^{-2}(\omega) - n_o^{-2}(2\omega)}{n_e^{-2}(2\omega) - n_o^{-2}(2\omega)}\,. \tag{13.7}$$

It should be noted that the angles between the direction of propagation of the waves and the crystal axes are for the waves inside the crystal. It may also be possible to achieve phase matching for a wave that is polarized so that it is partly an ordinary wave and partly an extraordinary wave and the frequency doubled wave is a pure extrordinary wave, $n_e(\omega,\theta) = n_e(2\omega)$. This is called the Type II method of phase matching. With phase matching it is possible in principle to convert the fundamental wave almost entirely into a second harmonic wave.

It should be noted that for Type I phase matching the fundamental wave and the second wave harmonic wave have orthogonal polarizations. One might think that the fundamental could not excite second harmonic generation since the polarizations are orthogonal. This is not correct. The anisotropic character of the crystal enables the fundamental wave to couple to the second harmonic wave even though they have different polarizations. The different polarizations of the fundamental and second harmonic waves enables one to separate the two waves by the use of a polarizer.

Although the concept of phase matching is straight forward it should not be thought that one can always obtain phase matching with any double refracting crystal. If the difference between the ordinary and extraordinary indices of refraction is not as large as the change in the index of refraction between ω and 2ω then phase matching is not possible.

In a birefringent crystal if the incident beams make an arbitrary angle with the optic axis then the ordinary and extraordinary waves do not propagate through the crystal in the same directions. This is called walk-off. If the incident beam is focused to increase the nonlinear effect then the walk-off can cause the fundamental and second harmonic beams to become separated after a distance called the aperture length. In order to avoid walk-off it is desirable to use a crystal such that phase matching occurs at an angle θ_{pm} near $\pi/2$. At a phase matching angle of $\pi/2$ there is no walk-off. For the 1.06 μm wavelength of the Nd:YAG laser, phase matching at nearly $\pi/2$ can be obtained with a lithium niobate crystal. Since the ordinary and extraordinary indices of refraction change at different rates as the temperature changes it is possible to fine tune the phase matching to be at almost exactly $\pi/2$ by controlling the temperature of the lithium niobate crystal.

There are many different crystals that are used for frequency doubling. Some important nonlinear crystals used for frequency doubling are ammonium dihydrogen phosphate (ADP), potassium dihydrogen phosphate (KDP), potassium titanyl phosphate (KTP), lithium niobate ($LiNbO_3$), barium titanate ($BaTiO_3$), and beta barium borate (BBO). These crystals and others have suitable properties in different wavelength regions and have different efficiencies. For example beta barium borate is especially useful for second harmonic generation in the ultraviolet region near 200 nm. With the correct choice of a crystal and the correct orientation of that crystal the doubling efficiency can be very high. For example for a Nd:YAG laser with an output of 1 J per pulse one can obtain 250 mJ per pulse of doubled light.

The major remaining problem is to determine for a given optical fundamental frequency, what crystal should be used and how to cut a crystal so that it is oriented to provide good frequency doubling. In addition to the phase matching conditions, the efficiency of the frequency doubling process at different wavelengths varies from one type of crystal to another. Normally nonlinear crystal suppliers can give advice on what crystal to use to double a particular frequency, and they have the proper X-ray equipment for determining the optic axis of the crystal. After they determine the crystal's optic axis they then cut the crystal for optimum doubling efficiency. Although the crystals are cut by the manufacturer it is usually necessary for the user to make small adjustments and therefore the crystals are normally mounted so that their angle with respect to the incident fundamental laser beam can be adjusted. It should also be mentioned that the indices of refraction of materials vary with the temperature so that the temperature of the crystal must be controlled. Many of the nonlinear crystals are hygroscopic and must be mounted in a manner to keep them free from water vapor.

Another type of phase matching is also possible in some types of nonlinear crystals including KTP and $LiNbO_3$. Phase matching is required when the crystal length is greater than the coherence length, so if the effective length of the crystal is kept less than the coherence length, *quasi-phase matching* is possible. Rather than use a short crystal, however, this method essentially divides the crystal into a series of short segments equal to the coherence length, with the orientation of the crystal reversed between each segment. As a result, the mismatched phase built up in each length is 'unwound' in the next segment. Such a crystal is said to be *periodically poled.* Certain crystals, such as $LiNbO_3$, can be periodically poled by applying very strong electric fields to the crystal while being fabricated which permanently alters the crystal structure.

While seemingly an exotic technology, devices that use second harmonic generation of the Nd^{3+} 1064 nm wavelength to create green 532 nm emissions have become commodity products. Semiconductor diode lasers ($\lambda \sim 808$ nm) are used to pump a Nd:YAG or Nd:YVO_4 crystal whose infrared output are then converted into bright green visible emissions by a KTP frequency doubling crystal. Such diode pumped solid state lasers (DPSS) are available to consumers as inexpensive green laser pointers with output powers of a few mW. DPSS Are also available as higher power laboratory class instruments (tens of Watts) commonly used to pump Ti:Sapphire lasers.

13.3 Frequency Summing, Differencing, and Tripling

Another process that is of great interest for generating coherent beams of high frequency light is the process of combining two laser beams of different frequency to form a beam whose frequency is equal to the sum frequency of the two incident beams. Frequency doubling can be viewed as the process of summing the frequencies of two beams that have the same frequency. The generation of a coherent beam with a frequency equal to the difference frequency between two beams is also of interest, although it is not used as often as frequency summing. In frequency summing or differencing one uses two different incident laser beams with angular frequencies, ω_1 and ω_2, and produces an output beam that contains coherent light with the angular sum or difference frequency $\omega_3 = \omega_1 \pm \omega_2$. Although it is not necessary that the two input laser beams to be summed or differenced be collinear, they are almost always selected to be so. This is because it is much easier to focus two beams into the same small volume if they are nearly collinear than if they intersect at a substantial angle. If one has two laser beams denoted by subscripts 1 and 2 incident on a nonlinear crystal the polarization of the crystal will contain terms of the form $\epsilon_0\, d_{ijk}\, \boldsymbol{E}_{1j}\boldsymbol{E}_{2k}$. If $\boldsymbol{E}_1$ varies as $\sin(\omega_1 t)$ and $\boldsymbol{E}_2$ varies as $\sin(\omega_2 t)$ then there are terms in the polarization of the nonlinear crystal that vary as

$$\sin(\omega_1 t)\sin(\omega_2 t) = \frac{\cos(\omega_1 - \omega_2)t - \cos(\omega_1 + \omega_2)t}{2} \tag{13.8}$$

so that the radiation generated by the accelerated electrons in the nonlinear crystalline solid will contain both the sum frequency $\omega_1 + \omega_2$ and the difference frequency $\omega_1 - \omega_2$. Thus it is possible to produce a beam of light that contains either the sum or difference frequency.

First consider the process of creating a beam with a frequency equal to the sum frequency of two incident beams. As was the case for frequency doubling the process of frequency summing demands good phase matching. This requires that

$$\boldsymbol{k}_1 + \boldsymbol{k}_2 = \frac{2\pi n(\omega_1)}{\lambda_1} + \frac{2\pi n(\omega_2)}{\lambda_2} = \boldsymbol{k}_3 = \frac{2\pi n(\omega_3 = \omega_1 + \omega_2)}{\lambda_3} \tag{13.9}$$

where λ_1, λ_2, and λ_3 are vacuum wavelengths. The phase matching condition for frequency summing can be rewritten as

$$\omega_1\, n(\omega_1) + \omega_2\, n(\omega_2) = \omega_3\, n(\omega_3) \tag{13.10}$$

or

$$n(\omega_3) = \frac{\omega_1\, n(\omega_1) + \omega_2\, n(\omega_2)}{\omega_1 + \omega_2}\,. \tag{13.11}$$

The two incident laser beams can be polarized in either the same direction so that they are either both ordinary waves or both extraordinary waves or they can be polarized so that one beam is an ordinary wave and the other is an extraordinary wave. The former is called Type I method of phase matching and the latter is called Type II method of phase matching.

For the Type I phase matching in a negative uniaxial crystal where both incident waves are ordinary waves the sum wave is an extraordinary wave. In a Type I phase matching situation for a negative uniaxial crystal the phase matching angle can be obtained from

$$n_e(\omega_3) = \left[\frac{\cos^2\theta_{\mathrm{pm}}}{n_o^2(\omega_3)} + \frac{\sin^2\theta_{\mathrm{pm}}}{n_e^2(\omega_3)}\right]^{-1/2} = \frac{\omega_1}{\omega_3}\, n_o(\omega_1) + \frac{\omega_2}{\omega_3}\, n_o(\omega_2)\,. \tag{13.12}$$

The expression for the θ_{pm} for Type I phase matching with a positive uniaxial crystal or for Type II phase matching with either positive or negative uniaxial crystals can also be derived.

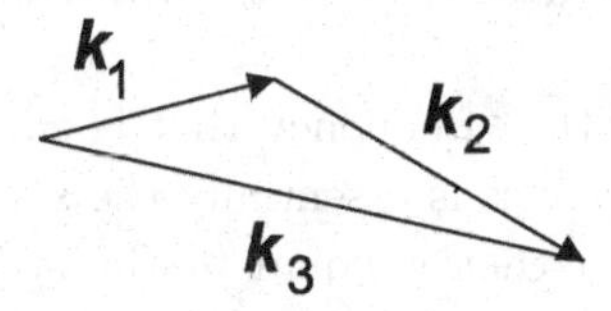

Fig. 13.2 The condition on the $\boldsymbol{k}$ vectors for phase matching when summing the frequencies of two laser beams. The three $\boldsymbol{k}$ vectors can be collinear.

One interesting feature that should be noted is that it is not necessary that both incident waves and the sum wave must travel at the same phase velocity. It is only necessary that $\boldsymbol{k}_3 = \boldsymbol{k}_1 + \boldsymbol{k}_2$ in order to satisfy the phase matching condition. This corresponds to the beat wave for the two incident waves moving through the nonlinear crystal at the same phase velocity as the outgoing sum wave. In general if the two incident waves are not collinear then the same condition applies but the two incident $\boldsymbol{k}$ vectors must be added as vectors to give the $\boldsymbol{k}$ vector for the sum wave. This is illustrated in Figure 13.2.

For producing a wave with a frequency that is the sum of the frequencies of the two incident waves it is of course necessary that $\omega_3 = \omega_1 + \omega_2$. From a photon view point this is just the conservation of energy. From the photon viewpoint the phase matching condition, $\boldsymbol{k}_3 = \boldsymbol{k}_1 + \boldsymbol{k}_2$, is just the conservation of linear momentum. For frequency doubling it is necessary to conserve both the energy and the linear momentum of the photons involved in the process.

The summing of two waves to produce a wave with a frequency equal to the sum of the incident wave frequencies is called up-frequency conversion. The principal use of up-frequency conversion is to convert infrared photons into visible photons. This is useful because the detectors for infrared photons are not very efficient whereas a photomultiplier can detect visible or near ultraviolet photons with a very high efficiency. For up frequency conversion one of the low frequency waves is an intense laser that has a frequency such that when it is added to the infrared frequency the sum is in the visible or near ultraviolet. This intense source is mixed in a nonlinear crystal with the infrared source. There is a component in the output at the sum frequency, that is at a wavelength in the visible or near ultraviolet that can be detected efficiently with a photomultiplier.

It is often desired to obtain a wave that is at three times the frequency of the incident wave. This tripled wave is usually obtained by first producing a second harmonic wave and then summing the second harmonic wave with the incident wave to form a wave at the third harmonic of the incident wave.

Producing a wave with a frequency that is equal to the difference frequency of two incident waves is essentially the same process as producing a wave with a frequency that is equal to the sum frequency of the two incident waves. One must only make sure that the index of refraction of the difference wave satisfies the phase matching condition for the difference frequency.

13.4 Parametric Amplifiers and Oscillators

Optical parametric amplifiers and oscillators (OPO) are a method of obtaining an intense coherent beam of light at wavelengths that are not readily generated in another manner. The name parametric amplifier or oscillator refers to the variation of some parameter of a nonlinear crystal to produce amplification or oscillation at optical frequencies. A time dependent electric field in a nonlinear crystal causes the index of refraction to vary as a function of the time. The index of refraction and hence the polarization of a nonlinear crystal is the parameter that is varied in a parametric amplifier or oscillator. This time dependent polarization can generate optical waves. Previous sections of this chapter discussed how two incident waves can be combined to form a wave with a frequency equal to the sum of the frequencies of the two incident waves. Parametric amplifiers and oscillators are closely related to this process, but run in reverse. A parametric amplifier or oscillator converts an input pump wave into two waves both at lower frequency. The two lower frequency waves are called the signal wave and the idler wave. The signal beam is typically the beam is desired for some use. If the angular frequency of the pump wave is ω_3, the angular frequency of the signal wave is ω_1, and the angular frequency of the idler wave is ω_2 then $\omega_3 = \omega_1 + \omega_2$. Phase matching is essential for a parametric amplifier or oscillator so that it is necessary that $\boldsymbol{k}_3 = \boldsymbol{k}_1 + \boldsymbol{k}_2$. Phase matching is typically in the forward direction for an OPO, i.e $\boldsymbol{k}_1$, $\boldsymbol{k}_2$ and $\boldsymbol{k}_3$ are colinear.

One might think from what has been said that a parametric amplifier or oscillator are exactly the same as the time reversal of the process of combining two waves into a wave with a frequency equal to the sum of the frequencies of the two incident waves. This is almost correct, but one must be careful in what is meant by this. When one has a single wave incident on a nonlinear crystal there is not just a single possible way to produce a pair of photons the sum of whose frequencies add up to the frequency of the incident wave. Instead there are many ways that the incident photon can split into two photons. This is somewhat analogous to the following. An atom can absorb a photon from a laser beam and make a transition to a higher level denoted by u. An atom in the excited level u can decay by the spontaneous emission of two photons. There is, however, no reason why the two photons spontaneously emitted should necessarily have particular frequencies. In fact two photon spontaneous emission has a wide spread in the frequencies of the emitted photons although of course the sum of the frequencies of the two photons multiplied by Planck's constant must be equal to the energy difference between the excited level, u, and the lower level, l.

In the situation where a laser beam is incident on a nonlinear crystal in a parametric oscillator there are many ways that the incident photon can be split up with the resulting spontaneous emission of two photons. The nonlinear crystal must have phase matching for the signal and idler beams to gain significant intensity. It should be clear that a stimulated emission process for the signal photons results in both signal and idler photons being produced. In order to obtain the substantial emission of coherent radiation at a particular frequency and into a particular mode from a parametric oscillator one must have enough stimulated emission that one reaches threshold for stimulated emission into the desired mode. For intensities below threshold only spontaneous emission is produced. Thus a parametric amplifier or oscillator has a threshold pump intensity for operation. At the threshold the stimulated emission into the signal beam is equal to the losses experienced by that beam. This is different from the situation for the process of combining two waves to form a wave whose frequency is equal to the sum frequency of the two incident waves. The process of summing two waves does not have a threshold. The process of summing two waves simply varies as the product of the intensities of the two waves.

Parametric oscillators are commonly used for the production of coherent light at frequencies where other lasers are not readily available. Figure 13.3 shows a schematic diagram for a possible OPO. The pump beam is incident on the nonlinear crystal through a dichroic mirror that is transparent for the pump beam but highly reflective for the signal beam. The dichroic mirror

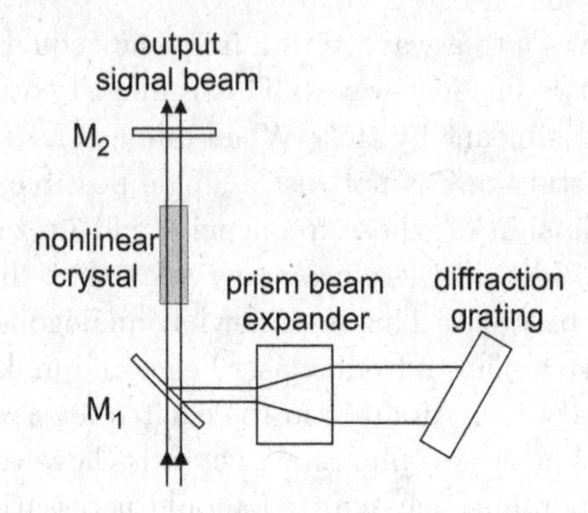

Fig. 13.3 A schematic diagram of an optical parametric oscillator. The mirror M_1 must pass the incident pump beam and reflect the signal beam. The mirror M_2 should couple some of the signal beam into the output beam.

forms the folding mirror in a folded optical cavity for the signal beam. The purpose of the optical cavity is to provide a long photon lifetime for the signal beam photons and thereby reduce the threshold pump power. The optical cavity contains a prism beam expander and a diffraction grating so that oscillation occurs at only a single wavelength. When the pump laser beam reaches threshold the very weak spontaneous emission at the signal frequency and in the single oscillating mode mode of the signal beam is stimulated and oscillation begins so that strong emission of light occurs at the signal frequency. As seen in Fig. 13.3 the output mirror of the optical cavity is partially reflecting for the signal beam in order to couple some of the signal beam out of the optical cavity.

The setup shown in Figure 13.3 is not a very useful for an optical parametric oscillator. It was only shown to illustrate the idea of an OPO in a simple way. Figure 13.4 shows a commonly used ring optical parametric oscillator. The mirrors are in what is called a bow tie configuration. In the ring oscillator, the beam is circulating around the optical cavity in only one direction. There is no device shown for selecting the wavelength. The wavelength selection is accomplished by varying the phase matching condition. This can be accomplished either by tilting the phase matching crystal or by changing its temperature or both. In Figure 13.3 the signal beam is the output beam.

Optical parametric oscillators are used most commonly in the wavelength range 600 nm - 3 μm. OPOs can be operated in either a pulsed mode or in a cw mode. The output power from a parametric oscillator can

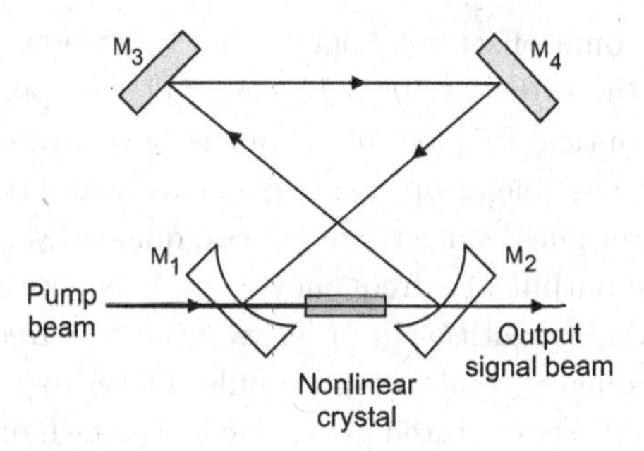

Fig. 13.4 A schematic diagram of a bow tie optical parametric oscillator. The dichroic mirror M_1 must pass the incident pump beam. Mirror M_2 couples some of the signal beam into the output. The cavity is resonant for the signal or both the signal and idler beams.

be high. They also have wide tuning ranges, commercial models of OPOs are available that produce tunable coherent radiation over an extremely wide wavelength range.

13.5 Frequency Combs

Frequency combs are remarkable sources of light. The Nobel Prize was awarded to Hall and Hänsch for the development of optical frequency combs. A frequency comb is a method of generating a light beam that contains light at frequencies that are exact harmonics of a low frequency standard. The term comb is used because a graph of the light intensity as a function of the frequency of the light looks like the teeth of a comb. Optical frequency combs can be discussed in either the time domain or in the frequency domain. The relationship between the time and frequency domains is simply that a signal in the frequency domain is obtained by taking the Fourier transform of the signal in the time domain. The discussion of the optical frequency comb is simplest in the frequency domain. The output signal from an optical frequency comb is a beam of light that contains an extremely large number of frequencies that are separated by equal spacings. This is very different from a conventional laser where the output light beam contains typically a single frequency. This is also very different from a typical broad bandwidth light source like an incandescant light which contains all frequencies in the output.

The output beam of a frequency comb contains frequencies given by $n\,f_R + \nu_0$ where n is an integer, f_R is the repetiton rate of the pulsed laser, and ν_0 is called comb offset frequency. The repetition rate of the pulsed laser can be on the order of 10^8 - 10^9 Hz. The output contains a great many individual optical frequencies all precisely separated by f_R. Nothing like this has been possible at optical frequencies before the invention of the laser. As one might guess since there are two quantities that determine the frequencies in the output of a frequency comb it is necessary to accurately control two different quantities in order to produce a frequency comb.

All optical frequency combs utilize a pulsed laser to generate the optical comb. A Ti:Sapph laser operated in a mode locked fashion can be used for a frequency comb. The Ti:Sappph mode locked laser produces optical pulses with a repetition rate, f_R. With modern developments a mode locked Ti:Sapph laser can produce pulses as short as a few fempto seconds (1 fs= 10^{-15}s). A pulse with a duration of only a few fempto seconds is so short that it contains only a few oscillations at the optical frequency. The pulses from a mode locked laser contain Fourier frequencies centered about the laser optical frequency which might be about 4×10^{14} Hz and extending over a range of about $\pm 3 \times 10^{13}$ Hz for a pulse duration of 5 fs. The range can be even larger for shorter duration pulses. The problem is how to develope a a series of comb like frequencies from the pulsed output of the mode locked Ti:Sapph laser.

One might wonder why the output of a mode locked Ti:Sapph laser is not a frequency comb automatically. The answer is that each individual pulse from the laser contains the same set of frequencies as every other pulse, but the phase of the electric field in each indivdual pulse is not the same as the phase in every other pulse. There are phase changes from pulse to pulse. Averaged over many pulses the output is more or less a a continuum of Fourier frequencies. In order to make the output of the laser into a frequency comb it is necessary to fix the phase of electric field in each pulse to be identical. If electric field in every pulse is the same then the laser output will be a frequency comb with a separation between the frequencies in the comb given by f_R. The method of setting the electric field phase is the second thing needed in addition to fixing the repetition rate in order to make a frequency comb. The most common way of accomplishing this is by self referencing. If the spread in the Fourier frequencies in the output from the mode locked laser is sufficiently large that it covers more than a factor of two in frequency then on can double a low frequency and beat it against the higher frequency that is twice the lower frequency. If one stabilizes the beat frequency then the phase of the electric field in each pulse is fixed to be the same. This requires a second method of stabilizing the path length

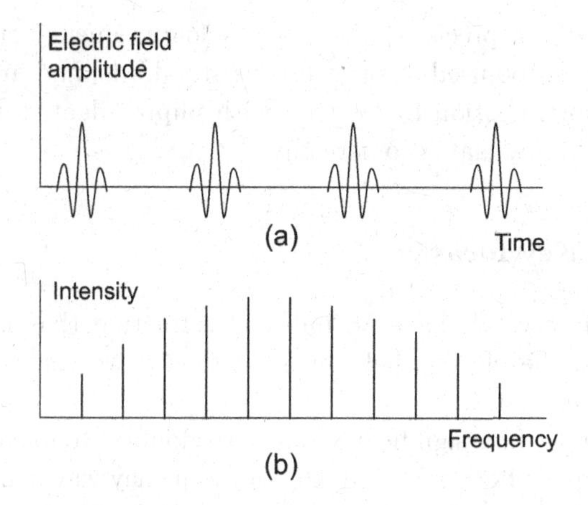

Fig. 13.5 (a) the electric field amplitude as a function of the time. The pulses are exactly equally spaced in time. (b) the output frequency comb.

in the ring laser. In summary it is necessary to have a method for fixing the path length in order to set f_R and a second independent method that fixes the beat frequency between a doubled low frequency in the comb a a higher frequency in the comb that is twice the frequency of the lower frequency without altering the repetition rate. If the spread in the frequenciees is not a factor of two then it is possible to beat two frequencies that are in the ratio 3 to 2 or other integral ratios. There are also methods to broaden the frequencies in a pulse i.e. to sharpen the pulses.

If the electric field in the series of pulses emitted by the laser is a perfect periodic function of the time then this periodic function of the time has a Fourier series that is a series of harmonic frequencies. In other words the electric field in the optical pulses contains only a series of harmonics of the low frequency f_R. The optical frequencies center at a harmonic of the low (radio) frequency standard near the line center of the laser transition and would have all the harmonics of the radio frequency standard ranging above and below the center frequency by about $\Delta\omega \sim 1/\tau$. In order for the frequency comb to be wide it is necessary for the pulse duration to be very short. The electric field of the pulses is illustrated schematically in Figure 13.5(a). The frequencies in the comb are illustrated in (b).

The discussion of a frequency comb has been very highly simplified and no details on how the comb is actually set up experimentally have been provided. The field is rapidly developing and any description would be out of date in the near future. It is truly remarkable that one can make optical frequencies that are precise multiples of a low frequency standard. Frequency combs can be used as optical time standards, and are also used to measure optical transition frequencies with unprecedented accuracy. Frequency combs are avialable comercially.

Summary of Key Ideas

- Anisotropic crystals have an index of refraction that is a non-linear function of the electric field in an optical wave that is propagating through the crystal.
- In order to produce significant non-linear doubled frequency intensity it is necessary for both the fundamental frequency wave and the doubled frequency wave to travel with the same phase velocity in the crystal. This is called *phase matching* (Sec. 13.2).
- For uniaxial crystal, the index of refraction varies for different polarizations of light. For ordinary waves, the index remains constant as the crystal is rotated about the optical axis. For extrodinary waves, the index varies between n_o and n_e depending upon θ,

$$\frac{1}{n_e^2(\theta)} = \frac{\cos^2\theta}{n_o^2} + \frac{\sin^2\theta}{n_e^2} \quad (13.6)$$

- For Type I phase matching, $n_o(\omega) = n_e(\theta, 2\omega)$.
- For Type II phase matching, $n_e(\theta, \omega) = n_e(2\omega)$.
- Frequency doubling (Sec. 13.1) is a widely used example of a nonlinear process. A common example of frequency doubling is the conversion of an infrared electromagnetic wave at 1064 nm from a Nd:YAG laser into visible light at 532 nm. It is also possible to triple the frequency of a laser beam and it is possible to form waves with frequencies that are sums and differences of the frequencies of two laser beams (Sec. 13.3).
- Optical parametric amplifiers and oscillators (Sec. 13.4) use nonlinear processes to create coherent beams at frequencies not easily obtained with a laser, such as infrared wavelengths (600-3000 nm). They work by converting an input pump beam into two waves (the idler and signal beams) both at lower frequencies, $\omega_{\text{pump}} = \omega_{\text{signal}} + \omega_{\text{idler}}$.
- Frequency combs use non-linear processes to produce light at frequency harmonics of a low (radio) frequency standard (Sec. 13.5). It enables one to measure the frequency of the low frequency signal and know the frequency of the harmonic optical waves with unprecedented accuracy.

Suggested Additional Reading

P. A. Franken, A. E. Hill, C. W. Peters, and G. Weinreich, "Generation of Optical Harmonics" *Phys. Rev. Lett.* **7**, 118 (1961). [First demonstration of the production of frequency doubled light using a non-linear crystal.]

R. W. Boyd, *Nonlinear Optics, 3rd Ed.*, Academic Press (2008).

N. Bloembergen, *Nonlinear Optics*, W. A. Benjamin (1965).

A. E. Siegman, *Lasers*, University Science Books (1986).

B. E. A. Saleh and M. C. Teich, *Fundamentals of Photonics*, John Wiley and Sons (1991).

William T. Silfvast, *Laser Fundamentals*, Cambridge University Press (1996).

M. E. Marhic, *Fiber Optical Parametric Amplifiers, Oscillators and Related Devices*, Cambridge University Press (2007).

B. J. Orr, Y. He, and R. T. White, "Spectroscopic applications of tunable optical parametric oscillators" (pp. 15-95), in *Tunable Laser Applications, 2nd ed.*, CRC Press (2009).

J. N. Eckstein, A. I. Ferguson, and T. W. Hänsch, "High-Resolution Two-Photon Spectroscopy with Picosecond Light Pulses" *Phys. Rev. Lett.* **40**, 847 (1978).

S. A. Diddams, D. J. Jones, J. Ye, S. T. Cundiff, J. L. Hall, J. K. Ranka, R. S. Windeler, R. Holzwarth, T. Udem, and T. W. Hänsch, "Direct Link between Microwave and Optical Frequencies with a 300 THz Femtosecond Laser Comb" *Phys. Rev. Lett.* **84**, 5102 (2000).

J. Ye and S. T. Cundiff, *Femtosecond Optical Frequency Comb: Principle, Operation and Applications*, Springer (2005).

Problems

1. Find an expression for the angle for Type II phase matching for a negative uniaxial crystal.

2. The wavelength of a Nd:YAG laser is 1.06 μm and the wavelength of the second harmonic is 530 nm. The nonlinear crystal lithium niobate ($LiNbO_3$) has indices of refraction of n_o=2.24 and n_e=2.16 at λ=1.06 μm and has indices of refraction of n_o=2.33 and n_e=2.23 at λ=530 nm. Calculate the Type I phase matching angle for second harmonic generation.

3. Sketch the variation of the ordinary and extrordinary indices of refraction for a situation where Type II phase matching is not possible in a negative uniaxial crystal.

4. Find an expression for the angle for Type I phase matching for frequency doubling in a positive uniaxial crystal used where the incident wave is an extraordinary wave.

5. Calculate the peak electric field for a 1 J pulse from Nd:YAG laser with a pulse duration of 8 ns. The laser is focused to a spot with a diameter of 10 μm.

6. How do you expect the second harmonic generation to change as when the lasers focal radius is decreased by a factor of 2?

7. What is the phase matching condition for two waves incident on a nonlinear crystal from opposite directions, where a third wave is generated?

8. What are the units for d_{ijk} in the MKS system?

9. Use the expression for $\boldsymbol{P}_i$ to show that the coefficients d_{ijk} must be zero for a crystal that has inversion symmetry.

10. Calculate the coherence length for second harmonic generation in a lithium niobate crystal for the wavelength of a Nd:YAG laser. The index of refraction of $LiNbO_3$ at λ=1.06 μm is n_o=2.24 and at λ=530 nm is n_o=2.33.

11. If the low frequency standard for a frequency comb operates at 5×10^8 Hz and the optical laser operates with a wavelength of 800 nm, what is the harmonic near the center of the frequency comb?

Chapter 14

Topics in Quantum Optics

For the most part, this book has avoided getting into the numerous applications of lasers. In this chapter, however, we explore some of the ways that lasers can be used to achieve ever increasing control of atoms. We start with the simple idea of using lasers to detecting the presence or absence of atoms. We then move onto using lasers to control the internal and external degrees of freedom of atoms. Some of these levels of control employ classical concepts such as the position and velocity of atoms. Other levels of control such as the entanglement require a quantum mechanical treatment. *Quantum optics* is the study of the interaction of light and matter at the microscopic scale where quantum effects are important. Much of this area involves an understanding of quantum mechanics beyond that expected of the reader. As a result, in this chapter we generally try to present the simplest treatment of the material that is consistent with a clear understanding.

The material covered in this chapter includes the use of laser beams for excitation of atoms or molecules into unique levels and application of this to optical pumping of atoms, the slowing of beams of atoms, trapping of atoms, and the use of trapped atoms or ions in quantum computers. Finally this chapter concludes with a discussion of the question as to what is a laser. This includes a discussion as to how one can determine if a beam of light coming through a hole in the wall was produced by a laser or not. This material will enable the reader to understand why people believe that many astrophysical objects produce maser beams. It will also enable the reader to understand lasers in comparison to other light sources such as thermal sources or synchrotron light sources.

14.1 Laser Induced Fluorescence: *Detecting Atoms*

The detection of atoms or molecules is a common operation in many areas of physics, chemistry and engineering. The light emitted by excited atoms provides a simple method of detecting atoms. Since the spectra of each atom is unique, the emission intensity can be proportional to number of atoms in the particular excited level of interest. It is often necessary to detect the density of atoms in the ground level or long-lived metastable levels. One method of detecting such atoms is to monitor the decrease in transmission of a laser beam as it passes through a sample. One variant of this laser absorption method is the Doppler-free saturated absorption spectroscopy experiment previously described in Chapter 5. One slight difficulty with this method is that it requires access to both sides of the sample, since the transmission through the sample is monitored. As a result, it can not be used to monitor a remote sample. Second, when the target density is very low it is necessary to measure a very small decrease in the transmitted laser power. These difficulties can be overcome by using laser induced fluorescence (LIF).

In a typical experiment utilizing LIF an intense tunable laser is used to illuminate a gas (Figure 14.1). The laser is tuned so that the wavelength of the laser corresponds to an absorption transition in some atom or molecule in the gas. The atoms in the lower level of the transition, denoted by l, absorb the laser photons and are thereby excited into a higher level, denoted by u. The level u then decays by spontaneous emission to various lower levels. The excited atoms are detected by observing the spontaneous radiation from the level u. Since some laser light may be scattered into the detector it is advantageous to detect the LIF from emission to some level

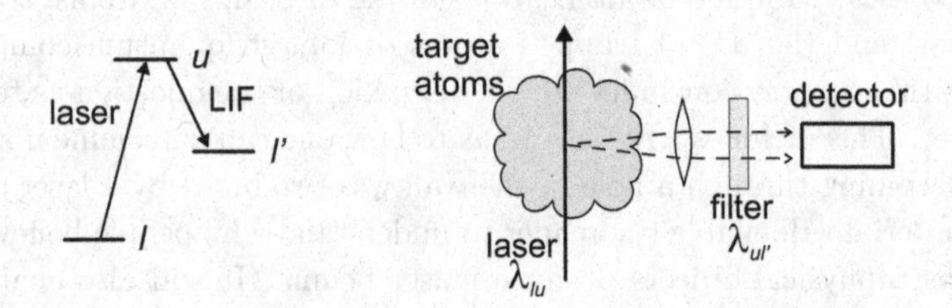

Fig. 14.1 Scheme for laser induced fluorescence. The laser excites atoms (or molecules) from level-l to level-u. Emissions out of level-u are observed on the $u \to l'$ transition. If $l = l'$, special care must be made to prevent scattered laser light from reaching the detector. If $l \neq l'$ and several photons per target atom are required, there needs to be some process for atoms in level-l' to return to the initial level l.

(l') other than the initial level, l, so that the LIF is emitted at a different wavelength ($\lambda_{ul'}$) than the laser wavelength (λ_{ul}). If this can be done then an interference filter or diffraction grating monochromator can be used to eliminate the scattered laser light. Sometimes, however, it is necessary to detect LIF by observing the spontaneous emission at the same wavelength as the laser wavelength. This requires great care to eliminate the effects of scattered laser light.

The sensitivity of detecting atoms by LIF can be very high. If the background is dark (i.e., if the gas sample is not radiating) then it may be possible to detect single atoms. To yield a measurable emission signal from a very small atomic sample, it is necessary that each atom absorb and re-radiate photons at a high rate. This means that it must not be possible for the atoms to decay into a level that does not absorb the incident light. It is not necessary that the atoms decay only into the initial level but only that, by some means, the lower level, l, is rapidly replenished. Against a bright background (i.e., if the gas is radiating at the wavelength to be detected) then LIF is not as sensitive as it is against a dark background. Nevertheless if one uses a pulsed tunable laser, the detection can be gated to detect LIF only during one or a few radiative lifetimes after the end of the laser pulse. LIF is also useful for the sensitive detection of molecules or free radicals.

The laser induced fluorescence signal is proportional to the number of atoms pumped into the excited level u times the branching fraction of the observed $u \to l'$ spontaneous emission decay channel. Absolute measurements require converting the measured optical signal (i.e., the current from a PMT or photodiode) into absolute units (i.e., photons/sec) and accounting for the limited solid angle of the detector. While simple in principle, absolute quantitative measurements require a careful understanding of all the physics involved in the LIF process. For example, for atoms (or molecules) the excited levels may be collisionally quenched before they radiate. This is particularly a problem at high gas densities or when the lifetime of the excited level is very long. The rate of excitation into level u is also sensitive to the frequency overlap between the laser's lineshape and $l \to u$ lineshape. For a narrowband laser, one typically integrates the observed signal as the laser's wavelength is swept over the width of the atomic transition. Saturation issues are also common (Chapter 5). In general, the number of excited atoms can be increased by increasing the pump laser intensity. At very high intensities, however, stimulated emission of the $u \to l$ transition limits the linearity of the LIF signal.

LIF measurements have the additional advantage that they can be spatially resolved. In a typical LIF experiment one observes the induced emissions along an axis perpendicular to the laser beam axis. Hence, the observed LIF signal arises from a small volume located at the intersection of the laser beam and the viewing region of the detector. This is in contrast to the laser absorption signal which is effectively a line of sight measurement along the entire path of laser through the sample.

14.2 Optical Pumping: *Exciting atoms into a particular state*

In addition to detecting atoms, lasers can be used to control atoms. Here, we consider the use of lasers for optical pumping. Although many different atoms can be optically pumped, in order to make the discussion concrete we consider the optical pumping of an idealized potassium (K) atom, which has no nuclear spin. The ground level of K is the $4^2S_{1/2}$ level and the lowest excited level of K is the $4^2P_{1/2}$ level. Each of these energy levels has two states one with $m = +1/2$ and the other with $m = -1/2$. The two lowest energy levels of K are shown in Figure 14.2. The resonant absorption wavelength from the $^2S_{1/2}$ level to the $^2P_{1/2}$ level is at $\lambda = 770$ nm. The transition rate connecting the various upper and lower states depend upon the polarization of the absorbed or emitted photon. Linearly polarized light drives transitions with $\Delta m = 0$ (sometimes called π transitions), while circularly polarized light drives $\Delta m = \pm 1$ (so called σ transitions).

If a K vapor is illuminated with σ^+ polarized light at 770 nm then only $\Delta m = +1$ absorption transition can be excited by absorption. Circularly polarized light at 770 nm can be produced by using a tunable Ti:Sapph laser followed by a linear polarizer and a quarter wave plate. When a σ^+ photon is absorbed by a K atom in the $m = -1/2$ state of the $^2S_{1/2}$ ground level it is excited to the $m' = +1/2$ state of the $^2P_{1/2}$ level. A K atom

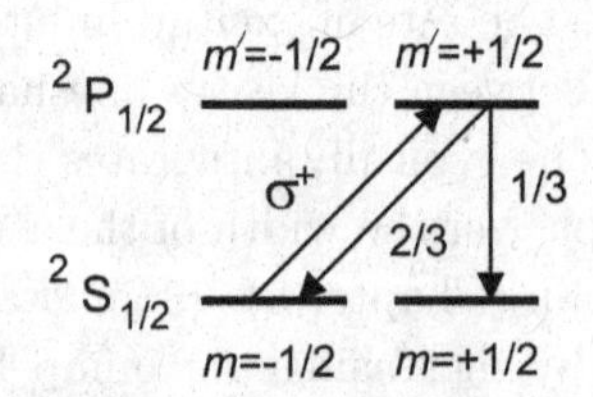

Fig. 14.2 The Zeeman energy levels of a K atom with absorption and spontaneous emission for optical pumping with a σ^+ polarized laser.

in the $m = +1/2$ state of the ${}^2S_{1/2}$ level can not absorb light since the selection rules for σ^+ light require $\Delta m = +1$ and the largest possible m' value for the ${}^2P_{1/2}$ level is 1/2 instead of 3/2. Once an atom is excited it decays back to the ground level by spontaneous radiation. The spontaneous radiation from the $m' = +1/2$ state of the ${}^2P_{1/2}$ level can be to either the $m = +1/2$ or the $m = -1/2$ states of the ground level. The branching ratios to the $m = +1/2$ and $m = -1/2$ states of the ground state are 1/3 and 2/3 respectively. Thus, while most excited $m' = +1/2$ ${}^2P_{1/2}$ atoms decay back to the 'original' $m = -1/2$ state of the ${}^2S_{1/2}$ ground level, a third of the atoms decay into the $m = +1/2$ state of the ${}^2S_{1/2}$ ground level which are transparent to the σ^+ laser beam. Since atoms are removed from the $m = -1/2$ state but not the $m = +1/2$ state of the ground level and since excited atoms decay back to both states of the ground level, with each absorption-emission cycle the number of atoms in the $m = -1/2$ state is reduced and the number of atoms in the $m = +1/2$ state is increased. Effectively atoms are "optically pumped" from the ${}^2S_{1/2}$ $m = -1/2$ state into the $m = +1/2$ state. If the relaxation rate from the $m = +1/2$ state to the $m = -1/2$ state of the ground level is small compared to the rate at which optical pumping process transfers atoms from the $m = -1/2$ state to the $m = +1/2$ state then a large population difference between the two ground level states is produced by optical pumping. The ground level is polarized by optical pumping. The polarization of the ground level is defined in terms of this population difference,

$$P = \frac{n_\uparrow - n_\downarrow}{n_\uparrow + n_\downarrow}, \tag{14.1}$$

where $n_\uparrow$ and $n_\downarrow$ are the populations of the $m = +1/2$ and $m = -1/2$ states respectively.

Laser optical pumping can produce large polarizations in a K vapor. In order to understand the possibilities better consider a Ti:Sapph laser beam with a diameter of 1 cm and with an output power of 4 W. At 770 nm, 4 W of light corresponds to 2×10^{19} photons/s. The absorption cross section for light at line center of the $4^2S_{1/2} \rightarrow 4^2P_{1/2}$ transition in K is equal to 9×10^{-12} cm^2. The absorption rate of a K atom in a 4 W laser beam at 770 nm is 0.4×10^8 s^{-1}. For an absorption rate this high it is possible to produce a high polarization even for a relatively high relaxation rate between the ground level m states. In order to maintain the polarization it is necessary to have an applied magnetic field parallel to the incident light beam. Although our discussion has focused on the an idealized K atom with no nuclear spin it is possible to optically pump a real K atom with nuclear spin and with the resulting hyperfine structure. In addition to K it is also possible to optically pump all the other alkali atoms as well as large number of other atoms and molecules. For example, very high density, high polarization targets of rubidium atoms can be achieved using high power diode lasers operating at 780 nm.

Optically-pumped spin-polarized targets have many uses. For example, they can be used to measure the Landé g-factors and hyperfine separations. This type of experiment can be done quite easily with LIF. After a target is optically pumped, the absorption of the laser beam is low and little light is emitted. Applying a rf field at the correct frequency induces transition between the $m = \pm 1/2$ states, depolarizing the target. This results in increased absorption and an increased LIF signal. The frequency required to drive the $m = +1/2 \rightarrow m = -1/2$ transition is equal to $\nu_{\rm rf} = g\,\mu_0\,B/h$ where g is the Landé g-factor of the ground level, μ_0 is the Bohr magneton, and B is the applied magnetic field. Optical pumping is also used to create polarized ion sources and targets for nuclear and high energy physics experiments. Spin polarized ^{129}Xe atoms have been used for magnetic resonance imaging (MRI) measurements of the lungs. Typically, a MRI image is produced by measuring the small polarization difference of a very large number of protons placed in a strong magnetic field. This does not work for the lungs which are mostly empty. Nevertheless, a significant MRI signal can be formed by using a low-density ^{129}Xe gas target with a very high spin polarization. This is produced by optically pumping a Rb vapor with a high-power diode laser array. Spin-exchange collisions between Rb and Xe atoms efficiently transfers the polarization to Xe atoms which can be harmlessly inhaled. A cold trap is used to condense the alkali Rb vapor.

14.3 Atom Trapping: *Controlling the motion of atoms*

In addition to controlling the internal degrees of freedom of an atom, lasers can also be used to control the external degrees of freedom of atoms. In this context, the external degrees of freedom are the atom's velocity and position. One of the most interesting current areas of research is the optical cooling and trapping of atoms by the use of tunable lasers. First we consider controlling the velocity distribution of a sample of atoms (Section 14.3.1), and then proceed to see how the position of atoms can also be controlled (Section 14.3.2).

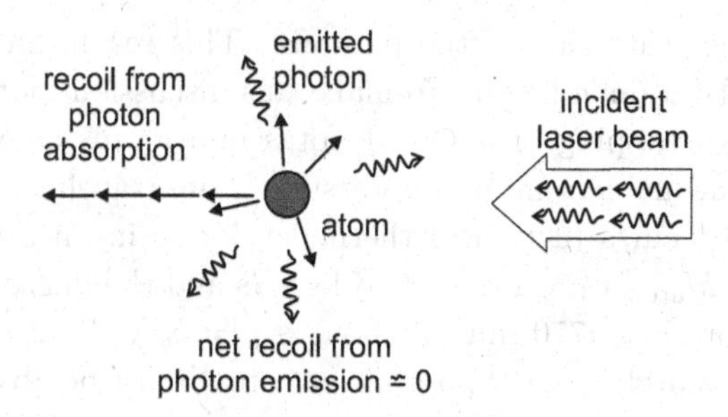

Fig. 14.3 Example of manipulating atoms with light. An atom sequentially absorbs and re-radiates four resonant photons coming from the laser beam on the right. Each time it absorbs a photon, the atom receives a kick to the left. The direction of photons emitted in the spontaneous emission process is essentially random, so the net recoil the atom receives in this step averages to zero in the atom's frame.

14.3.1 *Laser Cooling of Atoms*

14.3.1.1 *Manipulating Atoms with Light*

The basic principle for using lasers to control atoms is illustrated in Figure 14.3. An atom is illuminated by a laser beam with a wavelength that can drive a strong $l \rightarrow u$ resonance transition. Photons carry a momentum $\hbar \boldsymbol{k}$ where $\boldsymbol{k}$ is the wavevector (magnitude $2\pi/\lambda$) of the light. Each time an atom absorbs a photon, the photon's momentum is added to that of the atom's. When the atom decays via spontaneous emission, the emitted photon carries off some momentum and the atom receives a recoil. The direction of the spontaneously emitted photon is random. If the atom absorbs and re-emits a large number of photons, the recoil contributions from the emission process will average to zero in the atom's frame. The contributions from the absorption step, however, all occurs in the same direction. Hence, the laser will push the atom in the direction the laser beam is propagating. The momentum carried by individual photons is very small. But atoms are also relatively small and can, for strong resonance transitions, absorb and re-emit photons at a very rapid rate (i.e., $\sim 10^8$ s^{-1}).

14.3.1.2 *Stopping a Beam of Atoms*

Consider the process of using a laser to stop atoms emerging from a source. The atomic beam is illuminated by a tunable laser beam traveling in the opposite direction from the atomic beam. The tunable laser beam frequency is set to be equal to the resonant absorption frequency of the atoms in the beam. Following the discussion of Section 14.3.1.1, the laser exerts a force,

F_{scatt}, from the scattering of the photons. This results in slowing down the atoms in the atomic beam. To make this discussion more concrete, let us consider the stopping of beam of potassium atoms emerging from an oven. The linear momentum of a potassium atom traveling with a velocity of $v_0 = 4 \times 10^4$ cm/s (a typical thermal velocity in an atomic beam) is about $p_{\text{atom}} = M_{\text{atom}}\, v_0 = 2.8 \times 10^{-23}$ kg m/s and the linear momentum of a resonant photon (λ=770 nm) is about $p_{\text{ph}} = h/\lambda = 8.2 \times 10^{-28}$ kg m/s so that approximately 3×10^4 photons per atom must be absorbed in order to stop the atomic beam. The maximum rate at which photons can be absorbed is set by the maximum rate for spontaneous emission from the excited level. The maximum rate of spontaneous emission is approached when the laser intensity is equal to the saturation intensity (Chapter 5). In this situation an atom spends about half the time in the ground level and about half the time in the excited level so that the maximum spontaneous emission rate per atom is approximately $1/2\tau$ where τ is the spontaneous lifetime of the upper level. Although the laser intensity can be increased above the saturation intensity the spontaneous emission does not increase; only the stimulated emission increases. Absorption followed by stimulated emission does not slow down the atom since the stimulated photon has the same linear momentum as the original laser beam photon. Typically experiments are operated with a laser intensity of about 1/2 the saturation intensity. For potassium the lifetime of the excited level is τ=26 ns so that the maximum usable absorption rate is equal to 1.9×10^7 photon/s. This means that a time of about 2×10^{-3} s is required to stop the atomic beam. If the deceleration of the atoms in the atomic beam is uniform the minimum distance required to stop an atomic beam is equal to the average velocity of the beam, which is $v_0/2$, multiplied by the time require to stop the atomic beam. The minimum distance to stop an atom is therefore about 20 cm.

The discussion so far has overlooked an important point, the wavelength required to drive the resonance transition is Doppler shifted. For the laser to interact with the atomic beam the laser frequency must be equal to the Doppler shifted resonant frequency of the atoms in the beam. But if the velocity of the atoms is changing, so is the required wavelength. For an atomic beam moving with a velocity of 4×10^4 m/s the resonant frequency of an atom is shifted by a factor of 1.3×10^{-6} from its resonant frequency at rest. Thus as the atoms in the atomic beam are slowed either the laser frequency must decrease by a factor of about 1.3×10^{-6} in order that the atoms may continue to absorb the photons or the resonant atomic frequency must increase by the same factor. Both methods can be used.

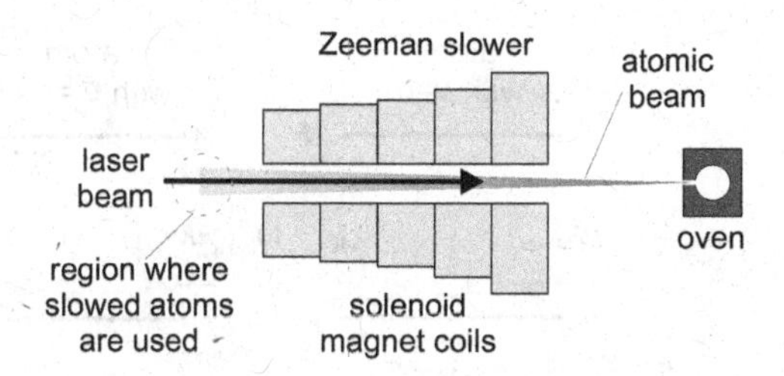

Fig. 14.4 Stopping an atomic beam using a Zeeman slower.

The first technique is called *chirp cooling* since the frequency of the laser is "chirped". Chirp cooling can be accomplished with the use of diode lasers where the frequency of the laser can be easily changed by varying the laser diode current. The second technique uses the Zeeman effect to shift the energy levels of the atom into resonance with the fixed frequency laser. Such a device is called a *Zeeman slower* and is illustrated schematically in Figure 14.4. The Zeeman shift of the $l \to u$ transition is equal to

$$\Delta E_{\text{Zeeman}} = (g_{J_u}\, m_{J_u} - g_{J_l}\, m_{J_l})\, \frac{\mu_B\, B}{\hbar}, \tag{14.2}$$

where g_{J_u} and g_{J_l} are the Landé g-factors of the upper and lower levels, μ_B is the Bohr magneton, and B is the magnetic field. A Zeman slower uses a spatially varying magnetic field. Decreasing the magnetic field decreases the Zeeman shift which offsets the varying Doppler shift as the atoms slow down. Note that the Zeeman shift is proportional to m_J. Since m_J can take on values between $+J$ to -J, the Zeeman slower is only resonant for one particular value of m_J. One either only stops the fraction of the atoms in the beam that initially have the proper m_J value, or optical pumping is used to put most or all the atoms in a single m_J sublevel so that the resonant frequency of all the atoms changes in the same way as they pass through the magnetic field.

14.3.1.3 *Optical Molasses*

Cold atoms have small velocity components in all directions. The slowing of an atomic beam does not eliminate the component of velocity transverse to the atomic beam axis since the recoil is not exactly zero and does not completely eliminate the component of velocity parallel to the atomic beam axis. Full laser cooling can be accomplished using the pairs of two counter propagating laser beams along each of three mutually orthogonal axes. The frequency of the laser beams must be set to be slightly lower than the atomic resonant frequency, or equivalently one can say that the laser is red detuned.

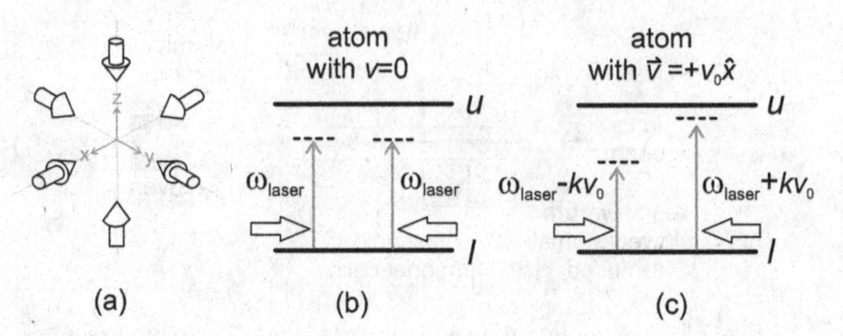

Fig. 14.5 (a) Schematic arrangement of laser beams for an Optical molasses. (b) Schematic energy level diagram indicating laser in red-detuned (lower energy). For a stationary atom, the detuning $\Delta = \omega_0 - \omega_{\text{laser}}$ is the same for the beam from the left and right, and it absorbs equally from both beams. (c) For an atom moving to the right, the laser beam from the right is Doppler shifted closer into resonance while the beam from the left is shifter further out of resonance, so the atom absorbs more photons from the right beam which eventually slows it down.

In this situation an atom moving in a given direction will have its resonant frequency Doppler shifted toward the resonant frequency as seen by the laser beam coming toward the atom in a direction opposite to direction of the atom's velocity and shifted away from resonance as seen by the laser beam propagating in the same direction as the atom's velocity. This results in the atom absorbing more photons from the laser beam incident opposite to the atom's velocity and fewer photons from the laser beam incident parallel to the atom's velocity since the absorption cross section is largest at the atom's resonant frequency. This method of cooling atoms is called *optical molasses* and illustrated schematically in Figure 14.5. The optical molasses force on an atom moving along one of the directions defined by a pair of counter propagating laser beams is given by

$$F_{\text{OM}} = F_{\text{scatt}}\left[\left(\omega_0 - 2\pi\frac{v}{\lambda}\right) - \omega\right] - F_{\text{scatt}}\left[\left(\omega_0 + 2\pi\frac{v}{\lambda}\right) - \omega\right] = -\alpha v \tag{14.3}$$

where F_{scatt} is the force on the atom due to the scattering of one of the two counter propagating laser beams, ω_0 is the atomic resonant frequency for an atom at rest, $2\pi v/\lambda$ is the Doppler shift of the atom's resonant angular frequency, ω is the laser's angular frequency, and $\alpha = 2dF_{\text{scatt}}/dv$. For an atom moving in an arbitrary direction the analysis is similar. In general the scattering force is such as to slow the atom and is directly proportional to the negative of the atom's velocity.

Even with the all the counter propagating beams of an optical molasses, the motion of atoms is not completely suppressed. The net result of continual absorption and emission cycles do not exactly average to zero velocity. When laser cooling was first considered it was expected that the lowest temperature, T, that could be reached would be of the order of

$$T = \frac{h}{4\pi\, k\, \tau} \tag{14.4}$$

where k is Boltzmann's constant and τ is the lifetime of the upper level. This is called the *Doppler cooling limit.* The Doppler cooling limit temperatures for rubidium and potassium atoms are both about 150 μK. It has been found, however, that it is possible to cool atoms to temperatures as low as a few μK if the counter propagating beams are correctly polarized. The explanation for this cooling depends upon multiple mechanisms involving optical pumping, AC Stark shifts of the atomic energy levels in the standing wave created by the counter propagating beams, and the polarization gradient in the standing wave. One form of this sub-Doppler cooling is called *Sisyphus cooling.* Essentially the standing wave creates regions of higher and lower potential energy. Each time an atom 'climbs' one of these 'hills' it looses a small amount of kinetic energy. Normally, it would regain this energy if it 'rolls' down the hill, however, atoms preferentially scatter photons when on the the top of these 'hills', which transfer them to a valley with the scattered photons carrying away a bit extra energy. The term *polarization gradient* cooling is used to describe another type of sub-Doppler cooling due to the optical pumping of atoms moving into regions of different polarization. Due to these sub-Doppler cooling mechanisms, atoms can be cooled to temperatures of a few μK, corresponding to atomic velocities of a few cm/s.

14.3.2 *Magneto-Optical Trap*

An *atom trap* requires both a dissipative force, $\boldsymbol{F} \propto -\boldsymbol{v}$, to dampen motion and a resorative force, $\boldsymbol{F} \propto -\boldsymbol{r}$ to confine atoms. The optical molasses described in Section 14.3.1.3 will cool atoms, but it does not confine them to a given region of space since their velocity is not reduced to zero. A given atom will diffuse slowly out of the region of the optical molasses. A *magneto-optical trap* (MOT) uses a minimum in the magnetic field to localize atoms to a small region within an optical molasses. The standard MOT consists of three pairs of mutually orthogonal counter propagating laser beams red-detuned from the resonance transition of the atom just as was used for an optical molasses. Two additional changes are required to convert this into an atom trap. First, the laser beams are circularly

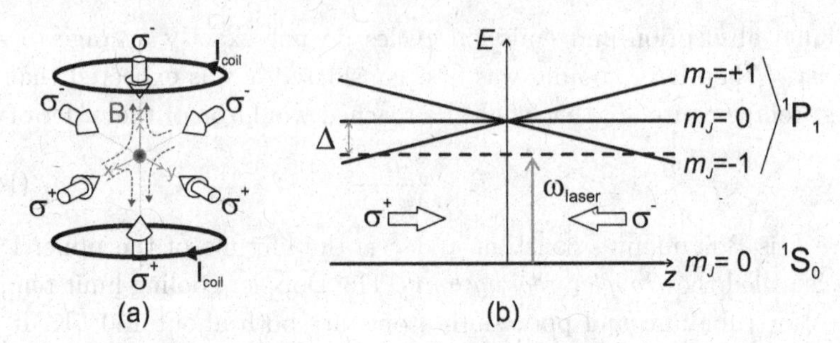

Fig. 14.6 (a) Schematic diagram of lasers and magnet coils in a magneto-optical trap. The dashed lines indicate the magnetic field directions along the x and z-axes. (b) Energy diagram for an idealized MOT with a J=0 ground state and $J = 1$ excited level.

polarized. Second, the trap utilizes a quadrapole magnetic field produced by a pair of coils similar to a pair of Helmholtz coils except that the current in the coils is run so that the magnetic field from the one of the coils is opposite in direction to the field from the other coil. In this situation the magnetic field on the axis of the coils and at the midpoint between the coils is zero, and the magnetic field increases in magnitude in all directions as one goes away from the midpoint between the two coils. As an atom moves from the center of the trap it moves away from the zero of the magnetic field. The Zeeman effect shifts the energy levels of the atom into resonance with one of the counter propagating beams, pushing the atom back towards the center of the trap.

In order to understand the operation of a MOT let us consider the simplified case of an atom with an $^1S_0$ ground level and a $^1P_1$ first excited level. The energy levels of this atom in a magnetic field are shown in Figure 14.6(b). As indicated in Figure 14.6(b) the frequency of all the laser beams is set to be slightly below the resonant absorption frequency from the $^1S_0$ ground level to the excited $^1P_1$ level. The laser beams are each circularly polarized as indicated in Figure 14.6(a). A circularly polarized photon has its spin angular momentum either parallel or anti-parallel to its direction of propagation depending on whether it is right of left circularly polarized. The dipole selection rules are such that σ^+ light drives transitions with $\Delta m_J = +1$ and σ^- light is required for $\Delta m_J = -1$. Let us consider a trap for which the magnetic field along the axis of the coils points away from the midpoint of the coils as shown in Figure 14.6(a). Consider an atom displaced from the trap center with $z > 0$. The Zeeman effect shifts the energy of the $m_J = -1$ sublevel closer into resonance with the red-detuned laser frequency, while the $m_J = +1$ sublevel is shifted further out of resonance with the laser. The optical absorption cross section for laser beam incident from the right in Figure 14.6(b) is larger than the absorption cross section for the beam from the left since the selection rules for the laser beam from the left results in a $\Delta m_J = -1$ transition in the atom whereas the beam from the right results in a $\Delta m_J = +1$ transition. Thus an atom experiences a resultant force back towards the center of the trap. Likewise an atom displaced in the $-z$ direction (or along any other axis) experiences a restoring force to the trap center. Due to the quadrapole field, the direction of B is parallel along the $z-axis$ and anti-parallel along the x and y-axes. Along the z-axis the laser beams along the axis of the coils must be polarized so that the photon spins are parallel to the direction of propagation of the laser beams. The circular polarization of the laser beams along the x and y-axes is thus reversed relative to that along the z-axis. If the currents in the coils are reversed then the magnetic fields are reversed and the circular polarizations of the laser beams must be reversed.

The total force on atoms in the MOT include both the Zeeman-shifted restoring force and the optical molasses damping force. Along the z-axis,

$$F_{\text{MOT}} = F_{\text{OM}} + F_{\text{Zeeman}} = -\alpha\, v_z - \beta\, z\,. \tag{14.5}$$

In principle the restoring force could lead to oscillatory motion, but in practice atom traps are highly over-damped. As a result, the trap consists of a small (~ 1 mm diameter), diffuse cloud of cold trapped atoms at the center of the trap. Depending upon the wavelength of the resonance transition, the trapped atoms appear as a bright diffusive cloud quite distinct from the fluorescence from untrapped atoms moving into the laser beams. The total number of atoms N in the trap is determined by the balance between the loading rate of atoms into the trap and the loss rate out of the trap. Many processes contribute to the loss rate. One loss mechanism is collisions with 'hot' untrapped atoms. A second is cold atom-atom collision processes between trapped atoms. The relative importance of these mechanisms depend upon the type of trapped atoms and the source of atoms being used to load the trap.

True magneto-optical traps are a bit more complicated than the simplified picture presented here. For example, rather than the simplified J=0 and J=1 levels considered in the simplified picture, one typically must include the effects of the spin of the nucleus and account for the hyperfine structure of the atom. For example, a typical ^{85}Rb trap operates on the $5^2S_{1/2}$ $(F = 3) \rightarrow 5^2P_{3/2}$ $(F' = 4)$ transition. Due to the requirement to continuously absorb and re-emit photons, atom traps

Table 14.1 Wavelengths of some of the atoms and ions that have been trapped.

Alkali	λ (nm)	Rare gas	λ (nm)	Other	λ (nm)	Ions	λ (nm)
Li	671	He*	389, 1083	Mg	285	Be^+	313
Na	589	Ne*	640	Ca	423	Mg^+	279
K	767	Ar*	811.5	Sr	461, 689	Ca^+	397
Rb	780	Kr*	811.3	Cr	425	Ba^+	493
Cs	852	Xe*	882	Ho	410.5	Hg^+	194
				Yb	399, 556	Yb^+	369

are only practical for closed two-level systems. Unfortunately, nature rarely complies rigorously with this requirement. For example, while the $5^2S_{1/2}$ $(F = 3) \to 5^2P_{3/2}$ $(F' = 4)$ transition is indeed a closed two-level system, occasionally, an atom will be excited to the $5^2P_{3/2}$ $(F' = 3)$ level due to absorption in the far wings of the line shape. This level can decay to the $5^2S_{1/2}$ $(F = 2)$ level which is approximately 3 GHz lower in energy than the $5^2S_{1/2}$ $(F = 3)$ level. Atoms 'accidentally' transferred to this level are no longer capable of being excited by the laser tuned within a few MHz to the cycling transition and will eventually fall out of the trap. Nevertheless, a second, repump laser can be used to transfer atoms out of this 'dark state' back onto the cycling transition. While discussed in terms of a Rb MOT, almost all atom traps require one or more repump lasers. The requirements for the repump laser are much looser than the cycling laser, a lower power laser is generally satisfactory and there is no need for counter propagating beams.

A number of different atoms can be trapped in a MOT, see for example Table 14.1. Magneto-optical traps with alkali atoms are particularly common. This is due to these atoms all having a strong $n^2S \to n^2P$ resonance transition with wavelengths in ranges that can pumped with a convenient choice of lasers. Rubidium MOTs are particularly common since the $5^2S_{1/2} \to 5^2P_{3/2}$ transition has a wavelength of 780 nm which can be produced using an external cavity diode laser. Another advantage of Rubidium is the simplicity of the Rb atom source. For many atoms this requires a high temperature oven to create an atomic beam. As mentioned in Section 14.3.1.3, it then requires a long distance to slow these hot atoms down to a velocity low enough to be captured by the MOT beams. This usually requires a Zeeman slower. Due to its high vapor pressure, however, a reasonably dense Rb atomic target can be formed by heating a sample of Rb metal to only a few degrees above room temperature. While the vast majority of these room temperature atoms are still too hot to be captured in the MOT, a small fraction of the atoms within the Maxwell-Boltzmann velocity distribution have a low enough velocity to be slowed within the $\sim$ 1 cm diameter laser beams used in a typical MOT. This is called a vapor-loaded trap. The source for ground state rare-gas atoms is also very simple. The wavelengths for the cycling transitions of the ground state resonance transitions, however, are in the vacuum ultraviolet region where tunable lasers are inconvenient. Nevertheless, rare-gas atom MOTs are quite common. The trick is to first to exite the atoms to a long-lived metastable level. For example, a neon MOT uses atoms in the Ne($2p^5 3s$) $J = 2$ 1s$_5$ metastable level (p. 236) as the lower level in the cycling transition.

Magneto-optical traps can be constructed in variety of ways besides the basic described verison. For example, a tetrahedral trap replaces the three pairs of crossed beams with four individual beams arranged as a tetrahedral. A 'mirror MOT' uses a mirror placed near the center of the trap to to reduced the number of required laser beams to three. Similarly, a pyramidal MOT can be created using a single input beam and four mirrors arranged as a type of retro-reflector. A 2D MOT eliminates one of the set of crossed beams to create a beam of slowed atoms which makes a useful source of cooled atoms to be loaded into other atoms traps.

Atom traps in general and MOTs in particular have a wide number of uses. Spectroscopic measurements with trapped atoms atoms is considerably simplified due to the absence of significant Doppler broadening. Another application is the study of the collisions between very slowly moving atoms. Atoms traps are also used to create Bose-Einstien condensates and Fermi condensates (p. 346) and are used in some types of quantum computers (Section 14.4). The performance characteristics required for these latter applications are beyond those obtainable directly with a magneto-optical trap and instead use some other designs to be discussed in Section 14.3.3. Nevertheless, these other traps still often use MOTs as a first stage in the cooling process.

14.3.3 *Other Types of Atom Traps*

In addition to the standard magneto-optical trap, there are a number of other types of atom traps with differing performance parameters. Here we present a brief overview of a few of these designs that are particularly relevant to lasers and quantum optics applications.

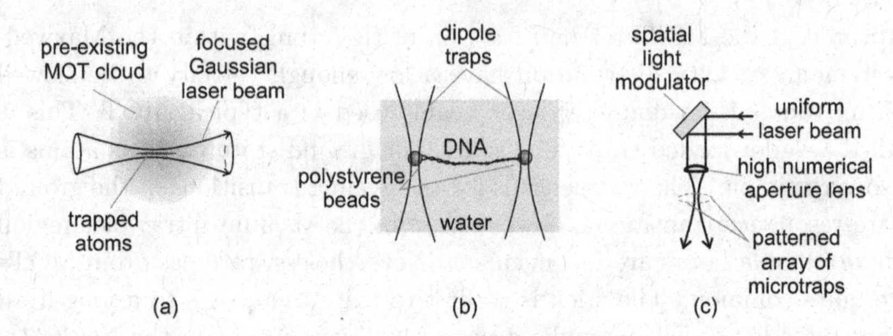

Fig. 14.7 Variants of dipole traps: (a) single beam dipole trap, (b) optical tweezers, (c) use of a spatial light modulator to create an array of microtraps.

14.3.3.1 *Dipole Traps*

A *dipole trap* uses the induced dipole moment of an atom to confine it at the focus of an intense laser beam. The wavelength of the laser is detuned far to the red of any resonance transitions of the atom. Indeed, another name used for a dipole trap is 'far off resonance trap' (FORT). For example, sodium atoms, which have resonance transition wavelengths of 589 nm, can be trapped in the focus of a Nd:YAG laser with a wavelength of 1064 nm. Even more impressive, rubidium atoms can be trapped using a CO_2 laser with a wavelength of 10.6 μm. The force is due to the gradient of the electric field times the electric dipole moment. Since the induced dipole moment of a neutral atom is proportional to the applied field, the force scales with the gradient of the electric field squared which is the same as the gradient of the intensity. Note that the dipole trap only provides a restoring force that pushes atoms towards the most intense part of the laser beam. A dipole trap can not cool atoms. Indeed, atoms in a dipole trap are very slowly heated by the momentum transfered from the occasional absorption of a photon from the far off resonance light. Typically, a MOT is used as a first stage of cooling and trapping. After the MOT has created a sufficient sample of cold atoms, the laser for the dipole trap is turned on and the MOT lasers are turned off. Alternatively, a dipole trap can be used for confinement in conjunction with an optical molasses for dampening.

Dipole traps can be used in many different configurations. Figure 14.7(a) illustrates a simple dipole trap formed at the focus of a single Gaussian laser beam. The intensity gradient is much higher in the radial direction than in the axial direction. As a result, the cloud of trapped atoms tend to be cigar shaped. Better confinement can be achieved by using the intersection of two lasers. Experimentally, it has been found that if one decreases the waist of the focused laser to only a few μm, the trap can only hold one atom at a time. If a second atom enters the trap, collision processes knock both atoms out of the trap. Such a single atom trap is called a *microtrap*.

Amazingly at the opposite extreme of holding individual atoms, dipole traps can also be used to hold microscopic particles. The term *optical tweezers* is used to describe this application of dipole traps. The particles can either be something like a small glass or plastic bead, or a biological sample such a cell. For example, optical tweezers have been used to measure the force required to pull a DNA strand apart by chemically bonding polystyrene beads to the two ends of a DNA molecule. Two dipole traps are used where the relative position of the beams (and thus the beads) can be varied by steering the lasers. Both the beads and the DNA molecules are all in an aqueous solution.

It is also possible to create arrays of trapped atoms by varying the intensity pattern from that of a simple TEM_{00} Gaussian beam. The first example of this was to use the standing wave interference pattern of two laser beams to create an *optical lattice* with a spacing of $\lambda/2$. Atoms can be trapped at each anti-node. A wide variety of schemes have been used to generate the intensity distributions. Figure 14.7(c) illustrates one such scheme for creating an arbitrary array of microtraps using a spatial light modulator. The light in the focal plane of a lens is the Fourier transform of the far field distribution of the light illuminating the lens. Thus, by controlling the intensity pattern of the laser before it is focused it is possible to create a large number of microtraps with almost any geometry in the focal plane. The spatial light modulator is essentially a mirror behind an array of liquid crystal shutters which can be programmed independently of one another. Nogrette *et al.* have created 2D arrays of up to 100 microtraps using such an arrangement. The maximum number of microtraps was due to the requirement that a minimum of 5 mW of 850 nm laser light was required for each microtrap. Arrays of single atoms are required for the construction of some proposed quantum computers using cold atoms. Note that for a red detuned laser the atoms are pulled into the maxima of the laser beam's intensity distribution. For a blue detuned laser, the dipole force pushes atoms into the minima. This is not useful for trapping with a single Gaussian beam, but when coupled with means to create 'dark spots' within a bright field, it is possible to trap atoms in the minima where the disturbance caused by the trapping lasers on the atoms will be minimized.

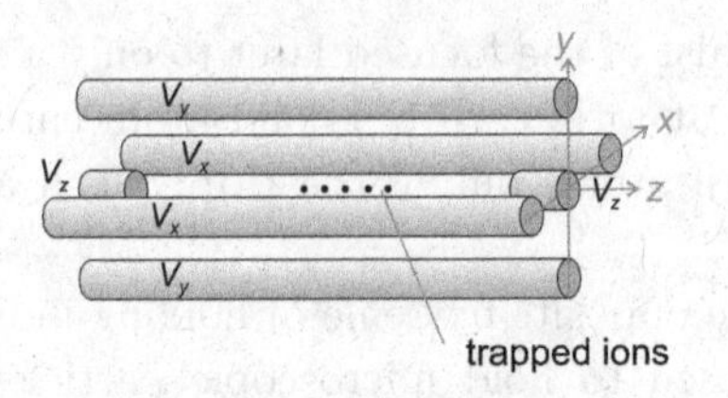

Fig. 14.8 Linear Paul trap for ions.

14.3.3.2 *Ion Traps*

Another possible implementation of a quantum computer uses a string of trapped ions. Ion traps use a combination of electric and or magnetic fields to confine ions. A *Penning trap* uses a static magnetic field and a set of electrodes to confine ions. A *Paul trap* uses only electric fields to confine ions. One can not create a local minimum in the electric field to trap ions using only static fields. Instead, in a Paul trap a time-varying electric field is used. The trap is surrounded with six electrodes with opposite pairs electrically connected. A large positive DC voltage is applied to the pair long the z-axis which prevent ions from escaping in this direction. The pairs along the x and y-axes are rapidly alternated (∼MHz) between positive and negative voltages. While the x-axis pair is positively charged, the positively charged trapped ions are repelled and are instead attracted towards the y-axis electrodes. Before the ions can escape along this route, however, the fields are reversed. The oscillatory motion of ions confined in the trap is called micromotion. A conventional Paul trap uses electrodes with a special hyperbolic shape. A linear Paul trap replaces the x and y-electrodes with rods as shown in Figure 14.8. This setup can be used to trap a series of ions along the z-axis.

Large voltages can be used with ion traps to create a very deep trap with excellent confinement. While the electric fields provide a large restoring force, they provide no cooling. Lasers, however, can be used to cool the ions to mK temperatures or lower. The last column of Table 14.1 (on page 340) lists the wavelengths of the transitions used to cool a number of different ions. While the general principle is similar to the laser cooling of atoms, the harmonic micromotion of ions results in a few subtle changes. Two slightly different schemes are used for the laser cooling of ions. The first, Doppler cooling, is essentially the same as that used to cool atoms using a red detuned laser. Unlike the six beams used in the optical molasses used to trap atoms only a single beam is required to cool ions. Due to the oscillatory motion, providing a drag force in only one direction is sufficient to eventually slow the motion in all directions. The second laser cooling techniques is called *sideband cooling.* A trapped ion has a series of vibrational energy levels similar to those of a molecule. The cooling laser is red detuned such that the frequency of the laser excites an ion from the ground state v'' vibrational level to the $v' = (v'' - 1)$ vibrational level of the excited state. In a molecule, the Frank-Condon principle can be used to estimate the transition rates into various vibrational levels of the ground state. For an ion trap, transitions with $\Delta v = 0$ are strongly favored. As a result, after one absorption and re-emission cycle, the vibrational energy of the ion is reduced i.e., the ion is colder.

It is interesting to note that the motions of a string of ions in a linear Paul trap are coupled together. As a result, it is only necessary to laser cool a single ion in the string to cool them all. The wavelengths of most of the ions listed in Table 14.1 are at a shorter wavelength than those used to trap atoms. Tunable lasers in the blue are less common, so typically a frequency doubled tunable red laser is used. The visible wavelength of the lasers allows one to directly observe individual ions from the resulting laser induced fluorescence.

14.3.3.3 *Magnetic Traps*

Cold atoms can also be confined in a magnetic trap. Since these type atom traps generally do not use a laser even for cooling, we will only briefly touch on them. The force on an atom is equal to the gradient of the atom's potential energy. As in a MOT, the magnetic field of a magnetic trap is designed so that the field increases in magnitude as an atom moves away from the trap center. For the proper m_J sub levels, the larger B-field corresponds to a larger potential energy and thus the atom is forced back towards the trap center. While the simple 2-coil quadrapole field of a MOT can be used for magnetic trapping, many magnetic traps use a more complicated arrangement of coils. Since the trapping force is proportional to the gradient of the magnetic field, only weak fields are needed when the dimensions of the trap are shrunk. As a result, it is possible to create millimeter sized miniature 'atom chip' magnetic traps using coils deposited onto a silicon wafer.

As in a dipole trap, the confining force in a magnetic trap is non dissipative with no cooling. Nonetheless, magnetic traps are used to create some of the coldest samples of trapped atoms via *evaporative cooling.* Typically one first loads a large number of atoms into a magnetic trap from a conventional MOT. The temperature of these atoms are typically on the order of mK or μK. The hottest of these atoms have enough energy to reach the outermost limits of the magnetic trap. At which point an rf-field is used to induce a $+m_J$ to $-m_J$ transition, so the 'hot' atoms no longer have the right sign of m_J to be trapped and exit the trap. Colder atoms do not have enough energy to venture away from the center of the trap and their $+m_J$ to $-m_J$ transition energies are not resonant with the applied rf-field. With the hottest atoms removed, the average energy of the remaining atoms is lower. At this point, one has a Maxwell-Boltzmann energy distribution with the high energy tail removed, the density of atoms at lower energies is unchanged. Collisions between atoms, however, eventually re-thermalize the distribution into a Maxwell-Boltzmann distribution with a lower temperature and increased density of really cold atoms. By slowly lowering the threshold for the removal of hot atoms, the temperature can be reduced and the density of the coldest atoms increased. Evaporative cooling reduces the number of atoms in the trap (often by a factor of ten or greater), but the temperature can be reduced to nK temperatures. While discussed in the context of magnetic traps, versions of evaporative cooling can also be used with a dipole trap.

The goal of evaporative cooling is generally to reduce the temperature of the trapped atoms and increase the density of atoms in the lowest possible energy levels. The wave functions of cold particles are more defuse, and quantum mechanical effects become manifest when the spacing between the individual atoms is less than than size of wave functions. The de Broglie wavelength of a particle with mass M and velocity v, is $\lambda_{\mathrm{dB}} = h/(Mv)$. Hence, as v is reduced, the wavelength increases and the particle is more delocalized. For a distribution of particles with a temperature T, the de Broglie wavelength is $\lambda_{\mathrm{dB}} = h/\sqrt{2\pi M kT}$. The distribution function of delocalized particles differs from the statistical Maxwell-Boltzmann distribution, depending upon the spin of the individual particles.

Particles can be classed into two categories based upon their spin. Particles with integral spin are called bosons while particles particles with half integer spin are called fermions. At low temperatures bosons and fermions behave very differently. Fermions have a Fermi-Dirac distribution which at low-temperatures fills up the lowest levels with two particles per level (i.e., one with spin up and the other with spin down). This results in the filling of the lowest $N/2$ levels at $T = 0$. In contrast, bosons can all go into the lowest energy level at $T = 0$. For trapped atoms with interger spins, a Bose Einstein Condensate (BEC) is expected to form when $n\lambda_{\rm dB}^3 > 2.6$. BECs of magnetically trapped ^{87}Rb, ^{23}Na, and ^{7}Li atoms were first created in 1995 with many additional atoms and isotopes since then. The first Fermi condensate of ^{40}K was created in 2003. A Bose condensate, has a large number of particles all in the lowest energy level and exhibits many interest phenomena. For example, if a Bose condensate is split in two and then brought back together a matter wave interference pattern is formed. This is possible because the de Broglie wavelength of all the particles is the same, similar to the narrow wavelength spread of a laser. It is possible to create a steady beam of atoms leaving a BEC (i.e. by inducing spin flips and allowing them to fall out of the trap) which are sometimes call 'atom lasers', although not really a laser.

14.4 Quantum Entanglement: *Control at the quantum level*

Entanglement is a key feature that distinguishes quantum mechanics from classical mechanics. Two particles are entangled if their wave function can not be written as a simple product of two single particle wave functions. In order to make this clear, let us consider two particles such as electrons, each with spin 1/2. These two particles can have a total spin angular momentum of S=1 or 0. The three $S = 1$ wave functions are given by $\alpha_1\alpha_2$ for $m = 1$, $\frac{1}{\sqrt{2}}(\alpha_1\beta_2 + \alpha_2\beta_1)$ for $m = 0$ and $\beta_1\beta_2$ for m= -1. The $S = 0$ wave function is given by $\frac{1}{\sqrt{2}}(\alpha_1\beta_2 - \alpha_2\beta_1)$ where the spin wave functions for the individual electrons with spin up along the z axis are given by α_1 and α_2 and the wave functions for spin down are given by β_1 and β_2. The wave functions for the total spin 1 and with $m = 1$ and $m = -1$ are simple product wave functions and are not entangled whereas the wave function for the spin 1 wave function and with m=0 and the spin 0 wave function can not be written as a simple product and these wave functions are entangled. Entanglement is a property of the system, but it is most interesting when trying to tease apart what is happening to individual components of the system. For example, while the electronic wave functions of all atoms other than hydrogen (which only has one electron) are entangled, this effect is usually called correlation rather than entanglement in this context since the individual electrons remain part of the same atom. Instead, the term entanglement is usually used to describe situations where the quantum system can be taken apart into discrete parts that classically one would think of as independent.

The use of two spin 1/2 particles to illustrate the difference between non-entangled wave functions and entangled wave functions is useful because of its simplicity. Photons are a bit different from electrons. Photons have an intrinsic spin angular momentum of 1. The projections of the photon intrinsic spin angular momentum along the direction of propagation of the photon must be either 1 or -1. Because the photon is a transverse wave, the projection of the photon angular momentum on the direction of propagation cannot be 0. Two photons can be entangled. In order to make this clear consider a relatively simple situation involving a positronium atom. Positronium is a hydrogen like atom consisting of an electron bound to a positron. Both the electron and the positron have spin 1/2. The positronium atom in the ground S level can thus have either a total spin angular momentum of 1 or 0. Of these, consider the case when $S = 0$. The positron and the electron can annihilate with the emission of photons. Based upon the conservations of momentum and energy, the annihilation results in two photons (each with energy of 511 keV) going in opposite directions in the frame where the positronium is at rest before annihilation. Since $S = 0$ before annihilation, we know the spins of the emitted photons must be in opposite directions, but we don't know if $s_1 = +1$ and $s_2 = -1$ (case a) or $s_1 = -1$ and $s_2 = +1$ (case b). Instead, the photons are entangled, The actual entangled wave functions of the two photons in this case are the sum and difference of (a) and (b).

The entanglement of the two particles (photons in this case) results in a remarkable and much discussed situation. The entangled wave function exists no matter how far apart the two particles are. If one makes a measurement of the polarization of one particle the wave function instantaneously collapses and the polarization of the other particle is instantaneously known to be opposite from the polarization of the photon whose polarization was measured. Einstein, Podolsky, and Rosen (EPR) published a paper in 1935 which discussed this situation with the conclusion that this was non-physical since they thought that the instantaneous collapse of the wave function implied an action that occurred faster than allowed by the speed of light and the theory of relativity since the entangled particles are separated in space. They felt that this indicated a flaw in quantum mechanics. Since the publication of this paper the subject of whether EPR was correct has been discussed many times. In 1964 Bell showed that quantum mechanics is inconsistent with EPR's ideas and that an assumption of locality is required for EPR to be correct. Bell further showed that there are inequalities in correlations implied by locality and that these correlations can be tested. If quantum mechanics is correct then the inequalities had to be violated. Bell's initial paper discussed two entangled spin 1/2 particles in a singlet state, however the ideas can be tested using entangled photons. Experimental tests of these inequalities have shown that EPR's ideas are incorrect and that quantum mechanics is indeed correct.

Entangled wave functions can be produced in several ways. One method to produce a pair of entangled photons is by the spontaneous parametric down conversion of a short wavelength photon into two photons of longer wavelength using a nonlinear crystal (Chapter 13). For spontaneous parametric down conversion to occur it is necessary for the sum of the energies of the two long wavelength photons to be equal to the energy of the short wavelength photon and it is necessary for the sum of the momenta of the two long wavelength photons to be equal to the momentum of the short wavelength photon. Spontaneous parametric down conversion has a very low probability for occurring (about $10^{-11} - 10^{-13}$). Without high power lasers for the incident beam it would not be possible to detect parametric spontaneous down conversion. The spontaneous emission of photons is sometimes considered as due to stimulated emission produced by vacuum fluctuations and the spontaneous down conversion is commonly treated as due to stimulated emission from vacuum fluctuations. Not all down conversion results in entangled photons, it is necessary to use a particular set up for the nonlinear crystal. In addition to photons, the wave functions of trapped atoms (and ions) can be entangled. Lasers are used to initially cool and trap the atoms (Section 14.3), entangle them, and to probe the state of the entanglement.

Entangled particles are the key component in *quantum computers*. Conventional computers store and process information as binary bits, 0 or 1. These are typically two different voltage levels in an electronic circuit. A conventional computer can solve essentially any computational task by breaking down the problem into a long series of basic logic operations which can in principle be processed using only single and dual bit Boolean algebraic steps. A quantum computer, in contrast, stores information as *qubits*, which use entangled wavefunctions that can take on values of '0' and '1' and complex superpositions of the two. For example, if a qubit were encoded using the spin of an atom with $s = 1/2$, a '1' might correspond to spin up (α_1) and '0' to spin down (β_1). A qubit would be written as $a\,\alpha_1 + b\,\beta_1$ where a and b are complex coefficients that satisfy $|a|^2 + |b|^2 = 1$. The quantum mechanical evolution of many entangled qubits can act as a quantum computer to solve complex problems. In principle, this can lead to a vast improvement in the time it takes to solve certain classes of problems, such as factoring large numbers, that are time-consuming with a conventional computer. The problem to be solved is determined by the initial configuration of the entangled qubits and their subsequent evolution as pairs of qubits are further entangled.

Various schemes have been suggested for building quantum computers, many of which use lasers. For example, one scheme uses a series of laser-cooled trapped ions in a linear Paul trap (Section 14.3.3.2) as the qubits. Another design uses an array of dipole microtraps (Section 14.3.3.1) each holding a single atom. Quantum dots and superconducting circuits are used as the qubits in some of the other promising schemes for quantum computers which do not use lasers. An optical quantum computer using entangled photons would be highly desirable, but it is difficult to create gates that process individual pairs of entangled photons owing to the weak coupling that exists between photons. Nonetheless, entangled photons may be useful for the long-range transportation of quantum entanglement. This is particularly true in the field of *quantum cryptology* which attempts to use the signatures of quantum entanglement to detect the presence of a third party eavesdropper in the communications between two parties. Both quantum computers and quantum cryptology are active areas of research in the field of quantum optics.

14.5 Squeezed Light: Overcoming the Heisenberg Limit

The Heisenberg uncertainty principle can be written in a number of ways. One way is

$$\Delta E \, \Delta t \geq \frac{\hbar}{2} \tag{14.6}$$

where ΔE is the uncertainty in the energy of a state and Δt is the uncertainty in the time needed to measure the energy. This is not a statement that experimental techniques or equipment for the measurement are inadequate or that it is fundamentally impossible to measure either the uncertainty in the energy or the time with great accuracy, but instead is a statement that in a quantum mechanical system it is impossible to measure both ΔE and Δt at the same time with an uncertainty in the product that is less than $\hbar/2$.

Consider the uncertainty principle as it applies to measuring the intensity of laser light inside an optical cavity. The energy of the light stored in the cavity is equal to the number of photons in the cavity (n) times the energy per photon ($h\nu = \hbar\omega$), so $\Delta E = \Delta n \, \hbar\omega$. The uncertainty in the time

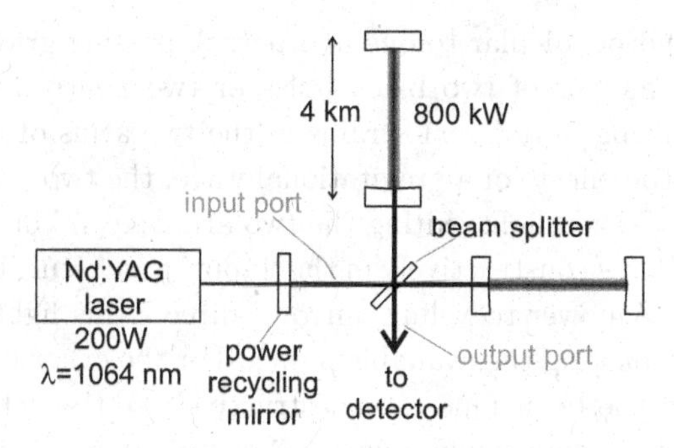

Fig. 14.9 Schematic design of Advanced LIGO. Each 4 km arm of the interferometer is a confocal Fabry-Perot cavity.

of the measurement can be related to the uncertainty in the phase of the photon's wavefunction, $\Delta\phi = \omega\,\Delta t$. Combining these two relations the uncertainty principle can be expressed as, $\Delta n\,\Delta\phi \geq 1$. It is possible, but not trivially easy, to produce an optical state with the minimum uncertainty so that $\Delta n\,\Delta\phi = 1$. Producing an optical state with this minimum uncertainty does not determine Δn or $\Delta\phi$ but determines only the product of the two uncertainties. The idea of squeezed states is that one can by novel means alter the uncertainty in either Δn or $\Delta\phi$ while leaving $\Delta n\,\Delta\phi = 1$. Thus it is possible to reduce Δn or $\Delta\phi$ to a small value while making the other quantity larger.

Normally the statistical fluctuation in the number of photons incident on a detector is determined by Poisson statistics. By squeezing the state it is possible to reduce the statistical uncertainty in the number of photons incident on a detector below the Poisson statistical uncertainty. Of course this means that the uncertainty in the phase is increased. It is also possible to reduced the uncertainty in the phase at the cost of increasing the uncertainty in the photon number.

Squeezed light can be generated by several methods. Squeezed states can be generated using optical parametric oscillators and in optical fibers. Also semiconductor lasers can generate amplitude-squeezed light when operated with a carefully-stabilized, low current. Squeezed light can be produced in atom-light interactions.

Squeezed states are potentially useful in many types of precision measurements. One long-sought use of squeezed states is in the detection of gravitational waves. The Large Interferometric Gravitational Observatory (LIGO) is a giant interferometer used to detect gravitational waves. As shown in Figure 14.9, the LIGO interferometer consists of two 4 km long arms oriented perpendicular to one another. A passing gravitational wave formed by the merger of two black holes or two neutron stars stretches spacetime, resulting in different strains in the two arms of the interferometer. Without the effects of a gravitational wave, the two arms have a zero path difference. When light exiting the two arms recombines at the beamsplitter, it interferes constructively in the 'input' port returning towards the Nd:YAG laser. A power recycling mirror redirects this light back into the interferometer boosting the available power. For the other beam, exiting in the output port the beams interfere destructively so the detector normally sees no signal other than random noise. A passing gravitational wave shifts the output from zero path difference, and a time varying signal is detected at the output port. LIGO is sensitive enough that it operates at or near the quantum limit for some frequencies.

The use of squeezed states enables the LIGO to use lower power lasers for the interferometer. Without squeezing, the photon counting signal-to-noise ratio of the 'dark fringe' signal scales as $1/\sqrt{P}$ where P is the power in the interferometer arms. Hence, higher power levels are desirable for reducing the Possion photon counting uncertainty. Unfortunately, the constant stream of photons also pushes on and heats the test mirrors which can shift the interferometer out of zero path difference. The random fluctuations caused by this photon pressure/heating scales as $\sqrt{P}$. Hence, the increase in this noise source eventually offsets the gain in the photon counting SNR obtained by increasing the laser power. Another way of describing the photon counting noise is to imagine that it is the 'signal' from vacuum fluctuations in the electromagnetic field that are injected into the interferometer from the output port. By replacing these random vacuum fluctuations with squeezed light, it is possible to reduce the noise level for a given laser power. Figure 14.10 shows a simplified version of the squeezed light injection system. A small amount of the primary 1064 nm Nd:YAG laser light is passed through a second harmonic non-linear crystal to generate 532 nm light phase-locked to the primary laser. This 532 nm light is then directed to a subthreshold optical parameter oscillator (OPO) that uses parametric down conversion to convert the 532 nm light back into 1064 nm photons. While this light is at the same wavelength as the starting beam, this process has created squeezed light.

Using squeezed light, it may be possible to improve the signal to noise ratio in LIGO by about a factor of two. This modest increase in the SNR nevertheless significantly expands the volume of space from which gravitational wave sources can be detected. In 2015 LIGO, without using squeezed

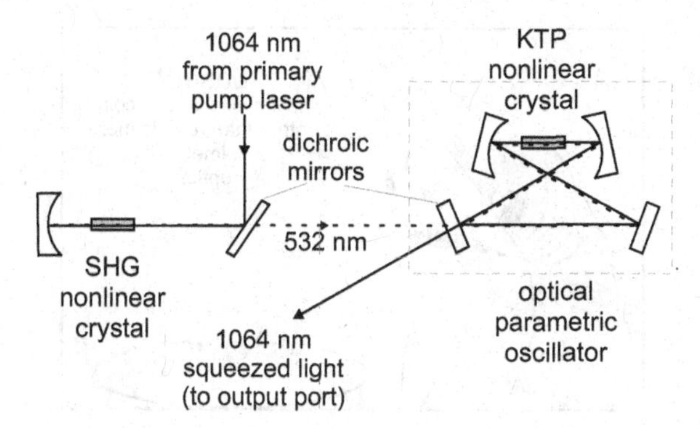

Fig. 14.10 Generation of squeezed light for improving the sensitivity of LIGO. Squeezed light at 1064 nm is generated by parametric down conversion of 532 nm light in an optical parameter oscillator. The OPO consists of a periodically-poled KTP crystal in a bow-tie cavity. The 532 nm light is obtained by second harmonic generation of light phase locked to the primary 1064 nm beam sent to the interferometer.

light, first detected gravitational waves from the merger of two black holes. It is expected that gravitational waves from neutron star mergers will also be observed.

14.6 What Makes a Laser Unique?

This may be a good place to address an interesting question. What is unique about a laser? Or in other words, how is a laser different from other light sources? Suppose that a He-Ne laser is inside a box and only the light beam emerges from the box. Is it possible for an external observer to determine whether there is a laser in the box or just a conventional light source? One can, after all, produce a very low divergence beam of light using a spatial filter and a large parabolic mirror using only conventional light sources. Also, one can produce a beam of light with a very narrow bandwidth and a large longitudinal coherence length using a large separation Fabry-Perot interferometer and conventional light source. Thus it is not enough to find that the light beam emerging from the box has a low divergence or a very narrow bandwidth to ascertain that the light source in the box is a laser.

14.6.1 *Equivalent Temperature*

There is, however, a way in which the laser differs from a conventional light source, its equivalent blackbody temperature is extraordinarily large. To see how this works, let us assume that the box contains a typical low-power

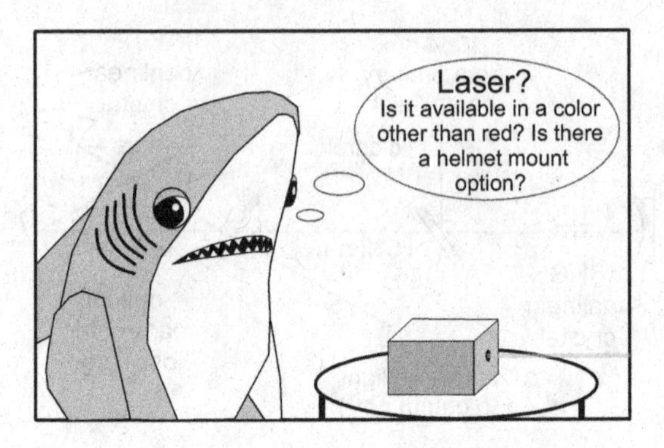

Fig. 14.11 A shark pondering one of the ultimate questions in life, 'Is it a laser?'

He-Ne laser, with 0.5 mW output at 632.8 nm and a 1 mm diameter beam. It is also assumed that the laser is running in a single longitudinal mode with a linearly polarized output beam. Consider the energy flux (mW) emitted from a blackbody with (i) the same area a as the laser beam, (ii) the same solid angle Ω as the laser, (iii) the same bandwidth $\Delta\nu$, and (iv) a sinlge polarization:

$$P_{\rm BB} = \frac{1}{2}\,\rho(\nu)\,\Delta\nu \left(\frac{\Omega}{4\pi}\right) c\,a = \frac{a\,h\,c\,\Omega\,\Delta\nu}{\lambda^3}\,\frac{1}{\exp\left(\frac{hc}{\lambda kT}\right)-1}\,. \tag{14.7}$$

Equating the blackbody power with that of the laser beam and solving for T,

$$T = \frac{h\,c}{\lambda\,k}\,\frac{1}{\ln\left(1+\frac{h\,c\,a\,\Omega\,\Delta\nu}{\lambda^3\,P_L}\right)} \approx \frac{\lambda^2\,P_L}{a\,\Omega\,\Delta\nu\,k}\,. \tag{14.8}$$

To estimate a typical laser bandwidth, it is assumed that the two mirrors of the laser have reflectivities of R_1=0.99 and R_2=1.00 and the length of the laser is 10 cm. In which case the photon lifetime in the laser is about $\tau_{\rm ph} = 6.6 \times 10^{-8}$ s, and the laser bandwidth is given by $\Delta\nu = 1/2\pi\tau_{\rm ph} = 2.4 \times 10^7$ Hz. Assuming the laser has a TEM_{00} output beam with the waist at the output coupler, w_0 = 0.5mm. So $a = \pi\,w_0^2 = 7.8 \times 10^{-7}$ m^2, and the solid angle of the beam is $\Omega \approx \lambda^2/\pi\,w_0^2 = 5.1 \times 10^{-7}$ sr. Inserting these numbers into Eq. (14.8) yields an equivalent blackbody temperature for the laser of $T = 5 \times 10^{12}$ K. This is significantly hotter than the center of the sun ($T = 1.5 \times 10^7$ K)! No known blackbody has temperatures this high and certainly no object on earth could have a temperature this high and fit inside a reasonably sized box. Compared to a blackbody, a laser concentrates its output into a very narrow bandwidth and small solid angle.

Many conventional light sources can not emit radiation greater than a blackbody since as the number density of the radiating body increases the emission goes up to the blackbody limit and then the linewidth begins to increase so that the radiation per unit frequency plateaus while the total radiation increases due to the linewidth of the radiation increasing. Thus a very high equivalent temperature over a very narrow bandwidth is a good indication that the box contains a laser. For the laser the equivalent blackbody temperature is very high for the narrow bandwidth of the laser and very much lower for wavelengths just outside the laser bandwidth. This method is used to identify maser or laser operation from astronomical sources.

There are, however, some other non-equilibrium light sources that have very high equivalent temperatures. Some examples of non-equilibrium sources with very high equivalent temperatures are fluorescent light sources, synchrotrons, and pulsars none of which is a laser. A synchrotron can produce highly collimated beams of light with narrow bandwidths from electrons circulating in a magnetic field (Chapter 12). The equivalent blackbody temperature for synchrotron radiation will be very high. It is unlikely, however, that a synchrotron could be made small enough to fit in a table top box. The same applies to a pulsar. One can, however, easily imagine a neon (or helium-neon) discharge light being in the box. Imagine, for a moment, a helium-neon discharge tube of the same dimensions as the He-Ne laser, but with simple windows in place of the cavity mirrors. To first order, the total power radiated on the 632.8 nm line from the discharge will be about the same as that from a similarly sized laser discharge tube, ~ 0.5 mW, but will be directed into 4π steradian instead of being concentrated into the narrow beam of the laser. The bandwidth is also greater, being that of a Doppler broadened line or about 10^9 Hz. With these differences, the equivalent temperature of a non-laser He-Ne discharge is *only* 8×10^3 K. While not quite as impressive as the equivalent temperature of a laser, this is still hotter than the surface of the sun (5800 K). Indeed, the 632.8 nm line is actually a very weak component of the neon emission spectrum, so an even higher temperature would be obtained using the stronger $2p_x \rightarrow 1s_y$ emission lines that produce the familiar orange-red glow of neon lights. Nevertheless, these temperatures are still orders of magnitude less than the equivalent temperature of a He-Ne laser. Thus a very high equivalent temperature over a narrow bandwidth with a low equivalent temperature outside this bandwidth is a good indication that the radiating light source is a laser and not a conventional or non-laser light source.

14.6.2 *Diode Lasers and LEDs*

Let us now consider the differences between a laser diode and an ordinary light emitting diode (LED). Diode lasers may seem much less 'laser like' than He-Ne lasers, since the output beam (without additional optics) is poorly collimated and the wavelength purity is much less. In both an LED and diode laser radiative recombination of injected electrons and holes produce photons. In an LED and a diode laser operating below the threshold current, spontaneous emission of these photons can occur in any direction. Further, not every electron-hole recombination event produces a photon. In a GaAs LED, only about half the pairs produce a photon. Hence, when spontaneous emission dominates (i.e., an LED or laser diode with $I < I_{\mathrm{th}}$), only a small fraction of electron-holes pairs produce a usable photon. When the current in the laser diode is raised above I_{th}, stimulated emission dominates spontaneous emission and the efficiency of the electron-hole to photon conversion process is raised substantially since both the likelihood of radiative recombination is increased and all of the emitted photons are added to output beam cavity mode. In Figure 8.15 on page 194 this increase is evident as the sharp rise in the slope of the power output above the threshold current. When a diode is lasing it has a very high effective temperature for the lasing modes. An LED does not have as high an effective blackbody temperature.

There is one more facet to this comparison of LEDs and diode lasers to discuss, a superluminescent LED is essentially a diode laser without the cavity mirrors. That is, the single pass gain along one axis is high enough that the spontaneous emissions are amplified (i.e., stimulated emission) before the light exits the device. This self amplified spontaneous emission (SASE) raises the output efficiency of the superluminescent LED in much the same way the power efficiency was raised in diode lasers. Nevertheless, without the cavity mirrors, the amplification process is incoherent and the output beam is not as well defined as it is in diode lasers.

In addition to their power efficiencies, LEDs and laser diodes also differ in their spectral output. The bandwidth of a ordinary red ($\lambda \approx 650$ nm) LED is about $\Delta\lambda_{\mathrm{LED}} \approx 25$ nm. This is also about the bandwidth of the gain medium in a red diode laser, but the actual emission wavelengths are restricted to the longitudinal modes closest to the peak of the gain curve, so $\Delta\lambda_{\mathrm{DL}} \approx 0.5$ nm. If the diode laser operates in a single longitudinal mode, $\delta\nu = 1/2\pi\,\tau_{\mathrm{ph}}$, where $\tau_{\mathrm{ph}} \approx (2\,n\,L)/(c\,T)$ (Eq. 3.6) which for typical values of L, and leads to $\Delta\lambda_{\mathrm{sm-DL}} \approx 0.05$ nm. The bandwidth of an external cavity tunable diode laser (Sec. 8.2.4) can be made even smaller due to the much longer cavity length.

14.6.3 *Optical Parametric Oscillator*

In Section 14.6.1 the criteria for deciding if an unknown beam is from a laser was to examine its equivalent temperature. A laser beam has a very high energy density in one or a few modes whereas other light sources including non-thermal sources such as a synchrotron beam usually have lower energy density per mode for similar total optical energy output, i.e. the optical energy is usually distributed over more modes than for a laser. The question might now be asked is an optical parametric oscillator (OPO) a laser? The output from an OPO is typically concentrated in one or a few modes and the OPO has a threshold for operation. In this sense the OPO acts like a laser. If an OPO were contained in a box it would be impossible to determine from the optical beam emerging from a hole if the box contained a laser or an OPO. Nevertheless, even if the output is coming from an OPO, the box will still most probably contain a laser being used to pump the OPO.

In this regard, how is an OPO different than a tunable Ti:Sapphire laser? Both devices are pumped by an external laser, have a threshold, and emit a beam at a user selected wavelength. The significant difference is that the OPO frequency depends on the frequency of the laser pump beam which is split into two beams of different frequencies in a crystal whereas for most lasers the output frequency is determined by the energy levels of the active laser medium. Once an atom is excited to a particular energy level the history of how it got there does not influence the emission process. Hence, the output wavelength of an optically pumped laser is generally decoupled from the wavelength of the pump laser. Note, however, this distinction does not work well with a free electron laser which lacks fixed energy levels and the output frequency is determined by the energy of the electron beam and the spacing of the wiggler magnets. Nonetheless, the direct coupling between the input and output beams in an OPO are what sets it apart. Indeed, this coupling is used to generate squeezed light (Sec. 14.5).

14.6.4 *Photon Correlation*

Some concepts from quantum optics present another way that one can tell if a light emerging from a box is from a laser or from some other source such as a discharge lamp. This approach is based upon measuring the distribution in the arrival times of individual photons in the beam. While a full mathematical treatment of this subject is beyond the level of this textbook (see for example the further reading), a qualitative understanding can be obtained by building upon the previously discussed topic of coherence.

There are orders of coherence. For first order coherence the electric field at a given time is highly correlated with the electric field at a different time. This means that the frequency phase and the wave vector at the given time, t, are in the same relationship to the same quantities at a different time, $t + \tau$, independent of the time, t. The electric field, however, can have different amplitudes at different times depending on how many photons are in the light beam. For example, consider the emissions from a Na discharge lamp. If the light is passed through a very narrow band pass filter it can be made to have a very narrow frequency spread. This leads to a high first order coherence. Nonetheless, the beam is expected to have times with a lower density of photons and times with a higher density of photons. This clumping of the photons is the result of Bose Einstein statistics for the photons when the emissions are not due to stimulated emission. The Bose Einstein statistics allow more than one photon to be in a mode at the same time due to the random statistical character of the photon emission. An interesting aspect of this is that Hanbury-Brown and Twiss used first order correlations of the photons from a star to measure the diameter of the star.

For second order coherence the intensity at a given time, t, is highly correlated with the intensity at the time, $t+\tau$ independent of the time t. For a cw laser above saturation when an atom is excited it is quickly stimulated to emit a photon. On the average as a photon is emitted into the oscillating mode a photon is lost from the mode and the intensity fluctuations are smoothed out. Due to the clumping of photons in a thermal optical source if one detects a photon then the probability that one will detect a second photon at a very short time after the first photon is detected is larger than the probability to detect a second photon at a longer time. Because of the smoothing of the intensity in a laser the probability of detecting a second photon after detecting a first photon is independent of the time. Thus by measuring the probability of detecting a second photon as a function of the time after detecting a first photon one can distinguish between a thermal optical source and a laser beam.

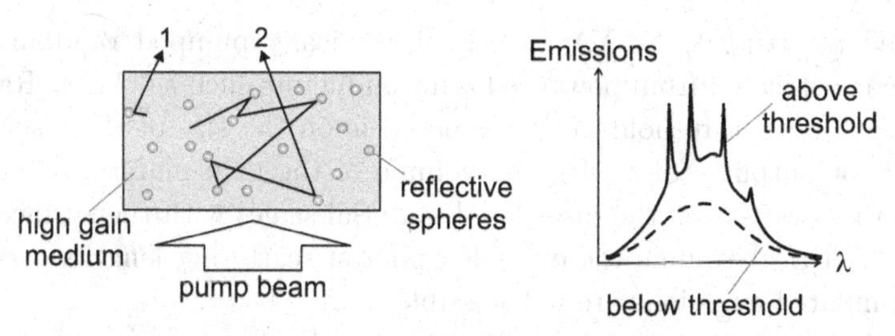

Fig. 14.12 Schematic of an optically pumped random laser using R6G dye as the gain material and TiO_2 reflective spheres. The paths of two photons of two sample photons are shown. For photon-2, the path length is long enough that stimulated emission is significant, an emission spectrum would show a peak at λ_2. Long path lengths are possible for larger sized lasers or higher concentrations of reflective spheres.

14.6.5 *Random Lasers*

Another system that explores of the boundaries of what 'is' and 'is not' a laser is the field of random lasers. A conventional laser consists of a gain medium inside an optical cavity. Astrophysical masers and and superluminescent LEDs have a high gain medium which permits amplified spontaneous emission without any cavity. A random laser falls somewhere between these two. It has a high gain material, along with a disordered array of light scattering elements. Light can reflect off these elements, extending the photon lifetime in the high gain material, much like a 'proper' optical cavity in a conventional laser. But the process is entirely random, with no predefined emission axis set by an external cavity as illustrated in Figure 14.12.

To illustrate how this system works, let use consider the particular random laser made up of a solution of high gain Rhodamine 6G (R6G) dye and a number of small, highly reflective TiO_2 spheres. The solution is excited by a pulsed, frequency-doubled Nd:YAG laser. When the concentration of reflective spheres is very low, only amplified spontaneous emission occurs. The output spectrum is rather broad, and the intensities are similar for all directions. As the concentration of TiO_2 spheres is raised, however, the path length (or alternatively the photon lifetime) for certain random paths which bounce off a few of the spheres will be longer. The random path acts as a laser cavity- with enhanced output and narrower spectral widths. Random lasers can also be made using a powder of Ti:sapphire, zinc oxide (ZnO) or gallium nitride (GaN) crystals. The powder serves as both the gain mechanism and disordered scattering element. Both the laser dye and the powered crystal approaches can be optically pumped with a frequency doubled (or tripled) Nd:YAG laser. Electrically pumped random lasers can be constructed from powdered semiconductors such as GaAs. Random lasers exhibit a threshold behavior based upon the size of the laser. The total light output scales with the volume of the gain material, while the photon loss rate from the disordered material scales with the surface area. Below a critical volume the extended photon scattering lengths necessary for stimulated emissions are not possible.

Besides testing the limits of what a laser is, random lasers may prove useful over conventional lasers in some applications. In principle they should be inexpensive to produce since no precision optical cavity components needs to be fabricated or aligned and the raw materials are relatively inexpensive. The randomness of the output beams is of course a major limitation for many applications, but will be useful for some applications such as the study of chaotic systems.

Summary of Key Ideas

- Laser Induced Fluorescence (LIF) is a very sensitive method for detecting small number of atoms or molecules (Sec. 14.1).
- Lasers can be used to control the velocity of atoms (Sec. 14.3.1). A Zeeman slower is used to slow an atomic beam (1D cooling). Optical molasses can be used to cool atoms to very low temperature (3D cooling). Sub-Doppler cooling allows cooling (i.e. velocities of a few cm/s) well below the Doppler cooling limit,

$$T = \frac{h}{4\pi\, k\, \tau}. \quad (14.4)$$

- Atom traps are used to confine cold atoms. A *Magnetic Optical Trap* (MOT) uses red-detuned lasers for both cooling and to provide a restoring force for confinement (Sec. 14.3.2). A *dipole trap* confines atoms (or microscopic objects) in the focus of an intense red-detuned laser beam (Sec. 14.3.3.1).
- Squeezing of quantum states can be used to produce states with a well defined number of photons or a well defined phase of the photons, but not both simultaneously. Squeezed light (Sec. 14.5) can be created via parametric down-conversion in an OPO.
- Determining whether a light beam is from a laser or another source of light is possible through by calculating the equivalent blackbody temperature (Sec. 14.6.1). In the active modes the effective temperature is

very high and it is low outside the bandwidth of the laser. Many non-thermal source have a high equivalent temperature, but not as high as a laser for the same power output.

Suggested Additional Reading

C. J. Foot, *Atomic Physics*, Oxford University Press (2005).

C. J. Foot, "Laser cooling and trapping of atoms" *Contemporary Physics* **32**, 369 (1991).

H. J. Metcalf and P. van der Straten, *Laser cooling of atoms and Trapping*, Springer (1999).

S. Chu, L. Hollberg, J. E. Bjorkholm, A. Cable, and A. Ashkin, "Three-dimensional viscous confinement and cooling of atoms by resonance radiation pressure" *Phys. Rev. Lett.* **55**, 48 (1985) [Optical molasses].

M. H. Anderson, J. R. Ensher, M. R. Matthews, C. E. Wieman, and E. A. Cornell, "Observation of Bose Einstein Condensation in a Dilute Atomic Vapor" *Science* **269**, 198 (1995).

A. Einstein, B. Podolsky and. N. Rosen, "Can Quantum-Mechanical Description of Physical Reality Be Considered Complete?" *Phys. Rev.* **47**, 777 (1935).

John Bell, "On the Einstein Podolsky Rosen Paradox" *Physics* **1**, 195 (1964).

S. J. Freedman and J. F. Clauser, "Experimental Test of Local Hidden-Variable Theories" *Phys. Rev. Lett.* **28**, 938 (1972).

A. Aspect, P. Grangier and G. Roger "Experimental Realization of Einstein-Podolsky-Rosen-Bohm Gedankenexperiment: A New Violation of Bell's Inequalities" *Phys. Rev. Lett.* **49**, 91 (1982).

M. A. Nielsen and I. L. Chuang, *Quantum Computation and Quantum Information*, Cambridge University Press (2000).

D. Walls, "Squeezed states of light" *Nature* **306**, 141 (1983).

R. E. Slusher, L. W. Hollberg, B. Yurke, J. C. Mertz, and J. F. Valley, "Observation of squeezed states generated by four wave mixing in an optical cavity" *Phys. Rev. Lett.* **55**, 2409 (1985).

L.-A. Wu, M. Xiao, and H. J. Kimble, "Squeezed states of light from an optical parametric oscillator" *J. Opt. Soc. Am. B* **4**, 1465 (1987).

The LIGO Scientific Collaboration, "Enhanced sensitivity of the LIGO gravitational wave detector by using squeezed states of light" *Nature Photonics* **7**, 613 (2013).

B. P. Abbott *et al.* (LIGO Scientific Collaboration and Virgo Collaboration), "Observation of gravitational waves from a binary black hole merger" *Phys. Rev. Lett.* **116**, 061102 (2016).

Meschede Dieter, *Optics, Light and Lasers, 2nd Ed.*, Wiley-VCH Verlag GmbH (2007).

R. Hanbury-Brown and R. Q. Twiss, "Correlation between photons in two coherent beams of light" *Nature* **177**, 27 (1956).

Problems

1. Describe why LIF is a more sensitive method for the detection of a small number of atoms than absorption spectroscopy.

2. Estimate the number of LIF photons scattered into a solid angle of 0.1 sr when a 10^{-3} J pulse of laser light at 589.6 nm corresponding to the Na $3^2S_{1/2} \to 3^2P_{1/2}$ transition ($A_{ij} = 6.1 \times 10^7$ s^{-1}) is incident on a Na cell with a density of 10^8 atoms/cm^3 and a cell volume of 1 cm^3. Assume the Na is Doppler broadened and has a linewidth of of 10^9 Hz. Assume that the cross section of the cell and the laser beam are both circular and have an area of 1 cm^2. You may ignore the reabsorption of the LIF by the Na atoms.

3. In two photon LIF, the energy of individual photons in the laser beam are only equal to half the energy needed to excite the $l \to u$ transition. Two photons must be absorbed by the sample. Two photon absorption is a second order processes like the non-linear effects discussed in Chapter 13. How can two photon LIF be used to construct a microscope to study the microscopic details of a sample? Can you think of any advantages of a microscope built using two photon LIF?

4. For an ideal Na atom with no nuclear spin, on the average how many photons are required to polarize the atom by optical pumping on the $3^2S_{1/2} \to 3^2P_{1/2}$ ($\lambda = 589.6$ nm) transition?

5. A Na atom at rest absorbs a photon of wavelength 589.6 nm. What is the recoil velocity of the Na atom?

6. Estimate the number of photons required to stop a Na atom that is moving with a velocity of 2×10^5 cm/s.

7. Strontium can be trapped in a MOT using two different choices of lines: (i) the strong $5s\,^1S_0 \to 5p\,^1P_1$ transition at 461 nm with $A_{ij} = 2.0 \times 10^8$ s^{-1}, and (ii) the weak $5s\,^1S_0 \to 5p\,^3P_1$ intercombination line at 689 nm $A_{ij} = 4.7 \times 10^4$ s^{-1}. (a) Calculate the Doppler cooling limit for a MOT operating at each wavelength. (b) To be captured into the MOT, an atom with an initial velocity v_0 must scatter enough photons to be stopped before it leaves the trap (i.e. it travels outside the the ~ 1 cm diameter region where the laser beams overlap). For each Sr wavelength,

calculate the number of photons required to stop an atom initially moving with $v_0 = 25$ cm/s. (c) Assuming the scattering rate is $\approx A_{ij}/2$, how long will it take to stop an atom using each wavelength?

8. MOTs are often used to create a sample of cold atoms to load into a magetic trap. (a) Explain why you can not create a magnetic trap using ground state Sr($5s\,^1S_0$) atoms. (b) When a strontium MOT is operated on the $5s\,^1S_0 \rightarrow 5p\,^1P_1$ transition at 461 nm, excited atoms will very occasionally decay into the metastable $5p\,^3P_2$ level via the $4d\,^1D_2$ intermediate level. Why does this allow magnetic trapping of Sr?

9. (a) Calculate the photon absorption cross section of the Rb($5^2S_{1/2}$ – $5^2P_{3/2}$) transition for a laser red-detuned 4 natural linewidths from the atomic transition. The transition has a wavelength of 780 nm and $A_ij = 3.8 \times 10^7$ s^{-1}. (b) The laser from part-a is used to create a Rb MOT containing 10^8 atoms. If the intensity of each of the six laser beams in the MOT is 0.5 mW/cm^2, what is the total photon scattering rate from the trapped Rb atoms?

10. (a) Calculate the photon absorption cross section of the Rb($5^2S_{1/2}$– $5^2P_{3/2}$) transition (see problem 9), for a photon at 1064 nm. (b) If a dipole trap uses a Nd:YAG laser ($\lambda = 1064$ nm) with an intensity of 10^7 W/cm^2 at the focus, what will be the scattering rate per Rb atom in the trap (neglecting all transitions other than the 780 nm line)?

11. Discuss your opinion on whether an OPO should be considered a laser.

12. Discuss your opinion on whether an "atom laser" (p. 347) should be considered a laser.

Solutions to Select Problems

Chapter 1

1.1 (a) $\lambda = c/\nu = 6 \times 10^{-5}$ cm
(b) $E = E_0 \sin(\omega t - kx)\hat{\jmath} = 0.1$ V/m $\sin(3.14 \times 10^{15} t - 1.05 \times 10^{7} x)\hat{\jmath}$.
Where $\hat{\imath}$, $\hat{\jmath}$, and $\hat{k}$ are unit vectors along the x, y, and z axes.
(c) $\vec{k} = (2\pi/\lambda)\hat{\imath} = 1.05 \times 10^{7}\ \mathrm{m}^{-1}\ \hat{\imath}$
(d) $B_0 = E_0/c = 3.33 \times 10^{-10}$ T. If $\vec{E}$ is in the y direction then $\vec{B}$ is in the z direction.

1.2 (a) $\ell_c = c\tau_c = 1.5 \times 10^{5}$ m $= 3 \times 10^{11}$ wavelengths
(b) $\Delta\nu = 1/(\pi\tau_c) = 6.37 \times 10^{2}$ Hz

1.3 (a) $\nu = c/\lambda = 6 \times 10^{14}$ Hz
(b) $W = h\nu = 3.98 \times 10^{-19}$ J $= 2.48$ eV

1.4 $W = h\nu = hc/\lambda$
$W(\text{in eV}) = \frac{hc}{e\lambda} = \frac{1239.85}{\lambda(\text{in nm})}$

1.5 (a) $W = I_\nu/c = 3.33 \times 10^{-10}\ \mathrm{J/m^3}$
(b) $W = \frac{1}{2}\epsilon_0 E_{\mathrm{RMS}}^2 + \frac{1}{2}B_{\mathrm{RMS}}^2/\mu^2 = \epsilon_0 E_{\mathrm{RMS}}^2$
so $E_{\mathrm{RMS}} = (W/\epsilon_0)^{1/2} = 8.68$ V/m
(c) $B_{\mathrm{RMS}} = E_{\mathrm{RMS}}/c = 2.89 \times 10^{-8}$ T
(d) $n_{\mathrm{ph}} = I_\nu/(h\,\nu\,c) = I_\nu/(h\,c^2/\lambda) = 8.4 \times 10^{9}\ \mathrm{photon/m^3}$

1.6 (a) The initial radius is $r_i = 10^{-2}$ m, and focused $r_f = 10^{-4}$ m.
$I_{\nu\ \mathrm{focus}} = (r_i/r_f)^2\, I_{\nu\ \mathrm{initial}} = 10^{3}\ \mathrm{W/m^2}$
(b) $W_{\mathrm{focus}} = (r_i/r_f)^2\, W_{\mathrm{initial}} = 3.33 \times 10^{-6}\ \mathrm{J/m^3}$
(c) $E_{\mathrm{RMS\ focus}} = (r_i/r_f)\, E_{\mathrm{RMS\ initial}} = 868$ V/m

1.7 $\lambda = (1239.85\ \mathrm{nm\,eV})/W = 590.4$ nm
$\nu = c/\lambda = 5.08 \times 10^{14}$ Hz

1.8 (a) $\lambda = \lambda_0/n = 333$ nm
(b) $\nu = c/\lambda_0 = 6 \times 10^{14}$ Hz, both inside and outside the material

1.9 $P_{\text{peak}} = P_{\text{avg}} \left(\frac{T}{\Delta t}\right) = 10^5$ W

1.10 No. The coherence time can not be greater than 5×10^{-9} s, but if the laser is not coherent for the entire pulse duration the coherence time can be less than 5×10^{-9} s.

Chapter 2

2.1 (a)$\lambda = 2L/n_x$, so $n_x = 2L/\lambda = 8 \times 10^5$
$\nu = c/\lambda = n_x\, c/2L = 6 \times 10^{14}$ Hz
$\Delta\nu = \Delta n_x\, c/2L = 7.5 \times 10^8$ Hz for $\Delta n_x = 1$.
$\frac{\Delta\nu}{\nu} = \frac{1}{n_x} = 1.25 \times 10^{-6}$
(b) $\nu = \sqrt{n_x^2 + n_y^2 + n_z^2}\ (c/2L)$
$\Delta\nu = \frac{n_x\,(c/2L)\,\Delta n_x}{\sqrt{n_x^2+n_y^2+n_z^2}}$
$\frac{\Delta\nu}{\nu} = \frac{n_x \Delta n_x}{n_x^2+n_y^2+n_z^2}$
For $\Delta n_x = 1$ and $n_x = n_y = n_z$, $n_x^2 + n_y^2 + n_z^2 = 3n_x^2$ and so $\Delta\nu/\nu = 1/3n_x$.
Since $\nu = c/\lambda = \sqrt{3}n_x\,(c/2L)$, $n_x = \frac{1}{\sqrt{3}}\frac{2L}{\lambda}$.
$\frac{\Delta\nu}{\nu} = \frac{\lambda}{\sqrt{3}\,2L} = 7.22 \times 10^{-7}$

2.2 $N_\nu = \left(e^{h\nu/kT} - 1\right)^{-1}$
Energy per photon, $h\nu = hc/\lambda = 2.48$ eV.
For $T = 5000$ K, $N_\nu = 3.13 \times 10^{-3}$ photons/mode.
For $T = 1000$ K, $N_\nu = 3.20 \times 10^{-13}$ photons/mode.
For $T = 100$ K, $N_\nu = 1.12 \times 10^{-125} \approx 0$ photons/mode.

2.3 $\bar{W} = h\nu\, N_\nu$
For $T = 5000$ K, $\bar{W} = 7.76 \times 10^{-3}$ eV.
For $T = 1000$ K, $\bar{W} = 7.94 \times 10^{-13}$ eV.
For $T = 100$ K, $\bar{W} \approx 0$ eV.

2.4 $\rho = \frac{8\pi\, h\, \nu^3}{c^3} N_\nu$
For $T = 5000$ K, $\rho = 2.6 \times 10^3$ eV/m^3 Hz.
For $T = 1000$ K, $\rho = 2.7 \times 10^{-7}$ eV/m^3 Hz.

For $T = 100$ K, $\rho \approx 0$ eV/m^3 Hz.

2.5 Number of modes/cm^3 $= \frac{8\pi\,\nu\,\Delta\nu}{c^3} = \frac{8\pi}{c\,\lambda^2}\,\Delta\nu = 4.7 \times 10^8$ modes/cm^3 within $\Delta\nu$.

2.6 From problem 2.5, for a volume of 10 cm^3, there are 4.7×10^9 modes. The total decay rate into all modes is $A_{ul} = 3 \times 10^8$ s^{-1}. Assuming decays into each mode are equally probable, the decay rate into a single mode is 6×10^{-2} s^{-1}.

2.7 $\sigma_{ul} = \frac{\lambda^2}{8\pi}\,A_{ul}\,g(\nu - \nu_0)$
At line center $g(0) \approx \frac{1}{\Delta\nu}$. So σ_{ul} at line center is
$\sigma_{ul} = \left(\frac{\lambda^2}{8\pi}\right)\left(\frac{A_{ul}}{\Delta\nu}\right) = 9.9 \times 10^{-12}$ cm^2
This is a large atomic cross section.

2.8 $\sigma_{lu} = \frac{g_u}{g_l}\,\sigma_{ul} = 2.98 \times 10^{-11}$ cm^2

2.9 If $g_u = g_l$, then $\sigma_{lu} = \sigma_{ul}$.
$\sigma_{ul} = (\pi\,r_e\,c\,f_{ul})/\Delta\nu = 1.6 \times 10^{-11}$ cm^2

2.10 $A_{ul} = (8\pi^2)(c/\lambda^2)\,r_e\,f_{ul} = 3.1 \times 10^7$ s^{-1}

Chapter 3

3.1 (a) $\sigma_{ul} = \frac{g_u}{g_l}\frac{\lambda^2}{8\pi}\,A_{ul}\,g(\nu - \nu_0) = \pi\,r_e\,c\,f_{ul}\,g(\nu - \nu_0)$
$A_{ul} = \frac{g_l}{g_u}\frac{8\pi^2\,r_e\,c\,f_{ul}}{\lambda^2} = 6.4 \times 10^7$ s^{-1}
So $\tau = \frac{1}{A_{ul}} = 1.6 \times 10^{-8}$ s
(b) $\sigma_{lu}(0) = \pi\,r_e\,c\,f_{ul}/\Delta\nu = 2.5 \times 10^{-12}$ cm^2
(c) $\alpha = n_l\,\sigma_{lu}(0) = 500$ cm^{-1}
(d) $I = I_0\,\mathrm{e}^{-n_l\,\sigma_{lu}(0)\,L} = 0.9 I_0$
$n_l\,\sigma_{lu}(0)\,L = 0.10$, $n_l = 8 \times 10^9$ Na atoms/cm^3

3.2 (a) $(p/V) = \frac{8\pi\,\nu^2\,\Delta\nu}{c^3} = \left(\frac{8\pi}{\lambda^2}\right)\left(\frac{\Delta\nu}{c}\right) = 8.43 \times 10^8$ modes per unit volume within $\Delta\nu$
(b) $\tau^{-1} = A_{ul}/p = A_{ul}/\left[(p/V)\,V\right] = 77$ s^{-1} for emission into a single mode

3.3 (a) $\Delta n_{\mathrm{ph}} = \left(\frac{p}{V}\right)\left(\frac{\tau_u}{\tau_{\mathrm{ph}}}\right) = \left(\frac{8\pi\,\nu^2\,\Delta\nu}{c^3}\right)\left(\frac{\tau_u}{\tau_{\mathrm{ph}}}\right)$
$= \left(\frac{8\pi\,\Delta\lambda}{\lambda^4}\right)\left(\frac{\tau_u}{\tau_{\mathrm{ph}}}\right) = 3.36 \times 10^{16}$ Xe_2^*/cm^3

(b) Energy is deposited in the Xe gas via electron-atom collisions, with the primary process being ionization (reaction i). The ionization energy of Xe is $I_{\text{Xe}} = 12.127$ eV.
$E_{\text{min dep}} = \Delta n_{\text{th}}\, I_{\text{Xe}}\, V = 6.1 \times 10^{-2}$ J/pulse.
(Alternatively, an even lower absolute minimum energy can be calculated by assuming Xe_2^* eximers are produced via reaction vi by Xe^* atoms produced by electron-impact excitation with an excitation threshold of 8.31 eV.)
(c) $P_{\text{min}} = E_{\text{min dep}}/t_{\text{pulse}} = E_{\text{min dep}}/(0.1 \times \tau_u) = 3 \times 10^7$ W/pulse
(d) $\sigma_{ul} = \frac{\lambda^2}{8\pi} A_{ul} \frac{1}{\Delta\nu} = \left(\frac{\lambda^4}{8\pi\,\Delta\lambda}\right)\left(\frac{1}{c\,\tau_u}\right) = 3.0 \times 10^{-18}\ \text{cm}^2$

3.4 (a) $n_{\text{th}} = \left(\frac{p}{V}\right)\left(\frac{\tau_u}{\tau_{\text{ph}}}\right) = \left(\frac{8\pi\,\nu^2\,\Delta\nu}{c^3}\right)\left(\frac{\tau_u}{\tau_{\text{ph}}}\right)$
$= \left(\frac{8\pi\,\Delta\lambda}{\lambda^4}\right)\left(\frac{\tau_u}{\tau_{\text{ph}}}\right) = 3.7 \times 10^{17}\ \text{He}_2^*/\text{cm}^3$
(b) $\sigma_{ul} = \frac{\lambda^2}{8\pi} A_{ul} \frac{1}{\Delta\nu} = \left(\frac{\lambda^4}{8\pi\,\Delta\lambda}\right)\left(\frac{1}{c\,\tau_u}\right) = 2.7 \times 10^{-19}\ \text{cm}^2$
(c) $p = \left(\frac{8\pi\,\nu^2\,\Delta\nu}{c^3}\right) V = \frac{8\pi\,\Delta\lambda}{\lambda^4} V = 2.45 \times 10^{17}$ modes in $\Delta\lambda$
(d) $A_{ul}/p = 4 \times 10^{-10}\ \text{s}^{-1}$
(e) $E_{\text{min dep}} = \Delta n_{\text{th}}\, I_{\text{He}}\, V = 29$ J/pulse.
(f) If $\sigma_{\text{ioniz}} \simeq \sigma_{ul} = 2.7 \times 10^{-19}\ \text{cm}^2$ or larger then the He_2^* laser can not work.

3.5 $\tau_{\text{ph}} = \left(\frac{n\,L}{c}\right)\left(\frac{1}{-\ln r_1 r_2}\right) \approx \frac{2n\,L}{c\,T_2} = 3.3 \times 10^{-7}$ s

3.6 $R_{ul} = A_{ul}\,(1 + N_\nu) = 200 A_{ul}$
$N_\nu = 199$ photons in the mode

3.7 If $1/\tau_{\text{ph}}$ is large enough then even with a population inversion the loss rate can exceed the stimulated emission rate.

3.8 $\Delta n_{\text{ph}} = \left(\frac{p}{V}\right)\left(\frac{\tau_u}{\tau_{\text{ph}}}\right) = \left(\frac{8\pi\,\Delta\lambda}{\lambda^4}\right)\left(\frac{\tau_u}{\tau_{\text{ph}}}\right)$
$\frac{\Delta n_{\text{ph}}(600\ \text{nm})}{\Delta n_{\text{ph}}(900\ \text{nm})} = \left(\frac{900}{600}\right)^4 = 5.1$

3.9 $\tau_{\text{ph}} = \frac{2nL}{c(1-R_2)}$
$R_2 = 1 - \frac{2nL}{c\,\tau_{\text{ph}}} = 0.998$

3.10 $\text{e}^{-\alpha L} = 0.2$
$\alpha = \frac{-\ln 0.2}{30\ \text{cm}} = 0.054\ \text{cm}^{-1}$
$\sigma_{lu} = \frac{\alpha}{n_l} = 5.4 \times 10^{-17}\ \text{cm}^2$

Chapter 4

4.1 (a) $\Delta\nu_L = \frac{1}{2\pi\,\tau_u} = 9.9 \times 10^7$ Hz
(b) $\Delta\lambda = \frac{\lambda}{\nu}\,\Delta\nu = \frac{\lambda^2}{c}\,\Delta\nu = 4.87 \times 10^{-13}$ cm $= 4.87 \times 10^{-5}$ Å

4.2 (a) $\Delta\nu_D = \left(7.2 \times 10^{-7}\right)\left(\frac{c}{\lambda}\right)\left(\frac{T}{A}\right)^{1/2} = 6.88 \times 10^{10}$ Hz
(b) $\Delta\nu_L$ was calculated in Problem 4.1(a). Using that, $\frac{\Delta\nu_D}{\Delta\nu_L} = 6.95{\times}10^2$

4.3 $\Delta\nu_L = \frac{1}{2\pi\,\tau_u} = \Delta\nu_D = \left(7.2 \times 10^{-7}\right)\left(\frac{c}{\lambda}\right)\left(\frac{T}{A}\right)^{1/2}$
Solving for T and using value of τ_u from problem 4.1,
$T = \left(\frac{\lambda}{c}\right)^2\left(\frac{1}{2\pi\,\tau_u}\right)^2\left(\frac{A}{5.18{\times}10^{-13}}\right) = 3.1 \times 10^{-3}$ K

4.4 (a) At STP (T=273 K), the number density of an ideal gas is $n = 2.69 \times 10^{19}$ molecules/cc, so $\Delta\nu_c = 3.63 \times 10^{10}$ Hz.
(b) Sodium has an atomic mass of 23.
$\Delta\nu_D = \left(7.2 \times 10^{-7}\right)\left(\frac{c}{\lambda}\right)\left(\frac{T}{A}\right)^{1/2} = 1.26 \times 10^9$ Hz
$n = \Delta\nu_D\,/\,(4.5 \times 10^{-10}$ Hz/cc$) = 2.8 \times 10^{18}$ cm^{-3}, which corresponds to P=0.104 atm = 79 Torr.
(c) $\Delta\nu_N = \frac{1}{2\pi\,\tau_u} = 9.9 \times 10^6$ Hz $= \Delta\nu_C = (4.5 \times 10^{-10}$ Hz/cc$)n$
$n = 2.2 \times 10^{16}$ cm^{-3}, so $P = 8.2 \times 10^{-4}$ atm $= 0.62$ Torr

4.5 The value of $v = \sqrt{\frac{3k\,T}{m}} = 4 \times 10^4$ cm/s.
$\Delta\nu_C = \frac{n\,\sigma\,v}{\pi} = (4.5 \times 10^{-10}$ Hz/cc$)n$, which yields $\sigma = 3.5 \times 10^{-14}$ cm^2
Approximating $\sigma = \pi b^2$, then $b = 1.1 \times 10^{-7}$ cm $= 21\ a_0$

4.6 $\tau_C = \frac{1}{n\,\sigma\,v} = 2.6 \times 10^{-11}$ s

4.7 $\Delta\nu_N = \frac{1}{2\pi}\left(\frac{1}{\tau_u} + \frac{1}{\tau_l}\right) = \frac{1}{2\pi}\left(\frac{\tau_u+\tau_l}{\tau_u\,\tau_l}\right) = 1.2 \times 10^7$ Hz

4.8 $g_D(\nu - \nu_0) = \frac{2\sqrt{\ln 2}}{\sqrt{\pi}\,\Delta\nu_D}\ \exp\left[-\frac{4\ln 2(\nu-\nu_0)^2}{\Delta\nu_D^2}\right]$
$g_C(\nu - \nu_0) = \frac{(\Delta\nu_C\,/\,2\pi)}{(\nu-\nu_0)^2+(\delta\nu_C\,/\,2)^2}$
$g_c(\nu - \nu_0) \geq g_D(\nu - \nu_0)$ when the Lorentzian packet are as large as the Doppler line shape. Graphing the left and right hand sides of this transcendental equation yields an approximate solution of $\nu - \nu_0 = 1.7 \times 10^9$ Hz $= 1.7\Delta\nu_C = 0.9\Delta\nu_D$.

4.9 (a) $N(J) = (2J + 1)\mathrm{e}^{-B\,J(J+1)hc/kT}$
$\frac{dN(J)}{dJ} = \mathrm{e}^{-B\,J(J+1)hc/kT}\left[2 - B(2J + 1)(2J + 1)\frac{hc}{kT}\right] = 0$
$(2J_{\max} + 1)^2 = \frac{2kT}{B\,hc}$

$J_{\max} = \frac{1}{2}\left[\left(\frac{2kT}{B\,hc}\right)^{1/2} - 1\right] = 0.589\left(\frac{T\,[\mathrm{K}]}{B\,[\mathrm{cm}^{-1}]}\right) - 0.5$
(b) For $B = 1.9$ cm^{-1} and $T = 325$ K, $J_{\max} = 7.2 \approx 7$.
(c) $J' = 7 \rightarrow J'' = 6$

4.10 (a) For λ=632.8 nm, T=400 K, and A=20 amu,
$\Delta\nu_D = (7.2 \times 10^{-7})\left(\frac{c}{\lambda}\right)\left(\frac{T}{A}\right)^{1/2} = 1.5$ GHz
(b) For λ=488 nm, T=3000 K, and A=40 amu,
$\Delta\nu_D = (7.2 \times 10^{-7})\left(\frac{c}{\lambda}\right)\left(\frac{T}{A}\right)^{1/2} = 3.8$ GHz
(c) $\Delta\nu = c \times (11\ \mathrm{cm}^{-1}) = 330$ GHz
(d) $\Delta\nu = c \times (6\ \mathrm{cm}^{-1}) = 180$ GHz

4.11 (a) For gas phase laser, λ=1.06 μm, T=1500 K, and A=144 amu,
$\Delta\nu_D = (7.2 \times 10^{-7})\left(\frac{c}{\lambda}\right)\left(\frac{T}{A}\right)^{1/2} = 0.65$ GHz.
For the solid state laser, $\Delta\nu = c \times (6\ \mathrm{cm}^{-1}) = 180$ GHz.
Hence, the linewidth of the solid state Nd^{3+} laser is 277× larger than the gas phase Doppler width.
(b) The column density of Nd^{3+} ions in the crystal is $n \times \ell = 10^{20}$ cm^{-2}. The number density of Nd atoms in the gas phase, $PV = NRT$, $n_0 = p/RT = 1.3 \times 10^{13}$ cm^{-3}. If 1% of the Nd is ionized to Nd^{3+}, then $n = 1.3 \times 10^{11}$ cm^{-3}. So $\ell = 7.8 \times 10^{8}$ cm $= 7800$ km. Better linewidth with the gas phase, but impractically large.

Chapter 5

5.1 $\Delta\nu_N = \frac{1}{2\pi}\frac{1}{\tau_u} = 1.0 \times 10^{7}$ Hz
$\sigma_{ul} = (\pi\, r_e\, c\, f_{ul}) / \left(\frac{2}{\pi\,\Delta\nu_N}\right) = 5.6 \times 10^{-10}$ cm^2
$I_{S'} = I_S(0) = \frac{h\nu}{2\sigma_{ul}\,\tau_u} = \frac{hc/\lambda}{2\sigma_{ul}\,\tau_u} = 189\ \mathrm{W/m^2} = 19\ \mathrm{mW/cm^2}$

5.2 $\alpha/\alpha_0 = 0.1 = \frac{1}{(1+I_\nu/I_S)^{1/2}}$, so $1 + I_\nu/I_S = 10^{+2}$
$I_n u = (10^2 + 1) I_S = 1.9 \times 10^4\ \mathrm{W/m^2} = 1.9\ \mathrm{W/cm^2}$

5.3 $\Delta\nu_{\mathrm{hole}} = \Delta\nu_N (1 + I_\nu/I_{S'})^{1/2} = 3.3\Delta\nu_N = 3.3 \times 10^{7}$ Hz

5.4 The value of $\alpha/\alpha_0 = 0.1$ independent of the saturation laser frequency.

5.5 $\Delta\nu_C = 10^{11}$ Hz
$\sigma_{ul} = (\pi\, r_e\, c\, f_{ul}) / \left(\frac{2}{\pi\,\Delta\nu_C}\right) = 5.6 \times 10^{-14}$ cm^2
$I_S(0) = \frac{h\nu}{2\sigma_{ul}\,\tau_u} = \frac{hc/\lambda}{2\sigma_{ul}\,\tau_u} = 1.9 \times 10^{6}\ \mathrm{W/m^2} = 190\ \mathrm{W/cm^2}$

5.6 $\frac{\alpha}{\alpha_0} = \frac{1}{1+I_\nu/I_S} = 0.1$
$1 + \frac{I_\nu}{I_S} = \frac{\alpha_0}{\alpha} = 10$
$I_\nu = 9I_S = 1.7 \times 10^3 \text{ W/cm}^2$

5.7 The gain or absorption coefficient has the same saturation independent of the frequency for a homogeneous line. Thus $\alpha/\alpha_0 = 10^{-1}$.

5.9 If $\Gamma_{u\ell} = \frac{2}{3}$, then $\Gamma_{ui} = \frac{1}{3}$.
$I_S^*/I_S = \left(\frac{2A_i}{\Gamma_{ui}\, A_u}\right) = 0.06$

5.10 The saturation intensity for an inhomogeneous line is the same as the saturation intensity for the homogeneous packets at line center. Therefore, the ratio of the saturation intensity with optical pumping to that without optical pumping is the same as the homogeneous case in Problem 5.9, 0.06. [Of course, the dependence of α on the saturation intensity is different for the homogeneous line versus inhomogeneous line.]

Chapter 6

6.1 $\tau_{\text{ph}} = \frac{2L}{c\,T_2} = 2 \times 10^{-7}$ s.
$P_e = T_2\, P_{\text{inside}}$, hence $P_{\text{inside}} = P_e/T_2 = 1$ W.

6.2 (a) $\nu = c/\lambda = 4.74 \times 10^{14}$ Hz
$\delta\nu_c = \frac{1}{2\pi\,\tau_{\text{ph}}} = 7.96 \times 10^5$ Hz
$\Delta\nu_{\text{laser}} = \left(\frac{2\pi\, h\nu}{P_e}\right)\, \delta\nu_c^2 \left(\frac{n_u}{\Delta n_{\text{th}}}\right) = 1.25 \times 10^{-4}$ Hz.
This assumes $n_u \approx \Delta n_{\text{th}}$.
(b) $\Delta\nu_{\text{laser}}/\nu = 2.6 \times 10^{-19}$

6.3 The number of photons/s striking the output mirror is equal to the power inside the laser (1 W) divided by the energy per photon which yields 3.2×10^{18} photons/s. The laser is 30 cm long, so at any instance in time there are laser photons going in each direction so that $N_\nu = 3.2 \times 10^{18}$ photons/s $\times \frac{2\times 30 \text{ cm}}{3\times 10^{10} \text{ cm/s}} = 6.4 \times 10^9$ photons.

6.4 $\Delta\nu_D = (7.2 \times 10^{-7}) \left(\frac{c}{\lambda}\right) \left(\frac{T}{A}\right)^{1/2} = 1.5 \times 10^9$ Hz
$\Delta n_{\text{th}} = \left(\frac{8\pi\, \nu^2\, \Delta\nu_D}{c^3}\right) \left(\frac{\tau_u}{\tau_{\text{ph}}}\right) = 4.4 \times 10^7$ atoms/cc
$N = \Delta n_{\text{th}} \times \text{Volume} = 1.3 \times 10^6$ atoms

6.5 For r=0.5, $N_\nu \simeq 0$
For r=1.0, $N_\nu = \sqrt{p} = 10^7$ photons
For r=2.0, $N_\nu = p(r-1) = 10^{14}$ photons
For r=5.0, $N_\nu = p(r-1) = 4 \times 10^{14}$ photons

6.6 For r=0.5, $n_u/\Delta n_{\text{th}} = 0.5$
For r=1.0, $n_u/\Delta n_{\text{th}} = 1$
For r=2.0, $n_u/\Delta n_{\text{th}} = 1$
For r=5.0, $n_u/\Delta n_{\text{th}} = 1$

6.8 $\phi = \frac{(h\nu/\epsilon V)^{1/2}}{(N_\nu\, h\nu/\epsilon V)^{1/2}} = \frac{1}{\sqrt{N_\nu}} = 1.26 \times 10^{-5}$ radians

Chapter 7

7.1 $g_1 = 1 - \frac{L}{R_1} = -\frac{1}{3}$
$g_2 = 1 - \frac{L}{R_2} = \frac{1}{3}$
$g_1\, g_2 = -\frac{1}{9}$, the cavity is not stable. Stability requires $0 \leq g_1\, g_2 \leq 1$.

7.2 $g_1 = 1 - \frac{L}{R_1} = 0$
$g_2 = 1 - \frac{L}{R_2} = 0$
$g_1\, g_2 = 0$, the cavity is on the borderline between stable and unstable. It is stable, but a small variation in the dimensions can cause it to become unstable. This is a confocal cavity.

7.3 $\frac{z_0^2}{L^2} = \frac{g_1 g_2(1-g_1 g_2)}{(g_1+g_2-2g_1 g_2)^2}$
In this case, $g_1 = g_2 \equiv g$, so $\frac{z_0^2}{L^2} = \frac{g^2(1-g^2)}{4g^2(1-g)^2} \to \frac{1}{4}$ as $g \to 0$
$z_0^2 = L^2/4$, so $z_0 = L/2 = 10$ cm
$w_0^2 = \frac{\lambda\, z_0}{\pi}$, so $w_0 = \sqrt{\lambda\, z_0/\pi} = 0.0126$ cm [waist]
$w(z) = \left[w_0 \left(1 + \frac{z}{z_0} \right)^2 \right]^{1/2} = 0.0178$ cm at z=10 cm [spot size = $1/e$ radius at mirrors]

7.4 The half angle to the beam radius is $\theta = \frac{\lambda}{\pi\, w_0} = 0.00126$ rad
$w(z) = \theta\, z$, $w(10\text{ m}) = 1.26$ cm, at the moon $w(0.38 \times 10^9\text{m}) = 480$ km
The $1/e$ diameter at z=10 m or at the moon is twice the value of $w(z)$.

7.6 (a) $z_0 = \frac{\pi w_0^2}{\lambda} = 1.57$ cm

(b) $w(z) = \left[w_0\left(1+\frac{z}{z_0}\right)^2\right]^{1/2} = 6.39 \times 10^{-2}$ cm

(c) $R = \left(z + \frac{z_0^2}{z}\right) = 20.12$ cm

(d) $\theta_{\text{far field}} = \frac{\lambda}{\pi w_0} = 3.18 \times 10^{-2}$ radians

7.7 Using a * to denote quantities after the lens,

(a) $\frac{1}{R^*} = \frac{1}{R} - \frac{1}{f} = -0.15\ \text{cm}^{-1}$

$R^* = -6.65$ cm. The minus indicates that the beam is converging.

(b) At the lens, $w = w* = 6.39 \times 10^{-2}$ cm.

$\frac{z_0^*}{z^*} = \frac{R^*}{\pi\, w*^2/\lambda}$

$w_0^* = w^*\left[1+\left(\frac{z*}{z_0^*}\right)^2\right]^{-1/2} = w^*\left[1+\left(\frac{R^*\,\lambda}{\pi\, w*^2}\right)^2\right]^{-1/2} = 1.65 \times 10^{-3}$ cm

$z_0^* = \frac{\pi\, w_0^{*2}}{\lambda} = 0.171$ cm

(c) As found above, the $1/e$ radius is $w_0^* = 1.65 \times 10^{-3}$ cm.

From $\frac{z_0^*}{z*} = \frac{R^*}{\pi\, w*^2/\lambda}$, The distance to the waist after the lens is

$z^* = z_0^*\,\frac{\pi w_0^{*2}}{\lambda\, R^*} = 6.6$ cm

7.8 $F = a^2/L\,\lambda$, thus $a = \sqrt{F\,L\,\lambda} = 0.03$ cm

7.9 For f=20 cm, the mirrors have R=40 cm.

$g_1 = g_2 = \left(1 - \frac{L}{R}\right)$

$0 \leq g_1\,g_2 = \left(1 - \frac{L}{R}\right)^2 \leq 1$

For the maximum value of L, $1 - L/R = -1$, so $L = 2R = 80$ cm

7.10

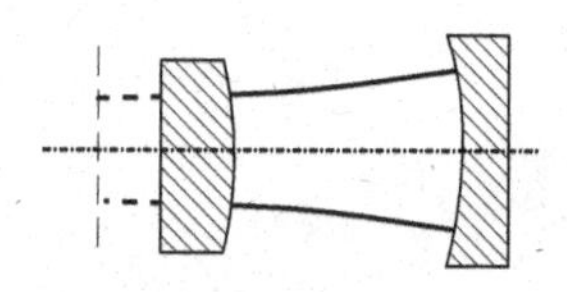

(a) The waist must be outside the cavity mirrors.

(b) It is not possible to construct a stable cavity with two diverging mirrors.

7.11 (a) $g_1 = \left(1 - \frac{L}{R_1}\right) = \frac{1}{3}$

$g_2 = \left(1 - \frac{L}{R_2}\right) = \frac{3}{5}$

$g_1\, g_2 = \frac{1}{5}$, so the cavity is stable.

(b) $\frac{z_1}{L} = \frac{R_2 - L}{R_1 + R_2 - 2L}$

$z_1 = L\left(\frac{R_2 - L}{R_1 + R_2 - 2L}\right) = 7.5$ cm. The waist is 7.5 cm from the mirror with R_1=15 cm.

(c) $z_0^2 = L^2\left[\frac{g_1\, g_2\,(1 - g_1\, g_2)}{(g_1 + g_2 - 2g_1\, g_2)^2}\right]$

$z_0 = 7.5$ cm [Rayleigh range]

$w_0 = \sqrt{\frac{z_0\,\lambda}{\pi}} = 1.09 \times 10^{-2}$ cm. [1/e radius]

$$z_1 = 7.5 \text{ cm}, \; w_1 = w_0\left[1 + \left(\frac{z_1}{z_0}\right)^2\right]^{1/2} = 1.54 \times 10^{-2} \text{ cm}$$

$$z_2 = L - z_1 = 2.5 \text{ cm}, \; w_2 = w_0\left[1 + \left(\frac{z_2}{z_0}\right)^2\right]^{1/2} = 1.14 \times 10^{-2} \text{ cm}$$

Chapter 8

8.1 (a) $\lambda = \frac{hc}{W} = 866$ nm

(b) $\Delta\lambda = (0.3 \text{ nm/°C})\,\Delta T$

$T' = T_0 + \Delta T = T_0 + \frac{\lambda' - \lambda_0}{0.3 \text{ nm/°C}} = -20$°C assuming room temperature is 20° C.

8.2 $\theta \cong \lambda/a = 0.8$ radians, or about 46°.

8.3 $N_0 = 1/\Delta x = 4 \times 10^7$ electrons/cm

$\lambda = \frac{2L}{N_0\, L/2} = \frac{4}{N_0} = 10^{-7}$ cm

8.4 $W = e\,E\,\ell$, so $\ell = W/(e\,E) = 0.33 \times 10^{-21}$ m

8.5 At room temperature $kT = 26 \times 10^{-3}$ eV.

$f = \frac{1}{1 + \exp[(E - E_F)/kT]} = e^{-27.5} \cong 10^{-12}$

8.6 $k = \frac{\pi}{a_0} = 1.26 \times 10^8$ cm^{-1}

8.9 $E_n = \frac{p_n^2}{2m} = \frac{h^2/\lambda^2}{2m} = \frac{h^2\, n^2}{8m\, L^2}$

$E_2 - E_1 = \frac{3h^2}{8m\, L^2} = 1.79 \times 10^{-21}$ J $= 1.12 \times 10^{-2}$ eV

8.10 $\Delta\nu_C = \frac{c}{2\,n\,L} = 4.12 \times 10^{11}$ Hz

Chapter 9

9.1

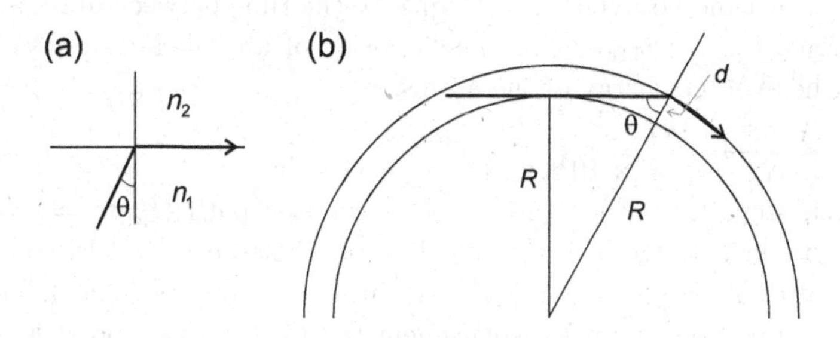

(a) $n_1 \sin\theta = n_2$
$\theta = \sin^{-1}(n_2/n_1) = 1.13\,\text{radians} = 64.5°$
(b) $\sin\theta = \frac{R}{R+d}$
$R = \frac{\sin\theta}{1-\sin\theta}\, d = 90\mu\text{m}$
This is rather small and most likely a fiber would break before it reached this radius of curvature.

9.3 $\Delta t = \frac{2\ell}{c} = 1.1 \times 10^{-4}$ s

9.6 (a) $P_{\text{peak}} = \frac{E_{\text{pulse}}}{\Delta t_{\text{pulse}}} = 10^8$ W
(b) $P_{\text{avg}} = \frac{E_{\text{pulse}}}{T} = \frac{E_{\text{pulse}}}{1/f} = 15$ W

9.7 (a) $n_{\text{ph}} = \frac{E_{\text{pulse}}}{hc/\lambda} = 1.3 \times 10^{18}$ photons
(b)$I_{\text{spot}} = \frac{P_{\text{peak}}}{\pi r^2_{\text{spot}}} = 3.2 \times 10^{19}\ \text{W/m}^2 = 3.2 \times 10^{15}\ \text{W/cm}^2$
$E_0 = \left(\frac{2\, I_{\text{spot}}}{c\,\varepsilon_0}\right)^{1/2}$ (see for example, Eq. 3.20)
$E_0 = 1.55 \times 10^{11}$ V/m. This is the magnitude of the electric field, the RMS field is $E_{\text{RMS}} = E_0/\sqrt{2} = 1.1 \times 10^{11}$ V/m.

9.8 $\Delta t = L/c = 10^{-8}$ s

9.10 (a) $(t_1 - t_2) = \frac{\ell}{c}(n_1 - n_2) = 2.67 \times 10^{-4}$ s
(b) Since $v = c/n$, the larger the value of n, the slower the velocity. Hence, the 800 nm pulse arrives first.

Chapter 10

10.1 For a random walk, consider the motion of the Ne atom as a series of steps. The time between steps is equal to the time between atom-atom collisions, $t_{\text{step}} = \ell_{\text{MFP}}/\langle v\rangle$. The velocity of atoms can be obtained from the average energy of the atoms,

$\langle E\rangle = \frac{1}{2}mv^2 = \frac{3}{2}kT$

$\langle v\rangle = \sqrt{3kT/m} = 6\times 10^4$ cm/s

In each step, the atom travels one mean free path ($\ell_{\text{MFP}} = 1/n\sigma$) in length in a random direction. The total distance traveled by the atom after N steps is $N\ell_{\text{MFP}}$, but since each step is in a different random direction, the net displacement from it's starting point is only $L = \sqrt{N}\ell_{\text{MFP}}$. Setting $L = r_{\text{tube}}$,

$N = \frac{r^2_{\text{tube}}}{\ell^2_{\text{MFP}}}$

$T = N\,t_{\text{step}} = \left(\frac{r^2_{\text{tube}}}{\ell^2_{\text{MFP}}}\right)\frac{\ell_{\text{MFP}}}{\langle v\rangle} = \frac{n\,\sigma\,r^2_{\text{tube}}}{\langle v\rangle} = 1\times 10^{-5}$ s

Note that this is only an order of magnitude calculation. A range of values are possible depending upon if one uses the average, most probable value, or RMS value in calculating $\langle v\rangle$. This approach also underestimates the true diffusion time *to the wall* of the capillary tube, since we assumed $L = r_{\text{tube}}$. But diffusion in the axial direction should not count towards the radial diffusion time. In kinetic theory, the diffusion constant is $D = \ell_{\text{MFP}}\,v/3$, with $L = \sqrt{Dt}$. This yields

$T = \frac{L^2}{D} = \frac{3r^2_{\text{tube}}}{\ell_{\text{MFP}}\,v} = 4\times 10^{-5}$ s.

10.2 For the 632.8 nm line (unprimed) to lase, it must have a lower threshold population difference, Δn_{th}, than the 3.39 μm line (primed). The minimum ratio is when the two Δn_{th} are the same.

$\Delta n_{\text{th}} = \frac{8\pi\nu^2\Delta\nu}{c^3}\frac{1}{A_{ij}\,\tau_{\text{ph}}} \leq \Delta n'_{\text{th}} = \frac{8\pi\nu'^2\Delta\nu'}{c^3}\frac{1}{A_{i'j'}\,\tau'_{\text{ph}}}$

For a Doppler broaden line, $\Delta\nu \propto \nu$, so

$\frac{\nu^3}{A_{ij}\,\tau_{\text{ph}}} \leq \frac{\nu'^3}{A_{ij'}\,\tau'_{\text{ph}}}$

$\frac{\tau_{\text{ph}}}{\tau'_{\text{ph}}} \geq \frac{A_{ij'}}{A_{ij}}\left(\frac{\nu}{\nu'}\right)^3$

$\frac{\tau_{\text{ph}}}{\tau'_{\text{ph}}} \geq \frac{A_{ij'}}{A_{ij}}\left(\frac{\lambda'}{\lambda}\right)^3 = 132$

10.3 (a) $\Delta\nu_D = (7.2\times 10^{-7})\left(\frac{c}{\lambda}\right)\left(\frac{T}{A}\right)^{1/2}$

for $\lambda = 632.8\,\text{nm}$: $\Delta\nu_D = 1.52\times 10^9$ Hz

for $\lambda = 3.39\,\mu\text{m}$: $\Delta\nu_D = 2.85\times 10^8$ Hz

(b) $\mu_B = 2.8\times 10^6$ Hz/Gauss

$\Delta\nu_B = \pm m_J\, g_J\, \mu_B\, B_0 = \pm 1.8 \times 10^8$ Hz for $m_J = \pm 1$.

(c) For $\lambda = 632.8\,\text{nm}$: $\Delta\nu_B = \pm 0.12\,\Delta\nu_D$

For $\lambda = 3.39\,\mu\text{m}$: $\Delta\nu_B = \pm 0.63\,\Delta\nu_D$

10.4 In general, competition among wavelengths which arise from the same upper level limits the laser output to the one with the lowest Δn_{th}. So only one visible output wavelength at a time, since all arise from the same $3s_2$ upper level. In principle, one could have a dual wavelength He-Ne laser with one wavelength using the $2s_2$ upper level (near-infrared) and the other using the $3s_2$ level (visible or infrared). Nevertheless, because the wavelengths are very different, it is difficult to have windows and mirrors work well with both wavelengths. For a very long laser spatial hole burning permits lasing on lines with the same upper level.

10.5 (a) $\Delta\nu_D = \left(7.2 \times 10^{-7}\right)\left(\frac{c}{\lambda}\right)\left(\frac{T}{A}\right)^{1/2} = 1.6 \times 10^9$ Hz

$\sigma_{lu} = \frac{g_u}{g_l}\left(\frac{\lambda^2}{8\pi}\right)\frac{A_{ul}}{\Delta\nu_D} = 4.8 \times 10^{-15}\,\text{cm}^2$

(b) $n = \frac{N}{V} = \frac{P}{RT} = 7.2 \times 10^{15}\,\text{cm}^{-3}$

$\ell_{\text{MFP}} = \frac{1}{\sigma\, n} = 0.3\,\text{mm}.$

ℓ_{MFP} is less than the tube radius, and most of the resonance transitions are reabsorbed before they reach the wall.

10.6 $\Delta\nu_C = \frac{c}{2L}$, thus $L = \frac{c}{2\Delta\nu_c} = 20.9$ cm

For a half angle θ, $\theta = \frac{\lambda}{\pi w_0}$, so $w_0 = \frac{\lambda}{\pi\,(\theta_{\text{full}}/2)} = 0.3$ mm. The spot size at the waist is $2w_0 = 0.6$ mm.

Note this is also the spot size at the output coupler, so the waist occurs at the output coupler. The output coupler is a plane mirror ($z_2 = 0$, $R_2 = \infty$). The high reflector is located at $z_1 = -L = -20.9$ cm. The radius of curvature of this mirror is

$-R_1 = z_1 + \frac{z_0^2}{z_1}$

where $z_0 = \frac{\pi\, w_0^2}{\lambda} = 20.9$ cm. This yields $R_1 = 118$ cm

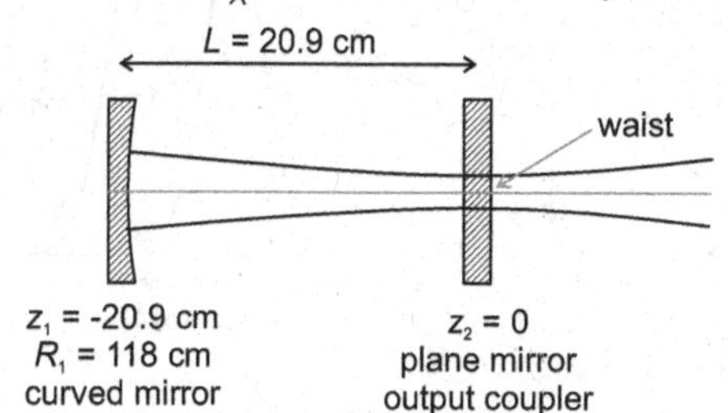

For a stable cavity, $0 \le g_1 g_2 \le 1$. For this cavity,
$g_1 g_2 = \left(1 - \frac{L}{R_1}\right)\left(1 - \frac{L}{R_2}\right) = (0.82)(1) = 0.82.$

Chapter 11

11.1 $\nu = c/\lambda = 2.83 \times 10^{13}$ Hz
$h\nu = 1.87 \times 10^{-20}$ J
n = number of photons/s $= P/h\nu = 5.35 \times 10^{22}$ photons/s

11.2 $\nu = c/\lambda = 10^{14}$ Hz
$h\nu = 6.6 \times 10^{-20}$ J
n = number of photons/s $= P/h\nu = 1.5 \times 10^{20}$ photons/s
The number of photons created in one hour is 5.4×10^{23} photons. One H_2 molecule is dissociated in the reaction that eventually produces one photon, so 5.4×10^{23} H_2, or 0.9 moles/hour H_2. Of course not every H_2 molecule will result in a photon in the laser mode, so on one hand this is a lower limit. On the other hand, the H atoms produced in the $F + H_2 \rightarrow HF^* + H$ reaction can recombine to form H_2 molecules. Assuming 100% recombination of the atomic hydrogen atoms, only 0.45 moles/hour H_2 would be required.

11.3 $\lambda = 248$ nm, $\nu = c/\lambda = 1.21 \times 10^{15}$ Hz
$h\nu = 7.98 \times 10^{-19}$ J
N = number of photons/pulse $= E_{\text{pulse}}/h\nu = 3.13 \times 10^{17}$ photons/pulse

11.4 (a) $\Delta\nu_D = (7.2 \times 10^{-7}) \left(\frac{c}{\lambda}\right) \left(\frac{T}{A}\right)^{1/2}$
For HeNe laser: T=400 K, M=20 amu, $\lambda = 632.8$ nm, $\Delta\nu_D = 1.5$ GHz
For HeCd laser: T=600 K, M=112 amu, $\lambda = 441.6$ nm, $\Delta\nu_D = 1.1$ GHz

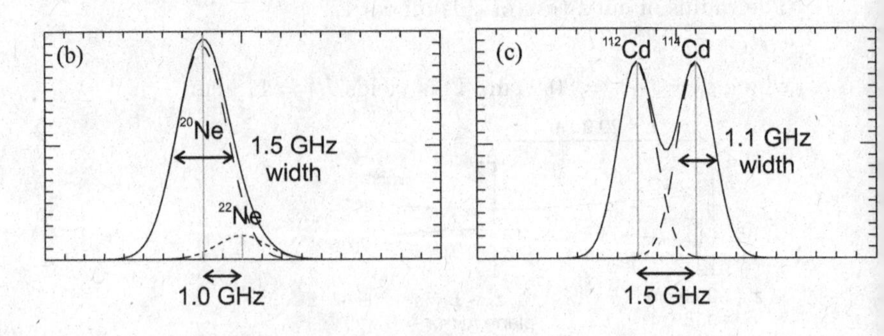

(d) This is inhomogeneous broadening since the light interacts with different atoms.

11.5 $\Delta\nu_C = \frac{c}{2L} = 7.5 \times 10^7$ Hz
Number of modes within Doppler width $= \frac{\Delta\nu_D}{\Delta\nu_C} = 40$ modes.
The use of a thick etalon and thin etalon within the cavity as well as a prism to separate lines other than the 514 nm line can produce lasing on a single mode.

11.6 Yes, spatial hole burning permits a laser with a homogeneously broadened line to lase on several longitudinal modes since the nodes are interacting with atoms at different locations.

11.7 Cu(^{2}D $\to$ ^{2}S) has $\Delta L = 2$ which is forbidden, hence the Cu(^{2}D) level is metastable. While the laser is on, atoms accumulate in the metastable level. The laser is pulsed to allow time for metastable atoms to be quenched back into ground state.

Chapter 12

12.1 The magnetic field is vertical. Since $\boldsymbol{F} = q\boldsymbol{v} \times \boldsymbol{B}$, the motion of the electron is in the horizontal plane. The polarization of the light is the same as the direction of the oscillating charge, so the polarization is horizontal.

12.2 At resonance,

$$\eta = \frac{1}{2} L \left(\frac{\omega^*}{2\gamma^2 c} - \frac{2\pi}{l} \right) = 0$$

$$\omega^* = \frac{4\pi\gamma^2 c}{l} = \frac{4\pi\gamma^2 c}{\lambda_u}$$

Since $\omega = 2\pi f = 2\pi c/\lambda$,

$$\frac{2\pi c}{\lambda^*} = \frac{4\pi\gamma^2 c}{\lambda_u}$$

$$\lambda^* = \frac{\lambda_u}{2\gamma^2}$$

12.3 (a) $K = 0.934\, B\, \lambda_u = 1.8$

$$\lambda^* = \frac{\lambda_u (mc^2)^2}{2E^2} \left(1 + \frac{K^2}{2}\right)$$

$$E = (mc^2)\sqrt{(1 + K^2/2)\ \lambda_u/(2\lambda^*)}$$

$$E(\lambda^* = 1.6\,\mu\text{m}) = 120 \text{ MeV}$$

$$E(\lambda^* = 2.2\,\mu\text{m}) = 100 \text{ MeV}$$

(b) $P_{\text{electrical}} = P_{\text{opt}}/\text{eff} = 670$ MW

12.4 $\frac{\Delta\omega^*}{\omega^*} = \frac{1}{N_w} = 0.05$
$\omega^* = \frac{2\pi c}{\lambda^*} = 1.89 \times 10^{12}$ Hz
$\Delta\omega^* = 9.4 \times 10^{10}$ Hz

12.5 $\Delta\nu = \frac{c}{L} = 0.25$ GHz

12.7 One method for creating a dual wavelength laser is to use a Littrow mounted diffraction grating where one order reflects λ_1 and a neighboring order reflects λ_2.
$n\,\lambda_1 = (n+1)\,\lambda_2 = 2d\,\sin\theta$
$n/(n+1) = \lambda_1/\lambda_2 = 0.96$. The best match using the lowest possible orders are with n=27 and $n+1$=28.
To keep the grating angle at a reasonable value ($\theta \leq 80°$), the grating spacing should be
$1/d = 2\sin\theta/(n\lambda_1) \approx 120$ grooves/mm.
Another possible design would create a broad band stimulated emission peak and use a pair of etalons that pass both lines but not other wavelengths. Or a design might use a 50-50 beamsplitter and two separately tuned gratings. Also note that a dye other than R6G might be required. R6G (normal tuning range 570-610 nm) while typically used to pump the Na 589.6 nm transition, may not work at 568.3 nm.

12.8 (a) P_{avg} =(energy/pulse)/(period) = 0.2 W
P_{peak} =(energy/pulse)/(pulse length) = 2×10^6 W
(b) Energy/photon = hc/λ
Energy/pulse$_{\lambda-2}$ = (λ_1/λ_2) Energy/pulse$_{\lambda-1}$ = 0.008 J

12.9 (a) The bandwidth of the laser should match the Doppler width.
$\Delta\nu_D = (7.2 \times 10^{-7})\left(\frac{c}{\lambda}\right)\left(\frac{T}{A}\right)^{1/2} = 1.7$ GHz
(b) $\Delta\nu_N = \frac{1}{2\pi\,\tau_u} = 10$ MHz

12.10 $\Delta\nu = \frac{1}{\pi\,\tau} = 6.4 \times 10^7$ Hz

Chapter 13

13.1 For type-II phase matching in a negative uniaxial crystal:

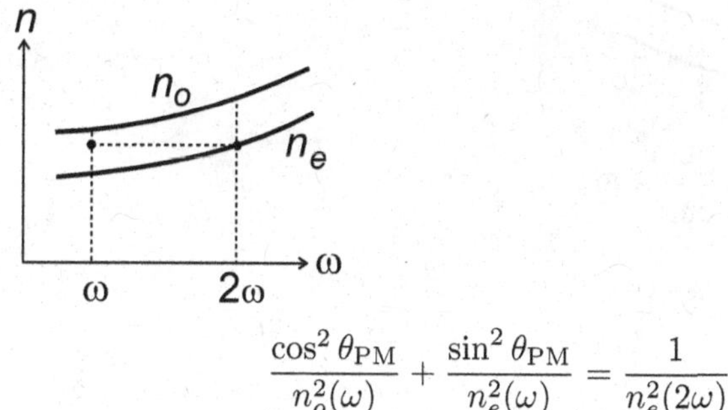

$$\frac{\cos^2\theta_{\text{PM}}}{n_o^2(\omega)} + \frac{\sin^2\theta_{\text{PM}}}{n_e^2(\omega)} = \frac{1}{n_e^2(2\omega)}$$

Solving for $\sin^2\theta_{\text{PM}}$ gives

$$\sin^2\theta_{\text{PM}} = \frac{n_e^{-2}(2\omega) - n_o^{-2}(\omega)}{n_e^{-2}(\omega) - n_o^{-2}(\omega)}$$

13.2 For type-I phase matching in $LiNbO_3$ (a negative uniaxial crystal):

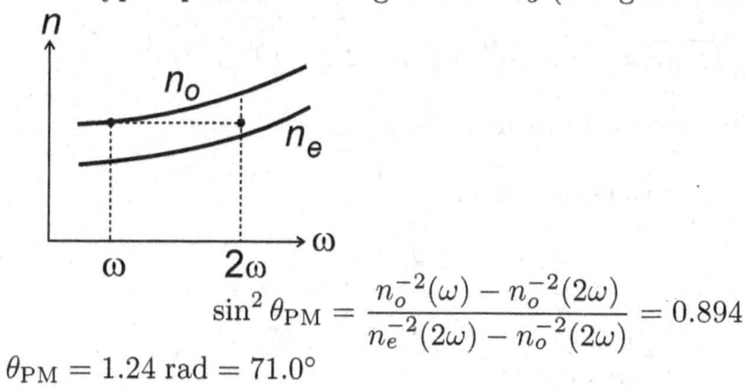

$$\sin^2\theta_{\text{PM}} = \frac{n_o^{-2}(\omega) - n_o^{-2}(2\omega)}{n_e^{-2}(2\omega) - n_o^{-2}(2\omega)} = 0.894$$

$\theta_{\text{PM}} = 1.24 \text{ rad} = 71.0°$

13.3 If $n_e(2\omega) > n_o(\omega)$ then Type-II phase matching is not possible.

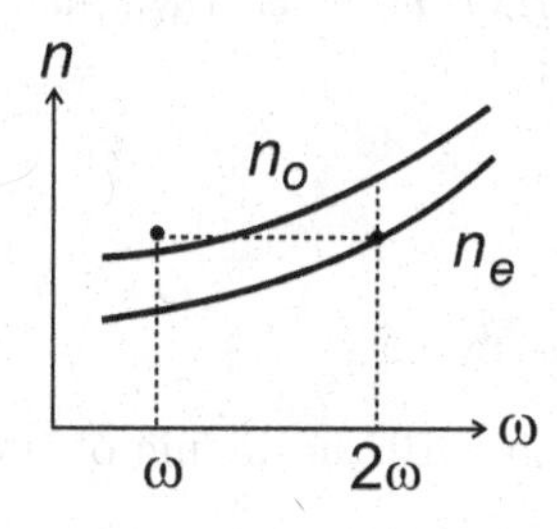

13.4 For type-I phase matching in a positive uniaxial crystal:

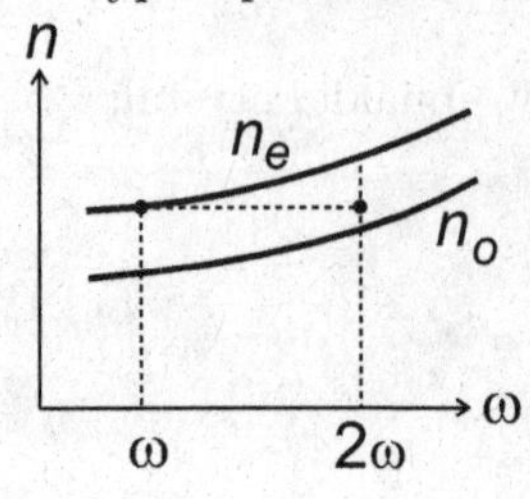

$$\frac{\cos^2\theta_{\mathrm{PM}}}{n_e^2(2\omega)} + \frac{\sin^2\theta_{\mathrm{PM}}}{n_o^2(2\omega)} = \frac{1}{n_e^2(\omega)}$$

Solving for $\sin^2\theta_{\mathrm{PM}}$ gives

$$\sin^2\theta_{\mathrm{PM}} = \frac{n_e^{-2}(\omega) - n_e^{-2}(2\omega)}{n_o^{-2}(2\omega) - n_e^{-2}(2\omega)}$$

13.5 $I = \frac{(E/\Delta t)}{\pi(d/2)^2} = 1.6 \times 10^{18}\ \mathrm{W/m^2}$
$I = \varepsilon_0\, E_{\mathrm{RMS}}^2\, c$
$E_{\mathrm{RMS}} = \sqrt{I/\varepsilon_0\, c} = 2.4 \times 10^{10}\ \mathrm{V/m} = 2.4 \times 10^{8}\ \mathrm{V/cm}$

13.6 The 2nd harmonic will increase by a factor of 4.

13.7 Using k values inside the material:

$$\overset{k_1}{\longrightarrow}\ \overset{k_2}{\longleftarrow}\ =\ \overset{k_3 = k_1 + k_2}{\longrightarrow}$$

$$\frac{2\pi\,\lambda_1\, n(\omega_1)}{c} - \frac{2\pi\,\lambda_2\, n(\omega_2)}{c} = \frac{2\pi\,\lambda_3\, n(\omega_3)}{c}$$

13.8 $\boldsymbol{P}_i = \varepsilon_0 \sum_{jk} d_{ijk}\, \boldsymbol{E}_j\, \boldsymbol{E}_k$
The units of $\boldsymbol{P}$ are C m^{-2}. ε_0 has units of C^2 N^{-1} m^{-2}. The units of electric fields can be expressed as N C^{-1} (i.e., $\boldsymbol{F} = q\boldsymbol{E}$). d_{ijk} has units of $\frac{P}{\varepsilon_0\, E^2}$, which are C/N.

13.10 $l_{\mathrm{coh}} = \frac{\lambda_0}{4[n(2\omega) - n(\omega)]} = 2.9\,\mu\mathrm{m}$

13.11 $\nu_{\mathrm{opt}} = \frac{c}{\lambda} = 3.75 \times 10^{14}$ Hz
harmonic number $= \nu_{\mathrm{opt}}/\nu_{\mathrm{low}} = 7.5 \times 10^5$
It is remarkable that one can have the 7.5×10^5 harmonic of the low frequency standard detected.

Chapter 14

14.1 LIF is measured against a dark, low-noise background. Absorption spectroscopy has noise from the light used for the absorption.

14.2 Assuming the laser bandwidth is much less than the Doppler width and is at line center,

$$\sigma_{lu}(\nu - \nu_0) = \frac{g_u}{g_l}\frac{\lambda^2}{8\pi} A_{ul}\, g(\nu - \nu_0) \approx \frac{g_u}{g_l}\frac{\lambda^2}{8\pi}\frac{A_{ul}}{\Delta\nu_D}$$
$$= 8.4 \times 10^{-16}\ \text{m}^2 = 8.4 \times 10^{-12}\ \text{cm}^2$$

If every absorption produces an LIF photon,
Number of LIF photons = Number photons incident × Number Na atoms/cm^2 × σ_{abs}

$$N_{total} = \frac{I}{h\nu} \times \frac{n_{Na} V}{A} \times \sigma_{abs} = \frac{I\lambda}{hc} \times \frac{n_{Na} V}{A} \times \sigma_{abs}$$
$$= 2.5 \times 10^{12}\ \text{LIF photons/pulse}$$

The LIF photons emitted into the solid angle Ω is

$$N_{detect} = N_{total}\left(\frac{\Omega}{4\pi}\right) = 2.0 \times 10^{10}\ \text{LIF photons/pulse}$$

This is very easily detected.

14.3 The two photon absorption occurs mostly when the beam is tightly focused. By focusing the beam inside a biological sample, one can determine the presence of particular molecules at that site inside a cell using LIF.

14.4 The branching fraction from $3^2P_{1/2}\ m = \frac{1}{2} \rightarrow 3^2S_{1/2}\ m = \frac{1}{2}$ is $\frac{1}{3}$ (see diagram on p. 330), so about 3 photons are needed.

14.5 $p_{ph} = \frac{h\nu}{c} = \frac{h}{\lambda} = 1.12 \times 10^{-27}$ kg m/s

$\Delta v = \frac{p_{Na}}{M_{Na}} = \frac{p_{ph}}{M_{Na}} = 0.029$ m/s $= 2.9$ cm/s

14.6 From problem 5, each photon slows the atom by $\Delta v = 2.9$ cm/s.

$N = \frac{v_{initial}}{\Delta v} = 6.9 \times 10^4$ photons

14.7 (a) $T = \frac{h}{4\pi k \tau} = \frac{h A_{ij}}{4\pi k}$

$\lambda = 461$ nm: $T = 7.6 \times 10^{-4}$ K $= 0.76$ mK

$\lambda = 689$ nm: $T = 1.8 \times 10^{-7}$ K $= 0.18\ \mu$K

(b) $\Delta v = \frac{p_{ph}}{M_{Sr}} = \frac{h}{M_{Sr}\lambda}$

$\lambda = 461$ nm: $\Delta v = 9.8 \times 10^{-3}$ m/s $= 0.98$ cm/s

$\lambda = 689$ nm: $\Delta v = 6.6 \times 10^{-3}$ m/s $= 0.66$ cm/s

$N = \frac{v_{initial}}{\Delta v}$

$\lambda = 461$ nm: $N = 26$ photons

$\lambda = 689$ nm: $N = 38$ photons

(c) $\Delta t = \frac{N}{(A_{ij}/2)}$

$\lambda = 461$ nm: $\Delta t = 2.6 \times 10^{-7}$ s $= 0.26\ \mu$s

$\lambda = 689$ nm: $\Delta t = 1.6 \times 10^{-3}$ s $= 1.6$ ms

So a trap using the $\lambda = 689$ nm line has a lower temperature, but it is harder to load since the scattering rate is lower.

14.8 (a) The ground state has $J = 0$, so $m_J = 0$ and the ground state energy is not effected by a magnetic field.

(b) The metastable level has $J = 2$, so atoms can have m_J values different from zero and thus are Zeeman shifted when they move in the spatially varying magnetic field which allows the atom to be trapped as long as it remains in the long-lived metastable level.

14.9 (a) $\sigma_{lu}(\nu - \nu_0) = \frac{g_u}{g_l} \frac{\lambda^2}{8\pi} A_{ul}\, g(\nu - \nu_0)$

$g(\nu - \nu_0) = \frac{\Delta\nu_N/2\pi}{(\nu-\nu_0)^2+(\Delta\nu_N/2)^2}$

For $\nu - \nu_0 = 4\Delta\nu_N$, $g(\nu - \nu_0) = [2\pi\,(16.25)\,\Delta\nu_N]^{-1}$

Since $\Delta\nu_N = \frac{1}{2\pi}\frac{1}{\tau_u} = \frac{A_{ij}}{2\pi}$

$\sigma_{lu}(\nu - \nu_0) = \frac{g_u}{g_l} \frac{\lambda^2}{8\pi} \frac{1}{16.25} = 3.0 \times 10^{-15}\ \text{m}^2 = 3.0 \times 10^{-11}\ \text{cm}^2$

(b) $N_{\text{scat}} = \left(\frac{I}{h\nu}\right) N_{\text{trap}}\, \sigma_{\text{abs}} = \left(\frac{I\,\lambda}{h\,c}\right) N_{\text{trap}}\, \sigma_{\text{abs}}$

The total intensity in the six beams is $I = 3$ mW/cm^2, so

$N_{\text{scat}} = 3.5 \times 10^{13}$ photons/s

14.10 (a) $\Delta\nu_N = \frac{A_{ij}}{2\pi} = 6.0 \times 10^6\ \text{s}^{-1}$

$\nu - \nu_0 = c\,(\lambda^{-1} - \lambda_0^{-1}) = 1.0 \times 10^{14}\ \text{s}^{-1}$

$g(\nu - \nu_0) = \frac{\Delta\nu_N/2\pi}{(\nu-\nu_0)^2+(\Delta\nu_N/2)^2} = 9.1 \times 10^{-23}$ s

$\sigma_{lu}(\nu - \nu_0) = \frac{g_u}{g_l} \frac{\lambda^2}{8\pi} A_{ul}\, g(\nu - \nu_0)$

$= 1.7 \times 10^{-28}\ \text{m}^2 = 1.7 \times 10^{-24}\ \text{cm}^2$

(b) $R_{\text{scat}} = \left(\frac{I}{h\nu}\right) \sigma_{\text{abs}} = \left(\frac{I\,\lambda}{h\,c}\right) \sigma_{\text{abs}} = 89$ photons/s for each atom

This is a much lower scattering rate per atom than the 3.5×10^5 photons/s for each atom calculated in problem 9. But that was for a laser tuned close to the atomic transition, while the rate in this problem is for a laser tuned *far* from the transition.

Fundamental Constants, Conversion Factors, and other Data

elementary charge	e	1.602×10^{-19} C
electron mass	m_e	9.110×10^{-31} kg 0.511 MeV/c^2
atomic mass unit	u	1.661×10^{-27} kg 931.5 MeV/c^2
speed of light	c	2.998×10^{8} m s^{-1} 30 cm/ns -or- $\sim$1 ft/ns
permittivity of free space	ε_0	8.854×10^{-12} C^2 N^{-1} m^{-2}
permeability of free space	μ_0	1.257×10^{-7} V s A^{-1} m^{-1}
gas constant	R	8.314 J mol^{-1} K^{-1} 62364 Torr cm^3 K^{-1} mol^{-1}
Boltzmann constant	k	1.381×10^{-23} J K^{-1} 8.617×10^{-5} eV K^{-1}
Planck's constant	h	6.626×10^{-34} J s 4.136×10^{-15} eV s
Avogadro number	N_A	6.022×10^{23} atoms mol^{-1}
Loschmidt number (STP)	N_L	2.687×10^{19} atoms cm^{-3}
Bohr magneton	μ_B	5.788×10^{-9} eV G^{-1}
Bohr radius	a_0	5.292×10^{-9} cm
Classical radius of the electron	r_e	2.818×10^{-13} cm
Rydberg energy	E_R	13.606 eV

$$
\begin{aligned}
1\ \text{eV} &= 8066.1\ \text{cm}^{-1} \\
&= 1.602 \times 10^{-19}\ \text{J} \\
&= 11605\ \text{K} \\
1\ \text{atm} &= 760\ \text{Torr}
\end{aligned}
$$

Periodic Table of the Elements

1	2	3	4	5	6	7	8	9	10	11	12	13	14	15	16	17	18
1 H 1.008																	2 He 4.00
3 Li 6.941	4 Be 9.012											5 B 10.81	6 C 12.01	7 N 14.01	8 O 16.00	9 F 19.00	10 Ne 20.1
11 Na 22.99	12 Mg 24.30											13 Al 26.98	14 Si 28.09	15 P 30.97	16 S 32.06	17 Cl 35.45	18 Ar 39.9
19 K 39.10	20 Ca 40.08	21 Sc 44.96	22 Ti 47.90	23 V 50.94	24 Cr 52.00	25 Mn 54.94	26 Fe 55.85	27 Co 58.93	28 Ni 58.70	29 Cu 63.55	30 Zn 65.38	31 Ga 69.72	32 Ge 72.59	33 As 74.92	34 Se 78.96	35 Br 79.90	36 Kr 83.8
37 Rb 85.47	38 Sr 87.62	39 Y 88.91	40 Zr 91.22	41 Nb 92.91	42 Mo 95.94	43 Tc (98)	44 Ru 101.1	45 Rh 102.9	46 Pd 106.4	47 Ag 107.9	48 Cd 112.4	49 In 114.8	50 Sn 118.7	51 Sb 121.8	52 Te 127.6	53 I 126.9	54 Xe 131
55 Cs 132.9	56 Ba 137.3	*	72 Hf 178.5	73 Ta 180.9	74 W 183.8	75 Re 186.2	76 Os 190.2	77 Ir 192.2	78 Pt 195.1	79 Au 197.0	80 Hg 200.6	81 Tl 204.4	82 Pb 207.2	83 Bi 209.0	84 Po (209)	85 At (210)	86 Rn (22
87 Fr (223)	88 Ra (226)	**	104 Rf (267)	105 Db (268)	106 Sg (269)	107 Bh (270)	108 Hs (277)	109 Mt (278)	110 Ds (281)	111 Rg (282)	112 Cn (285)	113 Nh (286)	114 Fl (289)	115 Mc (290)	116 Lv (293)	117 Ts (294)	118 Og (29

*	57 La 138.9	58 Ce 140.1	59 Pr 140.9	60 Nd 144.2	61 Pm (145)	62 Sm 150.4	63 Eu 152.0	64 Gd 157.2	65 Tb 158.9	66 Dy 162.5	67 Ho 164.9	68 Er 167.3	69 Tm 168.9	70 Yb 173.0	71 Lu 175.0
**	89 Ac (227)	90 Th (232)	91 Pa (231)	92 U 238.0	93 Np (237)	94 Pu (239)	95 Am (243)	96 Cm (247)	97 Bk (247)	98 Cf (251)	99 Es (252)	100 Fm (257)	101 Md (258)	102 No (259)	103 Lr (262)

Index

Printed in the United States
By Bookmasters